Project Planning & Design (PPD)

ARE 5.0 Exam Guide
(Architect Registration Examination)

ARE 5.0 Overview, Exam Prep Tips, Guide, and Critical Content

Gang Chen

ArchiteG®, Inc.
Irvine, California

Project Planning & Design (PPD) ARE 5.0 Exam Guide (Architect Registration Examination): ARE 5.0 Overview, Exam Prep Tips, Guide, and Critical Content

Copy Editor: Editing Team of ArchiteG, Inc.

ArchiteG, Inc.
http://www.ArchiteG.com

ISBN: 978-1-61265-042-5

PRINTED IN THE UNITED STATES OF AMERICA

What others are saying about *ARE Exam Guide series* and *ARE Mock Exam series …*

"Great study guide…"
"This was a great resource supplement to my other study resources. I appreciated the mock exam questions the most, and the solutions offer an explanation as to why the answer is correct. I will definitely check out his other ARE exam resources!

UPDATE: Got my PASS Letter!"
—Sean Primeaux

"Tried everything 4 times before reading this book and PASSED!"
"I had failed this exam 4 times prior to getting this book…I had zero clue as to what I was doing wrong. I read Ballast, Kaplan and random things on the forum but for the life of me couldn't pin point where I was missing it until I read THIS BOOK! Gang did an excellent job…I remember going through the ramp and reading Gang's book and saying Ohhhh like 4 or 5 times. I read his book several times until I became comfortable with the information. I went in on test day and it was a breeze. I remember walking out of there thinking I couldn't believe I struggled so much before. The tips in here are priceless! I strongly recommend this book…"
—hendea1

"Add this to your ARE study"
"This was a very helpful practice exam and discussion. I really appreciated the step-by-step review of the author's approach... As I studied it last before taking the test, Gang Chen's book probably made the difference for me."
—Dan Clowes ("XLine")

"Good supplemental mock exam"
"I found the mock exam to be very helpful, all of the answers are explained thoroughly and really help you understand why it is correct...Also the introduction and test taking tips are very helpful for new candidates just starting the ARE process."
—Bgrueb01

"Essential Study Tool"
"I have read the book and found it to be a great study guide for myself. Mr. Gang Chen does such a great job of helping you get into the right frame of mind for the content of the exam. Mr. Chen breaks down the points on what should be studied and how to improve your chances of a pass with his knowledge and tips for the exam.

I highly recommend this book to anyone…it is an invaluable tool in the preparation for the exam as Mr. Chen provides a vast amount of knowledge in a very clear, concise, and logical matter."
—Luke Giaccio

“Wish I had this book earlier”
“...The questions are written like the NCARB questions, with various types...check all that apply, fill in the blank, best answer, etc. The answer key helpfully describes why the correct answer is correct, and why the incorrect answers are not. Take it from my experience, at half the cost of other mock exams, this is a must buy if you want to pass...”
—Domiane Forte (“Vitruvian Duck”)

“This book did exactly like the others said.”
“This book did exactly like the others said. It is immensely helpful with the explanation... There are so many codes to incorporate, but Chen simplifies it into a methodical process. Bought it and just found out I passed. I would recommend.”
—Dustin

“It was the reason I passed.”
“This book was a huge help. I passed the AREs recently and I felt this book gave me really good explanations for each answer. It was the reason I passed.”
—Amazon Customer

“Great Practice Exam”
“... For me, it was difficult to not be overwhelmed by the amount of content covered by the Exam. This Mock Exam is the perfect tool to keep you focused on the content that matters and to evaluate what you know and what you need to study. It definitely helped me pass the exam!!”
—Michael Harvey (“Harv”)

“One of the best practice exams”
“Excellent study guide with study tips, general test info, and recommended study resources. Hands down one of the best practice exams that I have come across for this exam. Most importantly, the practice exam includes in depth explanations of answers. Definitely recommended.”
—Taylor Cupp

“Great Supplement!!”
“This publication was very helpful in my preparation for my BS exam. It contained a mock exam, followed by the answers and brief explanations to the answers. I would recommend this as an additional study material for this exam.”
—Cynthia Zorrilla-Canteros (“czcante”)

“Fantastic! ”
“When I first began to prepare for this exam; the number of content areas seemed overwhelming and daunting at best. However, this guide clearly dissected each content area into small management components. Of all the study guides currently available for this test - this exam not only included numerous resources (web links, you tube clips, etc..), but also the sample test was extremely helpful. The sample test incorporated a nice balance of diagrams, calculations and general concepts - this book allowed me to highlight any "weak" content areas I had prior to the real exam. In short - this is an awesome book!”
—Rachel Casey (RC)

Dedication

To my parents, Zhuixian and Yugen,
my wife Yongqing Fang,
and my daughters,
Alice, Amy, Angela, Athena and America.

Disclaimer

Project Planning & Design (PPD) ARE 5.0 Exam Guide (Architect Registration Examination) provides general information about Architect Registration Examination. The book is sold with the understanding that neither the publisher nor the authors are providing legal, accounting, or other professional services. If legal, accounting, or other professional services are required, seek the assistance of a competent professional firm.

The purpose of this publication is not to reprint the content of all other available texts on the subject. You are urged to read other materials, and tailor them to fit your needs.

Great effort has been taken to make this resource as complete and accurate as possible. However, nobody is perfect and there may be typographical errors or other mistakes present. You should use this book as a general guide and not as the ultimate source on this subject. If you find any potential errors, please send an e-mail to:
ArchiteG@Yahoo.com

Project Planning & Design (PPD) ARE 5.0 Exam Guide (Architect Registration Examination) is intended to provide general, entertaining, informative, educational, and enlightening content. Neither the publisher nor the author shall be liable to anyone or any entity for any loss or damages, or alleged loss or damages, caused directly or indirectly by the content of this book.

If you do not wish to be bound by the above, you may return this book to the publisher for a full refund.

Legal Notice

ARE Exam Guide series by ArchiteG, Inc.

Time and effort are the most valuable asset of a candidate. How to cherish and effectively use your limited time and effort is the key of passing any exam. That is why we publish the ARE Exam Guide series and the ARE Mock Exam series to help you to study and pass the ARE exams in the shortest time possible. We have done the hard work so that you can save time and money. We do not want to make you work harder than you have to. To save your time, we use a *standard* format for all our ARE 5.0 Exam Guide books and ARE 5.0 Mock Exam books, so that you can quickly skip the *identical* information you have already read in other books of the series and go straight to the *unique* "meat and potatoes" portion of the book.

The trick and the most difficult part of writing a good book is to turn something that is very complicated into something that is very simple. This involves researching and really understanding very complicated materials, absorbing the information, and then writing about the topic in a way that makes it very easy to understand. To succeed at this, you need to know the materials very well. Our goal is to write books that are clear, concise, and helpful.

Do not force yourself to memorize a lot of numbers. Read through the numbers a few times, and you should have a very good impression of them. You need to make the judgment call: If you miss a few numbers, you can still pass the exam, but if you spend too much time drilling these numbers, you may miss out on the big pictures and fail the exam.

The existing ARE prep books by others are either way too easy or way over-killed. They do NOT match the real ARE 5.0 exams at all.

We have done very comprehensive research on the official NCARB guides, many related websites, reference materials, and other available ARE exam prep materials. We match our mock exams as close as possible to the NCARB samples and the real ARE exams. Some readers had failed an ARE exam two or three times before, and they eventually passed the exam with our help.

All our mock exam books include a complete set of questions and case studies. We try to mimic the real ARE exams by including the same number of questions, using a similar format, and asking the same type of questions. We also include detailed answers and explanations to our questions.

There is some extra information on ARE overviews and exam-taking tips in Chapter One. This is based on NCARB *and* other valuable sources. This is a bonus feature we included in each book because we want our readers to be able to buy our ARE mock exam books together or individually. We want you to find all necessary ARE exam information and resources at one place and through our books.

All our books are available at
http://www.GreenExamEducation.com

How to Use This Book

We suggest you read *Project Planning & Design (PPD) ARE 5.0 Exam Guide (Architect Registration Examination)* at least three times:

Read once and cover all chapters, the Appendixes, the related *free* PDF files, and other resources. Highlight the information you are not familiar with.

Read twice focusing on the highlighted information to memorize. You can repeat this process as many times as you want until you master the content of the book.

After reviewing these materials, you can read the related mock exam book and take the mock exam, and then check your answers against the answers and explanations in the back, including explanations for the questions you answer correctly. You may have answered some questions correctly for the wrong reason. Highlight the information you are not familiar with.

Like the real exam, the mock exam will continue to use **multiple choice, check-all-that-apply,** and **quantitative fill-in-the-blank**. There are also two new question types: **hotspots, and case studies**

Review your highlighted information in both our ARE guidebook and mock exam book and take the mock exam again. Try to answer 100% of the questions correctly this time. Repeat the process until you can answer all the questions correctly.

Our books include most if not all the information you need to do the calculations, as well as step-by-step explanations. After reading our books, you will greatly improve your ability to deal with the real ARE questions and have a great chance of passing the exam on the first try.

Read our related mock exam book and take the mock exam at least two weeks before the real exam. You should *not* wait until the night before the real exam to take the mock exam. If you do not do well, you will go into panic mode and NOT have enough time to review your weaknesses.

Read for the final time the night before the real exam. Review *only* the information you highlighted, especially the questions you did not answer correctly when you took the mock exam for the first time.

Our books are very light so you can easily carry them around. These features will allow you to review our books whenever you have a few minutes.

The Table of Contents is very detailed so you can locate information quickly. If you are on a tight schedule you can forgo reading our books linearly and jump around to the sections you need.

All our books, including "ARE Mock Exams Series," "ARE Exams Guide Series," "LEED Mock Exams Series," and "LEED Exam Guides Series," are available at **GreenExamEducation.com**

Check out FREE tips and info at **GeeForum.com**, you can post your questions for other users' review and responses.

Table of Contents

Chapter Five Building Systems, Materials, & Assemblies (19-25%)

Appendixes

Back Page Promotion

Index

Chapter One

Overview of the Architect Registration Examination (ARE)

A. First thing first: go to the website of your architect registration board and read all the requirements of obtaining an architect license in your jurisdiction

See the following link:
https://www.ncarb.org/get-licensed/state-licensing-boards

B. Download and review the latest ARE DOCUMENTS AT THE NCARB website

1. Important links to the FREE and official NCARB documents

NCARB launched ARE 5.0 on November 1, 2016. The current version of the Architect Registration Examination (ARE 5.0) includes six divisions:

- Practice Management (PcM)
- Project Management (PjM)
- Programming & Analysis (PA)
- Project Planning & Design (PPD)
- Project Development & Documentation (PDD)
- Construction & Evaluation (CE)

All ARE divisions continue to use **multiple choice, check-all-that-apply,** and **quantitative fill-in-the-blank**. The new exams include two new question types: **Hotspots** and **case studies**.

There is a tremendous amount of valuable information covering every step of becoming an architect available free of charge at the NCARB website:
http://www.ncarb.org/

For example, you can find guidance about architectural degree programs accredited by the National Architectural Accrediting Board (NAAB), NCARB's Architectural Experience Program (AXP) formerly known as Intern Development Program (IDP), and licensing requirements by state. These documents explain how you can qualify to take the Architect Registration Examination.

We find the official ARE 5.0 Guidelines, ARE 5.0 Handbook, and ARE 5.0 Credit Model extremely valuable. See the following link:
http://www.ncarb.org/ARE/ARE5.aspx

You should start by studying these documents.

2. **A detailed list and brief description of the FREE PDF files that you can download from NCARB**
   The following is a detailed list of the FREE PDF files that you can download from NCARB. They are listed in order based on their importance.

   - All **ARE 5.0** information can be found at the following links**:** http://www.ncarb.org/ARE/ARE5.aspx

   - **ARE 5.0 Guidelines** includes extremely valuable information on the ARE overview, NCARB, registration (licensure), architectural education requirements, the Architectural Experience Program (AXP), establishing your eligibility to test, scheduling an exam appointment, taking the ARE, receiving your score, retaking the ARE, the exam format, scheduling, and links to other FREE NCARB PDF files. You need to read this at least twice.

   - **NCARB Education Guidelines** contains information on education requirements for initial licensure and for NCARB certification, satisfying the education requirement, foreign-educated applicants, the education alternative to NCARB certification, the Education Evaluation Services for Architects (EESA), the Education Standard, and other resources.

   - **Architectural Experience Program (AXP) Guidelines** includes information on AXP overview, getting started and creating your NCARB record, experience areas and tasks, documenting your experience through hours, documenting your experience through a portfolio, and the next steps. You need to read this document at least twice. The information is valuable.

     NCARB renamed the **Intern Development Program (IDP)** as **Architectural Experience Program (AXP)** in June 2016. Most of NCARB's 54-member boards have adopted the AXP as a prerequisite for initial architect licensure. Therefore, you should be familiar with the program.

     The AXP application fee is $100. This fee includes one free transmittal of your Record for initial registration and keeps your Record active for the first year. After the initial year, there is an annual renewal fee required to maintain an active Record until you become registered. The cost is currently $85 each year. The fees are subject to change, and you need to check the NCARB website for the latest information.

     There are two ways to meet the AXP requirements. The **first method** is **reporting hours**. Most candidates will use this method. You will need to document at least 3,740 required hours under the six different experience areas to complete the program. A minimum of 50% of your experience must be completed under the supervision of a qualified architect.

The following chart lists the hours required under the six experience areas:

Experience Area	Hours Required
Practice Management	160
Project Management	360
Programming & Analysis	260
Project Planning & Design	1,080
Project Development & Documentation	1,520
Construction & Evaluation	360
Total	**3,740**

Figure 1.1 The hours required under the six experience areas

Your experience reports will fall under one of **two experience settings**:
• **Setting A**: Work performed for an architecture firm.
• **Setting O**: Experiences performed outside an architecture firm.

You must earn at least **1,860 hours** in experience **setting A**.

Your AXP experience should be reported to NCARB at least every *six* months and logged within two months of completing each reporting period (the **Six-Month Rule**).

The **second method** to meet AXP requirements is to create an **AXP Portfolio**. This new method is for experienced design professionals who put their licensure on hold and allows you to prove your experience through the preparation of an online portfolio.

To complete the AXP through the **second method**, you will need to meet ALL the AXP criteria through the portfolio. In other words, you cannot complete the experience requirement through a combination of the **AXP portfolio** and **reporting hours**.

See the following link for more information on AXP:
https://www.ncarb.org/gain-axp-experience

- **Certification Guidelines** by NCARB (Skimming through this should be adequate. You should also forward a copy of this PDF file to your AXP supervisor.)

  See the following link which contains resources for supervisors and mentors:
  https://www.ncarb.org/gain-axp-experience/supervisors-mentors

- **ARE 5.0 Related FAQs (Frequently Asked Questions)**: Skimming through this should be adequate.

- **Your Guide to ARE 5.0** includes information on understanding the basics of ARE 5.0, new question types, taking the test, making the transition, getting ARE 5.0 done, and planning your budget. The document also contains FAQs, and links for more information. You need to read this document *at least twice*. The information is valuable.

- **ARE 5.0 Handbook** contains an ARE overview, detailed information for each ARE division, and ARE 5.0 references. This handbook explains what NCARB expects you to know so that you can pass the ARE exams. ARE 5.0 uses either **Understand/Apply (U/A)** or **Analyze/Evaluate (A/E)** to designate the appropriate cognitive complexity of each objective, but *avoids* the use of **"Remember,"** the lowest level of cognitive complexity (CC), or **"Create,"** the highest level of CC.

  This handbook has some sample questions for each division. The real exam is like the samples in this handbook.

  Tips:
  - ***ARE 5.0 Handbook*** *has about 180 pages. To save time, you can just read the generic information at the front and back portion of the handbook, and focus on the ARE division(s) you are currently studying for. As you progress in your testing, you can read the applicable division that you are studying for. This way, the content will always be fresh in your mind.*
  - *You need to read this document at least three times. The information is valuable.*

- **ARE 5.0 Test Specification** identifies the ARE 5.0 division structure and defines the major content areas, called **Sections**; the measurement **Objectives**; and the percentage of content coverage, called **Weightings**. This document specifies the scope and objectives of each ARE division, and the percentage of questions in each content category. You need to read this document *at least twice*. The information is valuable, and is the base of all ARE exam questions.

- **ARE 5.0 Prep Videos** include one short video for each division. These videos give you a very good basic introduction to each division, including sample questions and answers, and explanations. You need to watch each video *at least three times*. See the following link:
  https://www.ncarb.org/pass-the-are/prepare

- **The Burning Question: Why do we need an ARE anyway?** (We do not want to give out a link for this document because it is too long and keeps changing. You can Google it with its title. Skimming through this document should be adequate.)

- **Defining Your Moral Compass** (You can Google it with its title plus the word "NCARB." Skimming through this document should be adequate.)

- **Rules of Conduct** is available as a *free* PDF file at:
  http://www.ncarb.org/
  (Skimming through this should be adequate.)

C. The Intern Development Program (IDP)/Architectural Experience Program (AXP)

1. **What is IDP? What is AXP?**
IDP is a comprehensive training program jointly developed by the National Council of Architectural Registration Boards (NCARB) and the American Institute of Architects (AIA) to ensure that interns obtain the necessary skills and knowledge to practice architecture independently. NCARB renamed the **Intern Development Program (IDP)** as **Architectural Experience Program (AXP)** in June 2016.

2. **Who qualifies as an intern?**
Per NCARB, if an individual meets one of the following criteria, s/he qualifies as an intern:
a. Graduates from NAAB-accredited programs
b. Architecture students who acquire acceptable training prior to graduation
c. Other qualified individuals identified by a registration board

D. Overview of the Architect Registration Examination (ARE)

1. **How to qualify for the ARE?**
A candidate needs to qualify for the ARE via one of NCARB's member registration boards, or one of the Canadian provincial architectural associations.

Check with your Board of Architecture for specific requirements.

For example, in California, a candidate must provide verification of a minimum of five years of education and/or architectural work experience to qualify for the ARE.

Candidates can satisfy the five-year requirement in a variety of ways:

- Provide verification of a professional degree in architecture through a program that is accredited by NAAB or CACB.

  OR
- Provide verification of at least five years of educational equivalents.

  OR
- Provide proof of work experience under the direct supervision of a licensed architect.

See the following link:
https://www.ncarb.org/

2. **How to qualify for an architect license?**
Again, each jurisdiction has its own requirements. An individual typically needs a combination of about *eight* years of education and experience, as well as passing scores on the ARE exams. See the following link:
https://www.ncarb.org/become-architect/earn-license/state-licensing-boards

For example, the requirements to become a licensed architect in California are:

- Eight years of post-secondary education and/or work experience as evaluated by the Board (including at least one year of work experience under the direct supervision of an architect licensed in a U.S. jurisdiction or two years of work experience under the direct supervision of an architect registered in a Canadian province)
- Completion of the Architectural Experience Program (AXP)
- Successful completion of the Architect Registration Examination (ARE)
- Successful completion of the California Supplemental Examination (CSE)

California does NOT require an accredited degree in architecture for examination and licensure. However, many other states do.

3. **What is the purpose of ARE?**
   The purpose of ARE is NOT to test a candidate's competency on every aspect of architectural practice. Its purpose is to test a candidate's competency on providing professional services to protect the *health, safety, and welfare* of the public. It tests candidates on the *fundamental* knowledge of pre-design, site design, building design, building systems, and construction documents and services.

   The ARE tests a candidate's competency as a "specialist" on architectural subjects. It also tests her abilities as a "generalist" to coordinate other consultants' works.

   You can download the exam content and references for each of the ARE divisions at the following link:
   https://www.ncarb.org/pass-the-are/start

4. **NCARB has retired the rolling clock policy.**
   Per NCARB: The National Council of Architectural Registration Boards (NCARB) is retiring the Architect Registration Examination (ARE®) rolling clock policy and implementing a new score validity policy, effective May 1, 2023.

   See link below:
   https://www.ncarb.org/press/ncarb-eliminates-the-are-rolling-clock-policy-0

5. **How to register for an ARE exam?**
   See the instructions in the new ARE guideline at the following link:
   http://www.ncarb.org/ARE/ARE5.aspx

6. **How early do I need to arrive at the test center?**
   Be at the test center at least 30 minutes BEFORE your scheduled test time, OR you may lose your exam fee.

7. **Exam format & time**
   All ARE divisions are administered and graded by computer. The time allowances are as follows:

Division	Number of Questions	Test Duration	Appointment Time
Practice Management	80	2 hr 45 min	3 hr 30 min
Project Management	95	3 hr 15 min	4 hr
Programming & Analysis	95	3 hr 15 min	4 hr
Project Planning & Design	120	4 hr 15 min	5 hr
Project Development & Documentation	120	4 hr 15 min	5 hr
Construction & Evaluation	95	3 hr 15 min	4 hr
Total Time:		21 hr	25 hr 30 min

Figure 1.2 Exam format & time

English as a Second Language (ESL) Accommodations
NCARB provides two types of language support for eligible candidates who are non-native English speakers. These options are designed to assist individuals who may need extra support during the Architect Registration Examination (ARE):

- **Extended Exam Time**: Candidates may receive an additional 20% of the standard testing time for each exam section, giving them more time to complete the assessment.
- **Use of a Translation Dictionary**: Candidates testing in person at a PSI testing center may bring their own printed word-for-word translation dictionary. This dictionary must be free of any handwritten notes or annotations and will be reviewed by the test center staff before the exam begins. This option is *not* available for exams taken via online proctoring.

Remote proctoring was introduced in mid-December 2020. As of December 14, 2020, the number of questions and the time allotted for each exam division were adjusted to accommodate the new remote testing format:

Division	Number of Questions	Test Duration	Appointment Time
Practice Management	65	2 hr 40 min	3 hr 20 min
Project Management	75	3 hr	3 hr 40 min
Programming & Analysis	75	3 hr	3 hr 40 min
Project Planning & Design	100	4 hr 5 min	5 hr
Project Development & Documentation	100	4 hr 5 min	5 hr
Construction & Evaluation	75	3 hr	3 hr 40 min
Total Time:		19 hr 50 min	24 hr 20 min

Figure 1.3 Remote Exam format & time

NCARB suggests you arrive at the test center a minimum of 30 minutes before your scheduled appointment. You can have one flexible 15-minute break for each division. That is why the appointment time is 45 minutes longer than the actual test time for each division.

Practice Management has 80 questions and NCARB allows you 2 hours and 45 minutes to complete the exam, so you should spend an average of (2x60+45)/80=165/80= 2.06 minutes on each question.

Project Management and **Programming & Analysis** as well as **Construction & Evaluation** each have 95 questions and NCARB allows you 3 hours and 15 minutes to complete each exam, so you should spend an average of (3x60+15)/80=195/95= 2.05 minutes on each question.

Project Planning & Design as well as **Project Development & Documentation** each have 120 questions and NCARB allows you 4 hours and 15 minutes to complete each exam, so you should spend an average of (4x60+15)/120=255/120= 2.13 minutes on each question.

To simplify this, we suggest you spend about 2 minutes for each question in ALL divisions.

8. **How are ARE scores reported?**
   All ARE scores are reported as Pass or Fail. ARE scores are typically posted within seven to ten days. See the instructions in the new ARE guideline at the following link:
   http://www.ncarb.org/ARE/ARE5.aspx

9. **Is there a fixed percentage of candidates who pass the ARE exams?**
   No, there is *not* a fixed percentage of passing or failing. If you meet the minimum competency required to practice as an architect, you pass. The passing scores are the same for all Boards of Architecture.

10. **When can I retake a failed ARE division?**
    You can retake a failed division of the ARE sixty days after the previous attempt. You can only take the same ARE division three times within any twelve-month period.

11. **How much time do I need to prepare for each ARE division?**
    Every person is different, but on average you need about forty to eighty hours to prepare for each ARE division. You need to set a realistic study schedule and stick with it. Make sure you allow time for personal and recreational commitments. If you are working full time, my suggestion is that you allow no less than two weeks but NOT more than two months to prepare for each ARE division. You should NOT drag out the exam prep process too long and risk losing your momentum.

12. **Which ARE division should I take first?**
    This is a matter of personal preference, and you should make the final decision.

    Some people like to start with the easier divisions and pass them first. This way, they build more confidence as they study and pass each division.

Other people like to start with the more difficult divisions so that if they fail, they can keep busy studying and taking the other divisions while the clock is ticking. Before they know it, six months has passed, and they can reschedule if need be.

13. ARE exam prep and test-taking tips

You can start with Construction & Evaluation (CE) because it gives a limited scope, and you may want to study building regulations and architectural history (especially famous architects and buildings that set the trends at critical turning points) before you take other divisions.

Complete mock exams and practice questions and case studies, including those provided by NCARB's practice program and this book, to hone your skills.

Form study groups and learn the exam experience of other ARE candidates. The forum at our website is a helpful resource. See the following links:
http://GreenExamEducation.com/
http://GeeForum.com/

Take the ARE exams as soon as you become eligible, since you probably still remember portions of what you learned in architectural school, especially structural and architectural history. Do not make excuses for yourself and put off the exams.

The following test-taking tips may help you:

- Pace yourself properly. You should spend about two minutes for each question on average.
- Read the questions carefully and pay attention to words like *best, could, not, always, never, seldom, may, false, except,* etc.
- For questions that you are not sure of, eliminate the obvious wrong answer and then make an educated guess. Please note that if you do NOT answer the question, you automatically lose the point. If you guess, you at least have a chance of getting it right.
- If you have no idea what the correct answer is and cannot eliminate any obvious wrong answers, then do not waste too much time on the question and just guess. Try to use the same guess answer for all of the questions you have no idea about. For example, if you choose "d" as the guess answer, then you should be consistent and use "d" whenever you have no clue. This way, you are likely having a better chance at guessing more answers correctly.
- Mark the difficult questions, answer them, and come back to review them AFTER you finish all questions. If you are still not sure, go with your first choice. Your first choice is often the best choice.
- You really need to spend time practicing to become VERY familiar with NCARB's question types. This is because ARE is a timed test, and you do NOT have time to learn about the question types during the test. If you do not know them well, you will NOT be able to finish your solution on time.
- The ARE exams test a candidate's competency to provide professional services protecting the *health, safety, and welfare* of the public. Do NOT waste time on aesthetic or other design elements not required by the program.

ARE exams are difficult, but if you study hard and prepare well, combined with your experience, AXP training, and/or college education, you should be able to pass all divisions and eventually be able to call yourself an architect.

14. Strategies for passing ARE exams on the first try

Passing ARE exams on the first try, like everything else, needs not only hard work, but also great strategy.

- **Find out how much you already know and what you should study**

  You goal is NOT to read all the study materials. Your goal is to pass the exam. Flip through the study materials. If you already know the information, skip these parts.

  Complete the NCARB sample questions for the ARE exam you are preparing for NOW without ANY studying. See what percentage you get right. If you get 62% right, you should be able to pass the real exam without any studying. If you get 50% right, then you just need 12% more to pass.

  This "truth-finding" exam or exercise will also help you to find out what your weakness areas are, and what to focus on.

  Look at the same questions again at the end of your exam prep and check the differences.

  Note:
  We suggest you study the sample questions in the official NCARB Study Guide first, and then other study materials, and then come back to NCARB sample questions again several days before the real ARE exam.

  Per NCARB, with the launch of the updated Architect Registration Examination (ARE) 5.0 in December 2020, the new cutting scores are based on the following information:

  "**How many questions do I need correct to pass?**

  Each division of the ARE measures different content knowledge areas. The difference in knowledge areas and the relative difficulty of the questions that make up that content area vary between divisions; therefore, expectations around how many questions you will need to answer correctly also changes from division to division.

  - **Project Development & Documentation and Construction & Evaluation** require the lowest percentage of scored items to be answered correctly to pass. You need to answer between **57 – 62 percent** of scored items correctly on these divisions to pass.

  - **Practice Management and Project Management** require a slightly higher percentage of questions to be answered correctly to pass. You need to answer between **62 – 68 percent** of scored items correctly on these divisions to pass.
  - **Programming & Analysis and Project Planning & Design** require the highest percentage of questions to be answered correctly to pass. You need to answer between **65 – 71 percent** of scored items correctly on these divisions to pass."

  For detailed information, see the following link:
  https://www.ncarb.org/blog/what-score-do-you-need-to-pass-the-are

- **Cherish and effectively use your limited time and effort**

  Let me paraphrase a story.
  One time someone had a chance to talk with Napoleon. He said:
  "You are such a great leader and have won so many battles, that you can use one of your soldiers to defeat ten enemy soldiers."

  Napoleon responded:
  "That may be true, but I always try to create opportunities where ten of my soldiers fight one enemy soldier. That is why I have won so many battles."

  Whether this story is true is irrelevant. The important thing that you need to know is **how to concentrate your limited time and effort to achieve your goal. Do NOT spread yourself too thin**. This is a principle many great leaders know and use and is why great leaders can use ordinary people to achieve extraordinary goals.

  Time and effort are the most valuable asset of a candidate. How to cherish and effectively use your limited time and effort is the key to passing any exam.

  If you study very hard and read many books, you are probably wasting your time. You are much better off picking one or two good books, covering the major framework of your exams, and then doing two sets of mock exams to find your weaknesses. You WILL pass if you follow this advice. You may still have minor weakness, but you will have covered your major bases.

- **Do NOT stretch your exam prep process too long**
  If you do this, it will hurt instead of helping you. You may forget the information by the time you take the exam.

  Spend 40 to 80 hours for each division (a maximum of two months for the most difficult exams if you really need more time) should be enough. Once you decide on taking an exam, put in 100% of your effort and read the RIGHT materials. Allocate your time and effort on the most important materials, and you will pass.

- **Resist the temptation to read too many books and limit your time and effort to read only a few selected books or a few sections of each book in detail**

Having all the books but not reading them, or digesting ALL the information in them will not help you. It is like someone having a garage full of foods, and not eating or eating too much of them. Neither way will help.

You can only eat three meals a day. Similarly, you can ONLY absorb a certain amount of information during your exam prep. So, focus on the most important stuff.

Focus on your weaknesses but still read the other info. The key is to understand, digest the materials, and retain the information.

It is NOT how much you have read, but how much you understand, digest, and retain that counts.

The key to passing an ARE exam, or any other exam, is to know the scope of the exam, and not to read too many books. Select one or two really good books and focus on them. Understand the content and memorize it. For your convenience, I have underlined the fundamental information that I think is very important. You definitely need to *memorize* all the information that I have underlined.

You should try to understand the content first, and then memorize the content of the book by reading it multiple times. This is a much better way than relying on "mechanical" memory without understanding.

When you read the materials, always keep the following in mind:

- **Think like an architect.**
  For example, when you take the Project Development & Documentation (PDD) exam, focus on what need to know to be able to coordinate your engineer's work, or tell them what to do. You are NOT taking an exam for becoming a structural engineer; you are taking an exam to become an architect.

  This criterion will help you filter out the materials that are irrelevant and focus on the right information. You will know what to flip through, what to read line by line, and what to read multiple times.

  I have said this one thousand times, and I will say it once more:
  Time and effort is the most valuable asset of a candidate. How to cherish and effectively use your limited time and effort is the key to passing any exam.

15. ARE exam preparation requires short-term memory
You should understand that ARE Exam Preparation requires **Short-Term Memory**. This is especially true for the MC portion of the exam. You should schedule your time accordingly: in the early stages of ARE exam Preparation, you should focus on understanding and an **initial** review of the material; in the late stages of exam preparation, you should focus on memorizing the material as a **final** review.

16. Allocation of your time and scheduling

You should spend about 60% of your effort on the most important and fundamental study materials, about 30% of your effort on mock exams, and the remaining 10% on improving your weakest areas, i.e., reading and reviewing the questions that you answered incorrectly, reinforcing the portions that you have a hard time memorizing, etc.

Do NOT spend too much time looking for obscure ARE information because NCARB will HAVE to test you on the most **common** architectural knowledge and information. At least 80% to 90% of the exam content will have to be the most common, important and fundamental knowledge. The exam writers can word their questions to be tricky or confusing, but they have to limit themselves to the important content; otherwise, their tests will NOT be legally defensible. At most, 10% of their test content can be obscure information. You only need to answer about 62% of all the questions. So, if you master the common ARE knowledge (applicable to 90% of the questions) and use the guess technique for the remaining 10% of the questions on the obscure ARE content, you will do well and pass the exam.

On the other hand, if you focus on the obscure ARE knowledge, you may answer the entire 10% obscure portion of the exam correctly, but only answer half of the remaining 90% of the common ARE knowledge questions correctly, and you will fail the exam. We have seen many smart people who can answer very difficult ARE questions correctly because they are able to look them up and do quality research. However, they often end up failing ARE exams because they cannot memorize the common ARE knowledge needed on the day of the exam. ARE exams are NOT open-book exams, and you cannot look up information during the exam.

The **process of memorization** is like **filling a cup with a hole at the bottom**: You need to fill it faster than the water leaks out at the bottom, and you need to constantly fill it; otherwise, it will quickly be empty.

Once you memorize something, your brain has already started the process of forgetting it. It is natural. That is how we have enough space left in our brain to remember the really important things.

It is tough to fight against your brain's natural tendency to forget things. Acknowledging this truth and the fact that you cannot memorize everything you read, you need to focus your limited time, energy, and brainpower on the most important issues.

The biggest danger for most people is that they memorize the information in the early stages of their exam preparation but forget it before or on the day of the exam and still THINK they remember.

Most people fail the exam NOT because they cannot answer the few "advanced" questions on the exam, but because they have read the information and can NOT recall it on the day of the exam. They spend too much time preparing for the exam, drag the preparation process on for too long, seek too much information, go to too many websites, do too many practice questions and too many mock exams, and **spread themselves too thin**. They end up **missing the most important information** of the exam, and they fail.

The ARE Mock Exam series along with the tips and methodology in each of the books will help you find and improve your weakness areas, MEMORIZE the most important aspects of the test to pass the exam ON THE FIRST TRY.

So, if you have a lot of time to prepare for the ARE exams, you should plan your effort accordingly. You want your ARE knowledge to peak at the time of the exam, not before or after.

For example, if you have two months to prepare for a very difficult ARE exam, you may want to spend the first month focused on reading and understanding all of the study materials you can find as your **initial** review. During this first month, you can start memorizing after you understand the materials as long as you know you HAVE to review the materials again later to retain them. If you have memorized something once, it is easier to memorize it again later.

Next, you can spend two weeks focused on memorizing the material. You need to review the material at least three times. You can then spend one week on mock exams. The last week before the exam, focus on retaining your knowledge and reinforcing your weakest areas. Read the mistakes that you have made and think about how to avoid them during the real exam. Set aside a mock exam that you have not taken and take it seven days before test day. This will alert you to your weaknesses and provide direction for the remainder of your studies.

If you have one week to prepare for the exam, you can spend two days reading and understanding the study material, two days repeating and memorizing the material, two days on mock exams, and one day retaining the knowledge and enforcing your weakest areas.

The last one to two weeks before an exam is absolutely critical. You need to have the "do or die" mentality and be ready to study hard to pass the exam on your first try. That is how some people are able to pass an ARE exam with only one week of preparation.

17. **Timing of review: the 3016 rule; memorization methods, tips, suggestions, and mnemonics**

    Another important strategy is to review the material in a timely manner. Some people say that the best time to review material is between 30 minutes and 16 hours (the **3016** rule) after you read it for the first time. So, if you review the material right after you read it for the first time, the review may not be helpful.

    I have personally found this method extremely beneficial. The best way for me to memorize study materials is to review what I learned during the day again in the evening. This, of course, happens to fall within the timing range mentioned above.

    Now that you know the **3016** rule, you may want to schedule your review accordingly. For example, you may want to read new study materials in the morning and afternoon, then after dinner do an initial review of what you learned during the day.

OR
If you are working full time, you can read new study materials in the evening or at night and then get up early the next morning to spend one or two hours on an initial review of what you learned the night before.

The initial review and memorization will make your final review and memorization much easier.

Mnemonics is a very good way for you to memorize facts and data that are otherwise very hard to memorize. It is often arbitrary, or illogical but it works.

A good mnemonic can help you remember something for a long time or even a lifetime after reading it just once. Without mnemonics, you may read the same thing many times and still not be able to memorize it.

There are a few common types of mnemonics:
1) **Visual** mnemonics link what you want to memorize to a visual image.
2) **Spatial** mnemonics link what you want to memorize to a space, and the order of things in it.
3) **Group** mnemonics break up a difficult piece into several smaller and more manageable groups or sets, in which to memorize. One example is the grouping of the 10-digit phone number into three groups in the United States. This makes the number much easier to memorize.
4) **Architectural** mnemonics are a combination of visual mnemonics, spatial mnemonics and group mnemonics.

Imagine you are walking through a building several times, along the same path. You should be able to remember the order of each room. You can then break up the information that you want to remember and link this to several images, and then imagine you hang these images on walls of various rooms. You should be able to easily recall each group in an orderly manner by imagining you are walking through the building again on the same path and looking at the images hanging on walls of each room. When you look at the images on the wall, you can easily recall the related information.

You can use your home, office, or another building that you are familiar with to build an architectural mnemonic to help you to organize the things you need to memorize.

5) **Association** mnemonics allow you to associate what you want to memorize with a sentence, a similarly pronounced word, or a place you are familiar with, etc.
6) **Emotion** mnemonics use emotion to fix an image in your memory.
7) **First letter** mnemonics use the first letter of what you want to memorize to construct a sentence or acronym. For example, "**Roy G. Biv**" can be used to memorize the order of the 7 colors of the rainbow, it is composed of the first letter of each primary color.

You can use **association** mnemonics and memorize them as all the plumbing fixtures for a typical home, PLUS Urinal.

OR
You can use "Water S K U L" (**first letter** mnemonics) to memorize them:

Water Closets
Shower
Kitchen Sinks
Urinal
Lavatory

18. The importance of good and effective study methods
There is a saying: Give a man a fish, feed him for a day. Teach a man to fish, feed him for a lifetime. I think there is some truth to this. Similarly, it is better to teach someone HOW to study than just give him good study materials. In this book, I give you good study materials to save you time, but more importantly, I want to teach you effective study methods so that you can not only study and pass ARE exams, but also so that you will benefit throughout the rest of your life for anything else you need to study or achieve. For example, I give you samples of mnemonics, but I also teach you the more important thing: HOW to make mnemonics.

Often in the same class, all the students study almost the SAME materials, but there are some students that always manage to stay at the top of the class and get good grades on exams. Why? One very important factor is that they have good study methods.

Hard work is important, but it needs to be combined with effective study methods. I think people need to work hard AND work SMART to be successful at their work, career, or anything else they are pursuing.

19. The importance of repetition: read this book at least three times
Repetition is one of the most important tips for learning. That is why I have listed it under a separate title. For example, you should treat this book as part of the core study materials for the ARE exams and you need to read this book at least three times to get all of its benefits:

1) The first time you read it, it is new information. You should focus on understanding and digesting the materials, and also do an initial review with the **3016** rule.
2) The second time you read it, focus on reading the parts I have already highlighted AND you have highlighted (the important parts and the weakest parts for you).
3) The third time, focus on memorizing the information.

Remember the analogy of the memorization process as **filling a cup with a hole on the bottom**?
Do NOT stop reading this book until you pass the real exam.

20. The importance of a routine

A routine is very important for studying. You should try to set up a routine that works for you. First, look at how much time you have to prepare for the exam, and then adjust your current routine to include exam preparation. Once you set up the routine, stick with it.

For example, you can spend from 8:00 a.m. to 12:00 noon, and 1:00 p.m. to 5:00 p.m. on studying new materials, and 7:00 p.m. to 10:00 p.m. to do an initial review of what you learned during the daytime. Then, switch your study content to mock exams, memorization and retention when it gets close to the exam date. This way, you have eleven hours of exam preparation every day. You can probably pass an ARE exam in one week with this method. Just keep repeating it as a way to retain your architectural knowledge.

OR
You can spend 7:00 p.m. to 10:00 p.m. on studying new materials, and 6:00 a.m. to 7:00 a.m. to do an initial review of what you learned the evening before. This way, you have four hours for exam preparation every day. You can probably pass an ARE exam in two weeks with this preparation schedule.

A routine can help you to memorize important information because it makes it easier for you to concentrate and work with your body clock.

Do NOT become panicked and change your routine as the exam date gets closer. It will not help to change your routine and pull all-nighters right before the exam. In fact, if you pull an all-nighter the night before the exam, you may do much worse than you would have done if you kept your routine.

All-nighters or staying up late are not effective. For example, if you break your routine and stay up one-hour late, you will feel tired the next day. You may even have to sleep a few more hours the next day, adversely affecting your study regimen.

21. The importance of short, frequent breaks and physical exercise

Short, frequent breaks and physical exercise are VERY important for you, especially when you are spending a lot of time studying. They help relax your body and mind, making it much easier for you to concentrate when you study. They make you more efficient.

Take a five-minute break, such as a walk, at least once every two hours. Do at least thirty minutes of physical exercise every day.

If you feel tired and cannot concentrate, stop, go outside, and take a five-minute walk. You will feel much better when you come back.

You need your body and brain to work well to be effective with your studying. Take good care of them. You need them to be well-maintained and in excellent condition. You need to be able to count on them when you need them.

If you do not feel like studying, maybe you can start a little bit on your studies. Just casually read a few pages. Very soon, your body and mind will warm up and you will get into study mode.

Find a room where you will NOT be disturbed when you study. A good study environment is essential for concentration.

22. **A strong vision and a clear goal**
    You need to have a strong vision and a clear goal to master the architectural knowledge needed to become an architect in the shortest time. This is your number one priority. You need to master the architectural knowledge BEFORE you do sample questions or mock exams, except for "truth-finding" exams or exercises at the very beginning of your exam prep. This will make the process much easier. Everything we discuss is to help you achieve this goal.

    As I have mentioned on many occasions, and I say it one more time here because it is so important:

    The amount of architectural knowledge and information you can understand, digest, memorize, and firmly retain is what matters, not how many books you read or how many sample tests you have taken. The books and sample tests will NOT help you if you cannot understand, digest, memorize, and retain the important information for the ARE exam.

    Cherish your limited time and effort and focus on the most important information.

23. **Codes and standards used in this book**
    We use the following codes and standards:
    American Institute of Architects, Contract Documents, Washington, DC; ADA Standards for Accessible Design, ADA; Various International Codes by ICC. See Appendixes for more information.

24. **Where can I find study materials on architectural history?**
    Every ARE exam may have a few questions related to architectural history. The following are some helpful links to FREE study materials on the topic:
    http://issuu.com/motimar/docs/history_synopsis?viewMode=magazine

25. **Which sections of *Architectural Handbook of Professional Practice* (*AHPP*) should I read?**
    Architectural Handbook of Professional Practice (*AHPP*) is a very good resource for architectural practice, but it has over 1,000 pages, and can be overwhelming for people studying for the Architectural Registration Exams (ARE). To solve this problem, the publisher of *AHPP*, John Wiley & Sons, Inc., enlisted NarmourWright Design to set up a navigation document to help candidates studying for the ARE exams. Candidates for these exams can now just focus on reading the sections of the *AHPP* related to the division of

the ARE exam they are taking. Here is a link to the navigation documents for each division of the ARE exam: http://narmourwright.com/wp/index.php/wiley-publishing

Chapter Two

Project Planning & Design (PPD) Division

A. General Information

1. Exam content

The PPD division of the ARE has 120 questions which cover five different areas.

Sections	Target Percentage	Expected Number of Items
Section 1: Environmental Conditions & Context	10-16%	12-20
Section 2: Codes & Regulations	16-22%	20-27
Section 3: Building Systems, Materials, & Assemblies	19-25%	22-30
Section 4: Project Integration of Program & Systems	32-38%	38-46
Section 5: Project Costs & Budgeting	8-14%	9-17

Figure 2.1 Exam format & time

Note:
After December 13, 2020, the number of questions will be reduced. *See Figure 1.4.*

The exam content can be further broken down as follows:
Section 1: Environmental Conditions & Context (10-16%)

- Determine location of building and site improvements based on site analysis (A/E)
- Determine sustainable principles to apply to design (A/E)
- Determine impact of neighborhood context on the project design (U/A)

Section 2: Codes & Regulations (16-22%)

- Apply zoning and environmental regulations to site and building design (U/A)
- Apply building codes to building design (U/A)
- Integrate multiple codes to a project design (A/E)

Section 3: Building Systems, Materials, & Assemblies (19-25%)

- Determine mechanical, electrical, and plumbing (MEP) systems (A/E)
- Determine structural systems (A/E)
- Determine special systems such as acoustics, communications, lighting, security, conveying, and fire suppression (A/E)

- Determine materials and assemblies to meet programmatic, budgetary, and regulatory requirements (A/E)

Section 4: Project Integration of Program & Systems (32-38%)
- Determine building configuration (A/E)
- Integrate building systems in the project design (A/E)
- Integrate program requirements into a project design (A/E)
- Integrate environmental and contextual conditions in the project design (A/E)

Section 5: Project Costs & Budgeting (8-14%)
- Evaluate design alternatives based on the program (A/E)
- Perform cost evaluation (A/E)
- Evaluate cost considerations during the design process (A/E)

2. **Official exam guide and reference index for the Project Planning & Design (PPD) division**

NCARB published the exam guides for all ARE 5.0 division together as *ARE 5.0 Handbook*.

You need to read the official exam guide for the PPD division at least three times and become very familiar with it. The real exam is VERY similar to the sample questions in the handbook.

You can download the official *ARE 5.0 Handbook* at the following link:
https://www.ncarb.org/sites/default/files/ARE5-Handbook.pdf

Note: *We suggest you study the official ARE 5.0 Handbook first, and then other study materials, and then come back to Handbook again several days before the real ARE exam.*

B. The Most Important Documents/Publications for PPD Division of the ARE Exam

1. **Official NCARB list of formulas and references for the Project Planning & Design (PPD) division with our comments and suggestions**

You can find the NCARB list of references for this division in the Appendixes of this book and the *ARE 5.0 Handbook*.

The formulas will be available during the real exam. You should read through them a few times before the exam to become familiar with the content. This will save you a lot of time during the real exam, and will help you solve structural calculations and other problems.

Note:
While many of the MC questions in the real ARE exam ***focus on design concepts****, there are* ***some questions requiring calculations****. Therefore, it is absolutely necessary and critical*

for you to be very familiar with some of the basic and important equations, and to memorize them if possible. We have incorporated some of the most important equations into our PPD mock exam.

In the ARE exams, it may be a good idea to skip any calculation question that requires over 2 minutes of your time; just pick a guess answer, mark it, and come back to calculate it at the end. This way, you have more time to read and answer other easier questions correctly.

A calculation question that takes 20 minutes to answer will gain the same number of points as a simple question that ONLY takes 2 minutes.

If you spend 20 minutes on a calculation question earlier, you risk losing the time to read and answer ten other easier questions, which could result in a loss of ten points instead of one.

The following is the NCARB list of top references for this division. For a longer list of relevant reference materials, please see the reference matrix at the end of this book.

Publications
Architectural Graphic Standards
The American Institute of Architects
John Wiley & Sons, 11th edition (2007) and 12th edition (2016)

Building Construction Illustrated
Francis D. K. Ching
John Wiley & Sons, 6th edition, 2020

Heating, Cooling, Lighting: Sustainable Design Methods for Architects
Norbert Lechner
John Wiley & Sons, 4th edition, 2014

Mechanical & Electrical Equipment for Buildings (MEEB)
Walter T. Grondzik and Alison G. Kwok
John Wiley & Sons, 12th edition (2014)
and 13th edition (2019)

Note:
Many people think there are too many pages in NCARB's list of references to read. (about 4,500 pages in total)

My suggestion to you is to scan through all the NCARB listed materials, and focus on the parts that you are not familiar with. For example, if you are looking through MEEB, do not read the words, just look at the diagrams or figures. If you already understand them, then move on to the next page. If you do not understand, then stop and read the associated text. This will help you cut out many pages of reading from MEEB.

The following code and accessibility requirements have content covered in the Project Planning & Design division.

2010 ADA Standards for Accessible Design
U.S. Department of Justice, 2010

International Building Code (2018)
International Code Council, 2017

Focus on **Chapter 16**, particularly the sections on earthquakes and wind. Read them a few times, and have a general idea of the concepts. Do not force yourself to memorize all the details.

Read the following pages and become familiar with the Uniform and Concentrated Loads IBC table 1607.1:
***pg.* 285 & 286.**
OR
You can read the Uniform and Concentrated Loads IBC Table 1607.1 for FREE at the following link:
http://publicecodes.cyberregs.com/icod/ibc/2006f2/icod_ibc_2006f2_16_sec001.htm

Note:

- *The latest version of IBC is the 2018 version. For your convenience, we provide a link to the free online version of the IBC. The basic IBC content stays the same for many years. If the codes are updated by NCARB in the future, you just need to go to the root directory and find the latest version of the codes:*
  http://codes.iccsafe.org

- *You need to spend a large percentage (at least 20%) of your prep time on IBC Chapter 16 and become familiar with it. You do not need to force yourself to memorize the numbers and all the detail. Just reading it a few times and becoming familiar with the information should be adequate.*

AIA Contract Documents
None of the standard list of AIA Contract Documents related to the ARE have specific content covered in the Project Planning & Design division.

The following are some extra study materials that are useful if you have some additional time and want to learn more. If you are tight on time, you can simply look through them and focus on the sections that cover your weaknesses:

2. ***Manual of Steel Construction: Allowable Stress Design***; 9th Edition.
American Institute of Steel Construction, Inc. Chicago, Illinois, 1989

Look through the following pages and become familiar with the structural shapes, designations, dimensions, and properties:
***pg.* 1-9 thru 1-16, *pg.* 1-18 thru 1-32, *pg.* 1-40 thru 1-41, *pg.* 1-46 thru 1-52, *pg.* 1-94 thru 1-103.**

Make sure you understand what the designations stand for. For example, on pages 1-10 and 1-11, **W 40 x 298** means a W shape steel with a nominal depth of 40" (the actual depth is 39.69" per the Table on page 1-10), and a nominal weight of 298 lb. per ft. Use the diagram and Tables on pages 1-10 and 1-11 to look up the other detailed properties of a **W 40 x 298.** You do NOT need to remember any of these properties. You just need to know how to look them up and what they mean.

The most important information for an architect is the overall dimension of a structural member so that you can coordinate and make sure you have enough space to accommodate it. For example, you may need to find out if it will fit inside a wall or interstitial space.

You can also use the size of a structural member, the mechanical duct size, and the clearance space needed for a light fixture and/or fire sprinkler line to determine the interstitial height between floors.

Look through the following pages and become familiar with the beam nomenclature, diagrams, and formulas:
***pg.* 2-293 & 2-294, *pg.* 2-296, *pg.* 2-297, *pg.* 2-298, *pg.* 2-301, *pg.* 2-304, *pg.* 2-305.**

Look through the following pages and become familiar with bolts, threaded parts, and rivet tensions:
***pg.* 4-3 & 4-5.**

3. ***Steel Construction Manual***; 13th Edition. American Institute of Steel Construction, Inc. Chicago, Illinois, 2005
Read the following pages and become familiar with the round HSS dimensions and properties:
***pg.* 1-94 thru 1-98.**

4. **FREE information on truss and beam diagrams** can be found at the following link:
http://are.yuku.com/

5. **The FREE PDF file of FEMA publication number 454 (FEMA454),** *Designing for Earthquakes: A Manual for Architects*, is available at the following link:
http://are.yuku.com/

Note:

*You need to spend a large percentage (at least 30%) of your prep time on this **FEMA454** PDF file. Focus on **Chapters 4, 5, 8, and 9**. You may have **many real ARE questions** based on these chapters. This PDF book has many diagrams and photos and helps you understand what happened in buildings that failed during an earthquake.*

6. ***Wind Design Made Simple*** by ICC TRI-Chapter Uniform Code
   Note:
   You need to become familiar with this file. You do not need to force yourself to memorize the numbers and all the details, just reading it a few times and becoming familiar with the information should be adequate.

7. Arnold, Christopher. ***Building at Risk***, is available for FREE at the AIA website:
   http://www.aia.org/aiaucmp/groups/aia/documents/pdf/aiap016810.pdf

8. **Construction Specifications Institute (CSI) MasterFormat & *Building Construction***
   Become familiar with the new 6-digit CSI Construction Specifications Institute (CSI) MasterFormat as there may be a few questions based on this publication. Make sure you know which items/materials belong to which CSI MasterFormat specification section, and memorize the major section names and related numbers. For example, Division 9 is Finishes, and Division 5 is Metal, etc. Another one of my books, *Building Construction*, has detailed discussions on CSI MasterFormat specification sections.

Mnemonics for the 2018 CSI MasterFormat

The following is a good mnemonic, which relates to the 2018 CSI MasterFormat division names. Bold font signals the gaps in the numbering sequence.

This tool can save you lots of time: if you can remember the four sentences below, you can easily memorize the order of the 2018 CSI MasterFormat divisions. The number sequencing is a bit more difficult, but can be mastered if you remember the five bold words and numbers that are not sequential. Memorizing this material will not only help you in several divisions of the ARE, but also in real architectural practice

Mnemonics (pay attention to the underlined letters):
Good students can memorize material when teachers order. (divisions 1 to 8)
F students earn F's simply 'cause **forgetting** principles have **an** effect. (divisions 9 to 26)
C students **end** everyday understanding things without memorizing. (divisions 27 to 35)
Please make professional pollution prevention inventions **everyday**. (divisions 40 to 48)

0..Procurement and Contracting Requirements

1-Good.................................. General Requirements
2-Students............................. (Site) now Existing Conditions
3-Can......................................Concrete
4-Memorize............................Masonry
5-Material ....Metals

6-When................................... Woods, Plastics and Composites
7-Teachers.............................. Thermal and Moisture Protection
8-Order................................... Openings

9-F... Finishes
10-Students............................ Specialties
11-Earn.................................. Equipment
12-F's..................................... Furnishings
13-Simply............................... Special Construction
14-'Cause................................ Conveying Equipment
15... Plumbing + HVAC
16... Electrical + Lighting
20... Mechanical Support
21-Forgetting .... Fire Suppression
22-Principles.......................... Plumbing
23-Have................................ HVAC
25-An..................................... Integrated Automation
26-Effect............................... Electric

27-C...................................... Communication
28-Students............................ Electronic Safety & Security
31-End.................................. Earthwork
32-Everyday........................... Exterior Improvements
33-Understanding .... Utilities
34-Things............................... Transportation
35-Without Memorizing........ Waterways and Marine Construction

40-Please............................... Process Interconnections
41-Make................................ Material Processing and Handling Equipment
42-Professional...................... Process Heating, Cooling, and Drying Equipment
43-Pollution........................... Process Gas and Liquid Handling, Purification and Storage Equipment
44-Prevention......................... Pollution Control Equipment
45-Inventions......................... Industry-Specific Manufacturing Equipment
46... Water and Wastewater Equipment
48-Everyday......................... Electrical Power Generation

Note:
There are 50 CSI divisions beginning in 2004. The "missing" divisions are those “reserved for future expansion” by CSI, including divisions 17 to 19, 24, 29, 30, 36 to 39, 47 and 49. They are intentionally omitted from the list.

Chapter Three

Environmental Conditions & Context (10-16%)

NCARB Requirements
In this section, you'll use the site information gathered in the programming and analysis phase of a project to plan the site and environmental aspects of a project.

A. Determine location of building and site improvements based on site analysis (A/E) (4-8%)

NCARB Requirements
Based on views, wind, solar path, topography, adjacencies, planning concepts, and vulnerabilities and risks associated with climate change, you will need to locate a new built project on its site. This can include the orientation of one building, the arrangement of multiple buildings, the configuration of an addition to an existing building, or the layout of site improvements.

Architects must carefully evaluate a range of environmental and contextual factors when determining the optimal location for buildings and site improvements. This comprehensive site analysis forms the foundation for responsible, resilient, and sustainable design decisions. The process involves interpreting the site's natural and built conditions—including **climate, topography, hydrology, vegetation, surrounding development, infrastructure**, and **long-term sustainability goals**—to ensure that a proposed intervention supports both human use and environmental performance.

The goal of strategic building siting is to **maximize functionality, environmental performance, and climate resilience**, while **minimizing ecological disruption, infrastructure demand, and operational energy use**. This applies across project types—whether it's a single building, a large campus, or a renovation within an existing context.

Architectural Responsibilities Include:

- Assessing **views**, privacy, and visibility for key building functions
- Analyzing **prevailing wind** patterns for natural ventilation and comfort
- Optimizing **solar orientation** to support daylighting, passive heating, and PV systems
- Understanding **site slopes and natural drainage patterns**
- Addressing **adjacent land uses** and respecting spatial, acoustic, and visual relationships
- Planning for **circulation and access**, including ADA compliance and emergency vehicle routes
- Mitigating **noise, air pollution, and visual impacts** from surrounding infrastructure
- Identifying **climate risks** such as flooding, sea level rise, wildfire, drought, or extreme heat

This may apply to:

- **Single-building placement** on undeveloped or urban infill sites
- **Campus or multi-building site layouts**, such as schools, medical centers, or mixed-use developments
- **Additions or expansions** to existing buildings, including sensitive transitions and access points
- **Site improvements** such as parking, walkways, driveways, recreational space, stormwater features, and utility corridors

Key Considerations in Determining Building and Site Improvement Locations:

1. **Topography and Drainage**
   Topography—the shape and elevation of the land surface—plays a critical role in determining how and where buildings and site improvements should be located. A well-considered response to site elevation, slope, and natural drainage conditions can significantly reduce construction costs, mitigate environmental impacts, and improve long-term building performance and safety.

   Working with, rather than against, the land helps preserve ecological systems, reduce the need for retaining structures or excessive fill, and maintain effective **stormwater drainage**. Buildings should ideally be sited to integrate smoothly with existing contours and take advantage of natural landforms.

   Key Strategies:

   - **Site Structures to Minimize Grading and Earthwork Costs**
     Grading and excavation can be one of the largest hidden costs in a project. Minimizing cut and fill by placing buildings on flatter portions of the site or stepping the foundation to follow terrain helps preserve the site's natural integrity and reduce construction time, costs, and erosion risk.
   - **Avoid Steep Slopes and Place Buildings Along Contours When Feasible**
     Steep slopes pose challenges such as slope instability, limited accessibility, and complicated foundation systems. Where possible, align the building footprint parallel to existing contour lines. This minimizes slope disruption and supports easier integration of circulation paths, retaining walls, and outdoor spaces.
   - **Ensure Natural Drainage Patterns Are Maintained or Enhanced**
     Altering the natural flow of water can cause downstream flooding, erosion, or structural damage. Maintain or mimic original drainage patterns through thoughtful grading, vegetated swales, bioswales, rain gardens, or permeable paving. In areas prone to flooding, ensure finished floor elevations are set appropriately above expected stormwater levels.

   Design Implications:

   - **Foundation and Structural Considerations:**
     Sloped sites may require stepped footings, retaining walls, or pier foundations, all of which add complexity and cost. In some cases, split-level buildings or podium foundations are appropriate design solutions.

- **Accessibility and Circulation:**
  Ensure accessible routes (e.g., for ADA compliance) are feasible given the slope. Design pathways, ramps, and driveways to follow natural contours where possible to reduce grading and support ease of use.
- **Stormwater Management and Erosion Control:**
  Integrate sustainable drainage systems (e.g., French drains, infiltration basins, detention ponds) to slow runoff, promote infiltration, and reduce sediment loss. Employ erosion control measures such as stabilized slopes, mulch, or geotextiles during construction.
- **Landscape Integration:**
  Use retaining walls, terraced plantings, and contour-hugging design elements to blend built form with the landscape. This approach enhances aesthetics and ecological performance.

Examples:

- A **hillside residence in Northern California** uses a stepped floor plan that follows the site's natural slope. This minimizes excavation and blends the home into the terrain, while maintaining expansive views and stable foundation conditions.
- A **library in the Midwest** is sited at the high point of a gently sloping lot. The design uses natural swales and bioswales to direct rainwater away from the structure and into landscaped infiltration zones, reducing reliance on underground stormwater infrastructure.
- A **school campus in a flood-prone coastal area** elevates its structures above the base flood elevation. Classrooms are placed on a raised platform while playfields and parking areas are located in lower areas designed to accommodate temporary stormwater detention.

2. **Views and Visual Connections**

Designing with views in mind enhances both the **functionality** and **emotional experience** of a building. Visual connections to natural landscapes, urban features, or key architectural elements can increase the value of a property, improve user well-being, and strengthen spatial organization. A thoughtful approach to views includes both capturing **positive vistas** and **screening undesirable ones**, while balancing openness with privacy.

Views are a critical driver of building placement, window orientation, interior programming, and even landscape design. A successful site plan considers both long-distance vistas and close visual relationships to adjacent structures, pathways, and outdoor areas.

Key Strategies:

- **Frame Desirable Views (e.g., Mountains, Water Bodies, Urban Landmarks)**
  Orient windows, balconies, patios, and shared indoor spaces toward valuable natural or cultural assets. Use architecture—such as window framing, cutouts, or cantilevers—to focus and emphasize views. Design circulation spaces like hallways or staircases to reveal views incrementally, enhancing the spatial experience.
- **Position Private vs. Public Spaces Based on View Access**
  Assign spaces that benefit most from views—such as living rooms, conference areas, or lobbies—to the side of the building with the strongest visual access. More private or utilitarian spaces (e.g., restrooms, storage, mechanical rooms) can face less desirable views or be located along blank walls.

- **Use Landscape or Architectural Elements to Screen Unsightly Views**
  If the site faces highways, loading docks, neighboring buildings, or other visual detractors, introduce screening elements. These can include berms, trees, green walls, privacy fences, or translucent architectural panels. Screening should be integrated into the design language and not feel like an afterthought.

Design Implications:

- **Building Orientation and Massing:**
  The orientation of a structure may be dictated as much by visual access as by sun or wind. In high-value settings—like oceanfront or mountain-facing sites—view corridors can be prioritized over other siting considerations.
- **Window Placement and Sizing:**
  Maximize window openings where desirable views are available, using floor-to-ceiling glazing, clerestories, or corner windows. Conversely, minimize or obscure glazing on elevations facing less attractive elements.
- **Landscape Design:**
  Trees, plantings, and topographic changes can help focus attention on positive features while buffering the site from undesirable surroundings. For example, a row of evergreens might block the view of a parking lot, while a water feature draws the eye inward.
- **Zoning and Neighboring Context:**
  Understand and respond to zoning restrictions that protect view corridors, especially in dense urban or scenic areas. Also consider the visual privacy of neighbors—minimizing direct sightlines into adjacent homes or yards can improve relationships and property value.

Examples:

- A **residential tower in Seattle** is designed with units stacked to face Puget Sound and Mount Rainier. Large balconies and angled glazing enhance view exposure while minimizing the footprint's impact on surrounding buildings' views.
- A **visitor center near a national park** uses a long, horizontal window band to frame the mountain ridgeline. Interior seating areas, exhibit spaces, and circulation paths are aligned along this visual axis to maintain a constant connection to the landscape.
- A **corporate campus in an industrial zone** positions offices inward toward a landscaped courtyard with water features and sculptural plantings. Berms and vegetation are used along the property perimeter to screen surrounding warehouses and roads.

3. **Climate Resilience & Environmental Risks**
   Climate resilience is an increasingly critical factor in site selection and building design. As climate change leads to more extreme weather events, rising sea levels, and shifting environmental conditions, architects and planners must evaluate both **current vulnerabilities** and **future risks** when locating buildings and site improvements. The goal is to design for long-term safety, functionality, and environmental performance—even as climate patterns evolve.

   Siting decisions must avoid high-risk zones and incorporate data-driven strategies to **anticipate, mitigate, and adapt to environmental hazards**, ensuring that buildings remain habitable, insurable, and maintainable over their expected life span.

Key Strategies:

- **Avoid Siting in Floodplains, Wildfire-Prone Zones, or Areas Vulnerable to Sea Level Rise**
  Use caution when considering development in hazard-prone areas. In many cases, local regulations prohibit or strictly limit building in designated floodplains or fire zones. Even if construction is allowed, such siting can increase risk to life, property, and public infrastructure—and lead to long-term costs.
- **Use FEMA Flood Maps, Wildfire Risk Data, and Climate Forecasts**
  Reference authoritative data sources during site selection and planning:
  - **FEMA Flood Insurance Rate Maps (FIRMs)** for current and future floodplain delineation: You can use Fema.com to download flood map for your project location. Figure 3.1 is a FEMA Flood Insurance Rate Maps (FIRMs) for the White House.
  - **USDA Forest Service and CAL FIRE data** for wildfire risk zones
  - **NOAA, IPCC reports, and local climate adaptation plans** for long-range weather projections

  These tools can inform zoning setbacks, foundation elevation, materials, and access planning.
  Note: *NOAA stands for the* ***National Oceanic and Atmospheric Administration****. IPCC stands for the* ***Intergovernmental Panel on Climate Change****.*
- **Consider Future Risks from Increased Storm Intensity or Drought**
  Plan for intensifying rainfall, stronger hurricanes, longer droughts, and temperature extremes. This may influence stormwater capacity, foundation design, backup power requirements, and landscape choices (e.g., xeriscaping in arid zones). Even historically "safe" areas may face new risks over the next few decades.

Design Implications:

- **Foundation and Elevation Design:**
  In flood-prone areas, buildings may require raised foundations, pier supports, or flood vents. Critical systems (electrical panels, HVAC) should be elevated above base flood elevations.
- **Fire-Resistant Design:**
  In wildfire zones, specify ignition-resistant materials (e.g., metal roofing, fiber-cement siding), create defensible space around structures, and include ember-resistant vents and roofing assemblies.
- **Stormwater Management:**
  Oversize retention basins, bioswales, and permeable surfaces to accommodate heavier precipitation events. Green infrastructure can slow and filter runoff while adding ecological value.
- **Drought and Heat Adaptation:**
  Design drought-tolerant landscapes, collect rainwater, and minimize turf or water-dependent plantings. Use reflective roofing, shading devices, and ventilation strategies to mitigate extreme heat impacts.
- **Insurance, Permitting, and Code Compliance:**
  Building in high-risk zones often affects permitting timelines, insurance premiums, and

code requirements. Many jurisdictions have adopted resilience-focused building codes or overlay districts to reduce future disaster losses.

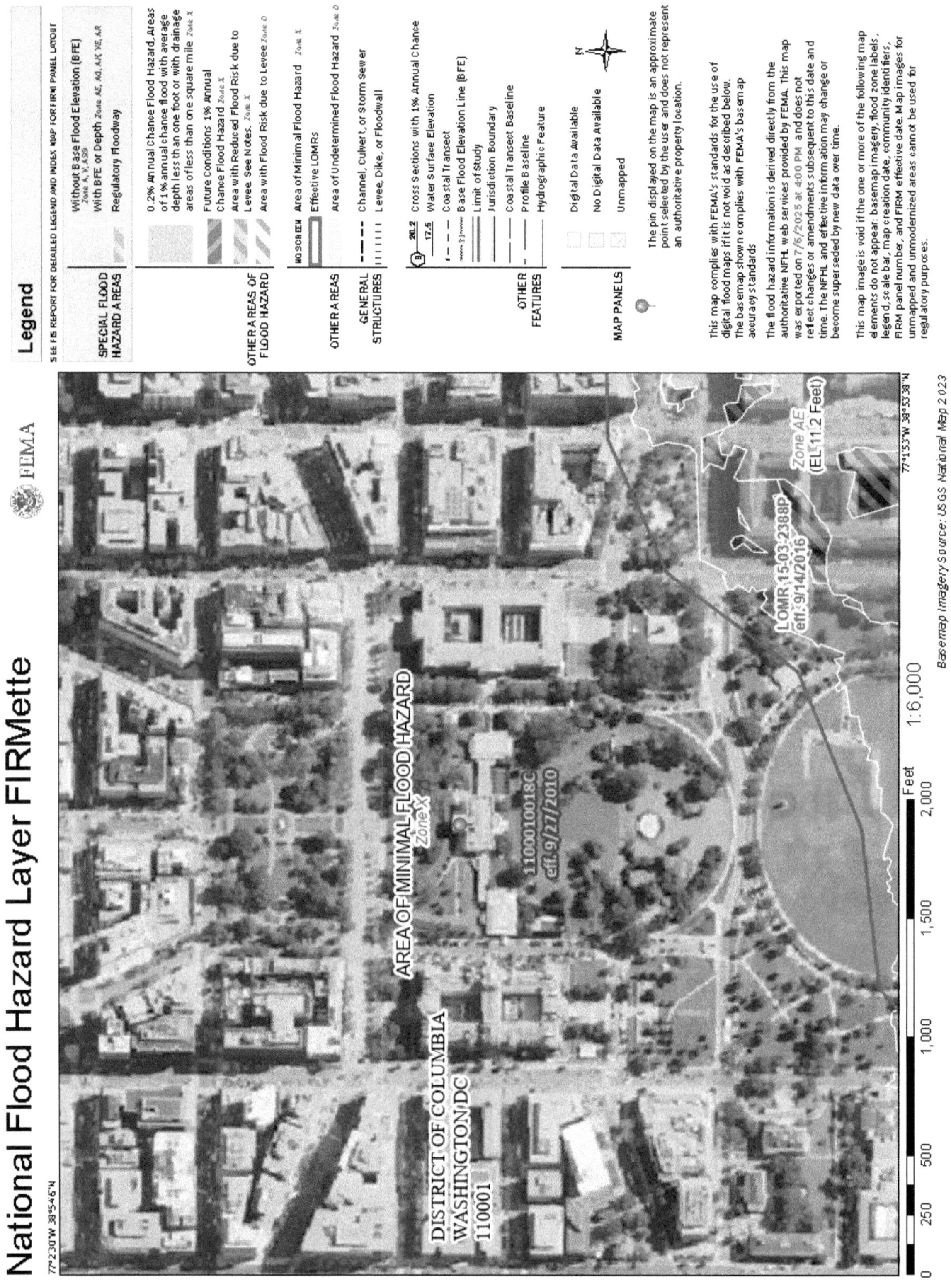

Figure 3.1 FEMA Flood Insurance Rate Maps (FIRMs) for the White House

Examples:

- A **coastal municipal building in Florida** is raised on concrete piers above the base flood elevation. All mechanical systems are located on the second floor, and the site includes bioswales and permeable pavers to manage runoff during storm surges.
- A **residential community in California's wildland-urban interface (WUI)** features fire-resistant siding, metal roofs, and a 100-foot defensible zone maintained by local vegetation management teams. Water tanks and fire hydrants are integrated into the site plan.
- A **new civic plaza in Phoenix** uses drought-tolerant native landscaping, passive shading structures, and high-albedo paving materials to reduce the urban heat island effect while remaining adaptable to rising temperatures and water scarcity.

4. **Contextual Adjacencies and Land Use Compatibility**

Successful site planning requires a careful analysis of **adjacent land uses**, **property lines**, and **community context** to ensure that new development fits sensitively within its surroundings. Good adjacency planning protects neighbors from negative impacts, fosters positive relationships, and can even enhance opportunities for **shared infrastructure**, **joint-use facilities**, and **mutual value**.

Placing buildings and site elements in harmony with existing and planned land uses helps avoid conflicts related to **privacy, noise, traffic, and scale** while supporting a sense of place, safety, and cohesion within the built environment.

Key Strategies:

- **Locate Buildings in Harmony with Neighboring Uses (e.g., Buffer Zones Between Residential and Commercial)**
  Consider compatibility of scale, function, and hours of activity. For instance, buffer residential areas from noisy or high-traffic commercial spaces with landscape screening, setbacks, or transitional land uses like office buildings or parks.
- **Respect Sight Lines, Setbacks, and Privacy for Adjacent Properties**
  Building massing, window placement, and roof terraces should be designed to minimize direct views into neighboring homes or yards. Honor zoning setbacks and height limits to avoid overbearing conditions. Fencing, planting, or changes in elevation can also enhance separation.
- **Consider Opportunities for Synergy or Shared Services**
  In compatible contexts, look for opportunities to share amenities or infrastructure—such as joint parking lots, loading docks, green spaces, or stormwater systems. Shared-use agreements between institutions or businesses can reduce land consumption and promote a more walkable environment.

Design Implications:

- **Zoning and Land Use Analysis:**
  Review zoning maps and comprehensive plans to understand existing and future land use designations. Projects may require zoning variances or conditional use permits if compatibility is in question.
- **Urban vs. Suburban Contexts:**
  In dense urban settings, adjacency may involve lot-line construction and shared walls,

requiring fire-rated assemblies and attention to noise transfer. In suburban or rural areas, greater setbacks and vegetative buffers are often expected.

- **Noise and Light Mitigation:**
  When siting near sensitive land uses (e.g., schools, hospitals, housing), design for reduced sound transmission and light spillover. Locate mechanical equipment away from property lines, and shield outdoor lighting to minimize glare.
- **Streetscape and Access Coordination:**
  Align driveways, crosswalks, and entrances with neighboring developments for visual consistency and safe circulation. Consider public realm impacts—such as pedestrian activity, visibility, and street presence—when placing entry points or landscape elements.

Examples:

- A **mixed-use development** includes retail on the street level with residential units above. A landscaped courtyard at the rear provides a buffer between the building and an adjacent single-family neighborhood, reducing noise and enhancing privacy.
- A **school campus expansion** places its new gymnasium adjacent to a neighboring public park, allowing shared use of athletic fields and parking through a municipal partnership.
- A **medical office complex** uses dense hedging, fencing, and a gradual building setback to shield a quiet residential neighborhood from after-hours traffic and service vehicle access, while aligning its main entrance with nearby transit stops to reduce car dependency.

5. **Access, Circulation, and Transportation**

Site access and circulation are foundational elements of a functional, safe, and sustainable development. Whether the project is a single building or a multi-use campus, successful site design must accommodate the **safe and efficient movement of people and vehicles**, while also supporting public transportation, accessibility, and emergency response needs.

Thoughtful circulation planning reduces **congestion**, improves **user experience**, and contributes to **long-term sustainability** by encouraging non-vehicular modes of transportation and reducing infrastructure strain.

Key Strategies:

- **Ensure Efficient and Safe Access for Vehicles, Pedestrians, and Bicycles**
  Organize site circulation to separate pedestrian pathways from vehicle routes where possible. Provide safe, ADA-compliant sidewalks, crosswalks, and bike lanes. Site design should reduce conflict points between different types of users, such as pedestrians crossing busy driveways or loading zones.
- **Maintain Clear Emergency Access Routes and Fire Lane Requirements**
  Fire lanes and emergency vehicle access routes must comply with local fire code requirements, typically providing unobstructed 20-foot-wide paths around buildings. Avoid obstructions such as landscaping, bollards, or parked vehicles in designated areas. Ensure turning radii and clearances are appropriate for fire trucks and ambulances.
- **Place Entries Near Transit Stops or Pedestrian Corridors When Possible**
  Entrances should be oriented toward transit stops, sidewalks, or pedestrian-friendly routes to reduce car dependency and enhance connectivity. This promotes walkability and aligns with principles of transit-oriented development (TOD), especially in urban or mixed-use areas.

Design Implications:

- **Site Layout and Driveways:**
  Driveways should be minimized and strategically placed to prevent traffic congestion and promote safety. Consider drop-off zones, delivery areas, and ride-share staging locations separate from main pedestrian entrances.
- **Parking and Bicycle Facilities:**
  Provide adequate and well-located vehicle and bicycle parking based on zoning requirements and user demand. Include covered bike racks, EV charging stations, and designated accessible parking spaces near primary entrances.
- **Pedestrian Safety and Comfort:**
  Design walkways with appropriate width, lighting, shade, and visibility. Use tactile indicators and curb ramps for ADA compliance. Buffer sidewalks from vehicle areas using landscaping, planters, or street furniture.
- **Transit Integration:**
  If the site is near a transit stop, coordinate with local transit agencies to align service routes, provide shelters or seating, and design pathways that support safe, direct access to and from the stop.
- **Wayfinding and Visibility:**
  Clear signage, lighting, and sight lines help users navigate the site efficiently and safely—especially for visitors, emergency responders, and delivery services.

Examples:

- A **community health clinic** includes a covered drop-off zone at the main entrance, a clearly marked fire lane around the building, and wide sidewalks that connect directly to a nearby bus stop and neighborhood trail system.
- A **university campus expansion** aligns its main pedestrian spine with an existing bike lane and transit corridor, promoting multimodal access. Car traffic is limited to the campus perimeter, with interior areas reserved for pedestrians and service vehicles.
- A **mixed-use development in a dense urban area** incorporates below-grade parking with elevator access near the street entrance, while the ground level is activated by wide sidewalks, shaded seating, and protected bike racks adjacent to a light rail station.

6. **Noise and Air Quality Buffers**

Protecting building occupants from excessive **noise pollution** and **poor air quality** is essential for ensuring health, comfort, and long-term building performance. In many sites—especially those near highways, rail lines, airports, or industrial zones—designers must take proactive steps to **minimize exposure to unwanted environmental impacts** through smart siting, landscaping, and architectural interventions.

These strategies are particularly important for sensitive uses such as **residences, schools, healthcare facilities**, and outdoor gathering spaces, where quiet, clean air is critical to user well-being.

Key Strategies:

- **Distance Buildings from Sources of Noise (e.g., Highways, Rail Lines, Industrial Areas)**
  Physical separation is the most effective method to reduce sound and airborne pollutants. Increase setback distances from major noise or emission sources whenever possible. In urban infill projects where space is limited, additional layers of mitigation may be required.
- **Use Berms, Trees, or Walls as Acoustic Buffers**
  Vegetated berms, dense hedgerows, sound walls, and site walls can reduce sound transmission from nearby roads or facilities. Multi-layered planting schemes can absorb and deflect sound while also filtering dust and particulates. Structural walls (e.g., masonry or concrete) are more effective for lower-frequency industrial or rail noise.
- **Consider Prevailing Wind Direction for Odor or Pollutant Control**
  Analyze wind patterns to avoid placing buildings downwind of pollution sources such as exhaust vents, gas stations, or waste facilities. Orient air intakes and outdoor air vents away from these areas, and use filters or positive air pressure where necessary.

Design Implications:

- **Site Layout and Building Orientation:**
  Place noise-sensitive spaces (e.g., bedrooms, classrooms) on the side of the building farthest from the noise source. Use less-sensitive spaces like corridors, bathrooms, stairwells, or storage areas as "buffer zones" between the noise and occupied areas.
- **Landscape and Terrain Features:**
  Incorporate earth berms, tree belts, green roofs, and vegetated screens to reduce particulate matter and lower ambient noise levels. Even modest elevation changes can improve air quality by promoting dispersion of ground-level pollutants.
- **Facade and Envelope Design:**
  Specify high-performance acoustic-rated glazing, insulated walls, and sealed openings to block airborne sound. Consider triple glazing and acoustic louvers in particularly noisy environments. Avoid operable windows on the noise-facing side where possible.
- **Mechanical System Placement:**
  Locate HVAC intakes, vents, and air filtration systems strategically to reduce intake of outdoor pollutants or odors. Use enhanced filtration (e.g., MERV 13 or higher) in buildings near high-traffic or industrial zones.

Examples:

- A **new elementary school near a busy expressway** is sited with a 200-foot landscaped setback that includes mounded earth berms and three rows of evergreen trees. The classroom wing is placed on the quiet side of the building, while administrative and utility spaces face the roadway.
- An **affordable housing complex adjacent to a light rail line** uses double-layered concrete walls with sound-absorbing insulation. The site design incorporates a landscaped courtyard in the center, creating a quiet communal space shielded from external noise.
- A **healthcare facility near an industrial zone** orients its air intakes and operable windows away from prevailing downwind directions. Rooftop equipment is protected by acoustic screens, and all indoor air is filtered using HEPA systems.

7. **Zoning and Regulatory Setbacks**
   - Comply with required front, rear, and side setbacks.
   - Respect height restrictions, easements, and open space requirements.
   - Use zoning overlays to guide land use and density.

Zoning codes and land use regulations define the legal parameters for how land can be used and developed. Understanding and applying **zoning and regulatory setbacks** is essential during site planning to ensure **code compliance**, protect **adjacent property rights**, and maintain neighborhood **scale and character**.

Setbacks, height limits, and open space requirements directly influence the **footprint, massing, and orientation** of new buildings. In some areas, zoning overlays or special districts introduce additional design guidelines or restrictions to achieve specific planning goals.

Key Strategies:

- **Comply with Required Front, Rear, and Side Setbacks**
  Setbacks define the minimum required distance between a structure and property lines, sidewalks, streets, or adjacent buildings. These ensure light, air, privacy, and fire separation between neighboring properties. Setbacks vary by zoning district and may also apply differently to accessory structures (e.g., garages or sheds).
- **Respect Height Restrictions, Easements, and Open Space Requirements**
  Maximum height limits help control building scale, preserve views, and reduce shadow impacts. Easements—such as those for utilities, drainage, or access—must remain unbuilt or accessible. Open space requirements may dictate a percentage of the site to remain landscaped or permeable, especially in residential or environmentally sensitive zones.
- **Use Zoning Overlays to Guide Land Use and Density**
  Zoning overlays may add special provisions to base zoning regulations. These can relate to historical preservation, transit-oriented development (TOD), affordable housing incentives, or environmental protection. Overlay districts may influence building height, floor area ratio (FAR), use types, parking minimums, or façade design.

Design Implications:

- **Building Envelope and Massing:**
  Setbacks and height limits define a "buildable envelope" within which the entire structure must fit. Design teams must explore creative massing solutions—such as step-backs, courtyard forms, or sunken levels—to work within these legal constraints while achieving design goals.
- **Site Utilization:**
  Setbacks and open space rules can significantly reduce the usable area of a parcel, especially on small or irregular lots. Efficient space planning and compact footprints become critical in urban infill projects.
- **Permit and Entitlement Strategy:**
  Projects that exceed height limits, encroach into setbacks, or deviate from use classifications may require variances, conditional use permits (CUPs), or zoning amendments. These can add time, cost, and risk to the project approval process.

- **Contextual Fit:**
  Complying with zoning helps new development align with the scale and rhythm of the surrounding neighborhood. For example, matching street-facing setbacks and rooflines in a historic district supports visual continuity and helps reduce opposition from community stakeholders.

Examples:

- A **multifamily housing project** on a corner lot in a medium-density zone must maintain a 20-foot front setback and a 10-foot rear setback, limiting how far the building can extend toward the street or property line. The design incorporates a stepped courtyard along the side setback to meet open space requirements and provide tenant amenities.
- A **mixed-use development** in a downtown overlay zone benefits from relaxed parking requirements and additional building height in exchange for including ground-floor retail and affordable housing units. This zoning incentive allows for a taller, more active building than otherwise permitted in the base zone.
- A **new commercial building** on a suburban arterial must avoid building over a 12-foot utility easement at the rear of the property. Designers shift the structure forward and add landscaping within the easement to satisfy open space requirements without losing buildable area.

8. **Infrastructure and Utility Access**

Access to utilities such as **water, sewer, electricity, gas, and telecommunications** is a critical factor in site planning and development feasibility. Early coordination with utility providers and consideration of infrastructure layout can greatly influence both **initial construction costs** and **long-term operational efficiency**.

Strategic placement of buildings, site improvements, and utility routes helps avoid costly trenching, reduces environmental disruption, and ensures code-compliant connections to essential services.

Key Strategies:

- **Consider Proximity to Water, Sewer, Power, and Telecom Service Connections**
  Assess the site's access to public utility infrastructure during the feasibility and schematic design phases. Sites lacking nearby service connections may require off-site improvements or easements, increasing both cost and approval time.
- **Site Structures to Reduce Trenching Distances and Utility Costs**
  Orient buildings and service entrances to align efficiently with nearby utility sources. Shorter service runs reduce the need for long trenching through paved or landscaped areas and limit site disturbance. Grouping utilities in shared corridors can simplify installation and maintenance.
- **Coordinate with Existing Utility Easements and Rights-of-Way**
  Utility easements must be left unbuilt and accessible for maintenance. These may affect where buildings, parking areas, fences, or trees can be located. Work with civil engineers and surveyors to identify and avoid conflicts with underground or overhead utilities early in design.

Design Implications:

- **Foundation and Site Grading Conflicts:**
  Foundations should be located clear of major utility lines. Deep trenching or infrastructure crossings under foundations may trigger costly special construction (e.g., sleeving, encasement, or rerouting).
- **Access and Maintenance Requirements:**
  Utilities—especially stormwater management systems, electrical transformers, and telecom hubs—require clear and often driveable access for maintenance vehicles. Allow adequate clearance, surfacing, and turning radii near these features.
- **Coordination with Civil and MEP Engineers:**
  Early collaboration with the civil and MEP (mechanical, electrical, plumbing) teams helps define efficient routes, connection points, and metering locations. This avoids late-stage conflicts between building placement and utility runs.
- **Cost and Permitting Impacts:**
  Long or complex utility extensions can trigger developer-funded improvements, especially in undeveloped or rural areas. These costs may affect feasibility. In some jurisdictions, utility availability is tied to permitting or development capacity limits (e.g., sewer moratoriums).

Examples:

- A **commercial office project** is sited near an existing water main and underground power line. The main electrical room is placed adjacent to the service entry to minimize conduit runs and avoid routing beneath the slab. A telecom pull box is included at the driveway edge for future fiber access.
- A **multi-building housing development** uses a central utility corridor beneath a landscaped pedestrian spine. This shared trench reduces the overall disturbance area and simplifies access for maintenance. Utility easements are coordinated with open space and walking paths.
- A **warehouse facility** must avoid building over a large underground storm drain easement. Designers shift the structure slightly and use the easement zone for surface parking, which maintains required clearances and makes efficient use of non-buildable land.

9. Cultural, Historic, or Environmental Preservation

Site planning must account for **existing natural, historical, and cultural features** that hold legal, ecological, or community value. Sensitive or significant features—such as wetlands, archaeological sites, old-growth trees, or historic structures—should be **preserved, respected, or thoughtfully integrated** into the design. Failing to do so can trigger legal challenges, community opposition, or costly project delays.

Beyond compliance, honoring a site's **natural heritage and cultural narrative** strengthens its connection to place, enhances sustainability, and builds trust with surrounding communities.

Key Strategies:

- **Avoid Disrupting Archaeological Sites, Tree Groves, Wetlands, or Historic Resources**
  Early site surveys should identify protected features and evaluate their significance under local, state, or federal criteria. Projects should avoid or minimize impacts by relocating

buildings, reducing footprints, or designing around sensitive zones. In some cases, preservation is mandatory, such as with designated historic landmarks or federally protected wetlands.

- **Integrate Buildings with the Cultural Identity or History of the Site**
  Consider the site's social, architectural, or ecological legacy in the design concept. Use materials, forms, or landscape elements that reflect the local vernacular or honor historical uses. This adds authenticity and increases community support for the project.
- **Follow Local, State, or Federal Environmental Review Processes (e.g., NEPA, CEQA)**
  Projects in certain jurisdictions or on public land may require formal environmental review to assess and disclose impacts on historical, cultural, or ecological resources. These processes—such as the **National Environmental Policy Act (NEPA)** or **California Environmental Quality Act (CEQA)**—can affect project scope, timeline, and approval pathways.

Design Implications:

- **Site Planning and Massing Adjustments:**
  Preservation may require reconfiguring site layouts to leave heritage trees, wetlands, or building foundations undisturbed. This can affect circulation routes, utility locations, and building orientation.
- **Adaptive Reuse Opportunities:**
  Where historic buildings are present, adaptive reuse may be more cost-effective—and publicly supported—than demolition. This approach also contributes to sustainability goals through embodied carbon savings.
- **Landscape and View Corridor Integration:**
  Viewsheds to or from historic structures or natural features (e.g., a grove of oaks or a native burial mound) may need to be preserved or framed. Landscape design can create respectful transitions between new and old elements.
- **Permitting and Documentation Requirements:**
  Documentation, mitigation plans, and consultation with preservation agencies or tribal authorities may be required. This could include historic resource surveys, cultural assessments, or environmental impact reports (EIRs).

Examples:

- A **university expansion** includes an open-air plaza that preserves the footprint of a 19th-century trading post uncovered during excavation. Interpretive signage and salvaged foundation stones are incorporated into the landscape design to tell the site's story.
- A **residential development** is redesigned to preserve a protected grove of native oak trees by clustering buildings on the site's flatter, previously disturbed portion. The trees become a central natural amenity and contribute to stormwater management.
- A **visitor center in a national park** is designed with low-profile rooflines, natural stone cladding, and timber detailing to harmonize with the surrounding historic district. The project undergoes a NEPA environmental assessment to ensure protection of wildlife corridors and tribal cultural resources.

10. Future Expansion and Flexibility

Planning for **future growth and adaptability** is a core principle of resilient and sustainable site development. Whether for a single facility or a multi-phase campus, anticipating how a site may evolve over time allows for smoother expansions, reduced future costs, and continued functionality as user needs change.

By integrating flexibility into the layout of buildings, infrastructure, and circulation systems, designers can support long-term value while avoiding expensive retrofits or disruptive site overhauls.

Key Strategies:

- **Reserve Areas for Future Building Additions, Parking Expansion, or Open Space**
  Designate and protect zones on the site for future construction, whether for additional floors, wings, auxiliary structures, or parking. Avoid placing permanent landscaping, utilities, or structures in areas that may be developed later.
- **Design Circulation and Infrastructure with Scalability in Mind**
  Size utilities, drives, and service routes to accommodate future phases. Leave space for future utility stubs, duct banks, or road extensions. Plan emergency vehicle access and service zones so that expansion won't compromise life-safety systems or maintenance access.
- **Orient Structures and Improvements to Allow Phased Development**
  Arrange buildings, entrances, and site features in a way that facilitates logical, cost-effective growth. Use modular design principles and "soft edges" where future connections or additions might be made.

Design Implications:

- **Massing and Structural Systems:**
  Structural grids, floor plates, and circulation cores should be designed to support future horizontal or vertical expansion. For example, elevator and stair cores can be sized and located to serve an eventual addition.
- **Access and Utility Infrastructure:**
  Provide oversized utility lines or conduit sleeves to accommodate future loads. Design roads and walkways to connect logically to areas of anticipated growth, avoiding dead ends or redundant infrastructure later.
- **Zoning and Entitlements:**
  Master planning for future expansion may require special zoning approvals or development agreements that allow phased implementation. It's important to ensure that initial phases don't exhaust allowed FAR, impervious coverage, or parking minimums.
- **Landscaping and Environmental Impact:**
  Temporary landscaping or permeable open space can be used in areas planned for future development, maintaining site aesthetics while keeping options open. Care should be taken to avoid excessive site disturbance twice.

Examples:

- A **medical clinic** is designed with knock-out panels on one wing to allow a future imaging suite to be added without modifying the existing structure. Utility stubs and extra conduit runs are included in the initial build.
- A **K-12 school campus** completes Phase 1 with core facilities (gym, cafeteria, admin, and 12 classrooms), while grading and infrastructure are already extended to a vacant zone for a future classroom wing as enrollment grows.
- A **tech company headquarters** includes a parking lot built with removable permeable pavers in an area slated for future R&D space. Utility easements and stormwater systems are pre-sized for the ultimate build-out.

B. Determine sustainable principles to apply to design (A/E) (2-6%)

NCARB Requirements

An architect must use sun, wind, temperature, precipitation, and other climatic data to determine building orientation and shape, building envelope design, site features, and active and/or passive building systems. Sustainable principles also include considering recycled content, identifying opportunities to minimize waste when selecting building materials, and other strategies to increase the project's resilience and reduce its negative impact on the environment.

As part of the site design and planning process, architects must determine the most appropriate location for a building or site improvement based on a comprehensive site analysis. This task involves synthesizing various factors including environmental, spatial, social, and regulatory influences.

When determining location, consider:

- **Views**
- **Prevailing wind direction**
- **Solar orientation and path**
- **Topography**
- **Site adjacencies and access**
- **Urban or rural planning concepts**
- **Climate-related vulnerabilities and risks (e.g., flooding, wildfire)**
- **Existing site infrastructure**
- **User needs and programmatic requirements**

This applies whether you're orienting a single structure, arranging multiple buildings, designing an addition to an existing facility, or planning external site improvements such as parking, landscaping, or circulation paths.

Key Considerations & Examples

1. Solar Orientation

Solar orientation is one of the most impactful and cost-effective strategies in sustainable architecture. It refers to positioning and shaping buildings in accordance with the sun's daily and seasonal movements to maximize energy efficiency, thermal comfort, daylighting, and photovoltaic potential. Proper orientation can significantly reduce heating, cooling, and lighting loads while enhancing occupant well-being.

Core Objectives:

- **Maximize Passive Solar Heating, Daylighting, and Photovoltaic Potential**
  Orient structures and fenestration to harness the sun for thermal and electrical energy. South-facing façades are ideal for passive heating and solar panels. Passive design features like thermal mass, insulated glazing, and sunspaces can lower heating demands in cooler climates.
- **Orient Major Glazing Toward the South (Northern Hemisphere)**
  South-facing windows receive reliable sunlight year-round. In winter, low-angle sun warms interior spaces; in summer, roof overhangs and shading elements prevent overheating. This consistent solar access also supports indoor plantings, learning environments, and health-focused design (e.g., biophilia and circadian rhythm support).
- **Minimize East/West-Facing Glass to Reduce Glare and Overheating**
  East and west façades experience harsh, low-angle sunlight that causes glare, heat gain, and increased cooling loads. Limit glazing here, or mitigate impacts with vertical louvers, reflective glazing, deciduous trees, or deep recesses.

Design Implications:

- **Building Massing and Layout**
  The most effective passive solar designs use an **east-west building orientation**, with primary façades facing north and south. This allows for strategic placement of occupied spaces along the south, mechanical and storage spaces along the north, and deep floor plates that still allow daylight penetration.
- **Shading Devices**
  Use **horizontal shading** (e.g., overhangs, brise-soleil) for south-facing windows and **vertical shading** (e.g., fins, screens, exterior louvers) for east and west exposures. Consider movable or operable shading systems to adapt to seasonal sun angles and occupant preferences.
- **Photovoltaic Integration**
  Solar panels should be installed at the optimal tilt and orientation for the geographic location. Roof design must accommodate unobstructed solar access and account for potential shading from trees, equipment, or adjacent buildings. Coordination with structural systems and mechanical equipment zones is essential.
- **Daylighting Strategy**
  Proper orientation allows for daylight harvesting systems that reduce artificial lighting demand. Use clerestory windows, light shelves, and high-reflectance finishes to bring natural light deep into floor plates. Well-designed daylighting enhances occupant comfort, productivity, and well-being.

Examples:

- A **net-zero energy school** is designed with a long, narrow floor plan aligned on an east-west axis. Classrooms face south with operable clerestory windows and deep overhangs to optimize daylighting and natural ventilation. Rooftop photovoltaic panels on a sloped south-facing roof meet the building's entire electrical demand.

- A **multifamily housing development** in a temperate climate uses a courtyard arrangement that opens to the south. Living rooms and balconies are positioned for solar access, while stairwells and corridors line the north side to reduce conditioned space and glazing on less effective façades.
- A single-family residential home in a cold climate is oriented with its long axis running **east–west**, so that the **major living spaces** face **south**. Large, triple-glazed south-facing windows allow for passive solar heat gain in the winter. A properly sized roof **overhang** prevents overheating during the summer months by blocking the higher-angle sun. On the north side, smaller windows reduce heat loss and exposure to cold prevailing winds.
- An **urban office building** with limited footprint rotates its core functions to the north side and uses a glass curtain wall on the south, with dynamic shading systems and solar panels integrated into the façade. This improves daylight use while reducing glare and solar gain during peak hours.
- In a **desert climate**, the building is oriented to minimize direct solar exposure and uses shaded courtyards to stay cool.

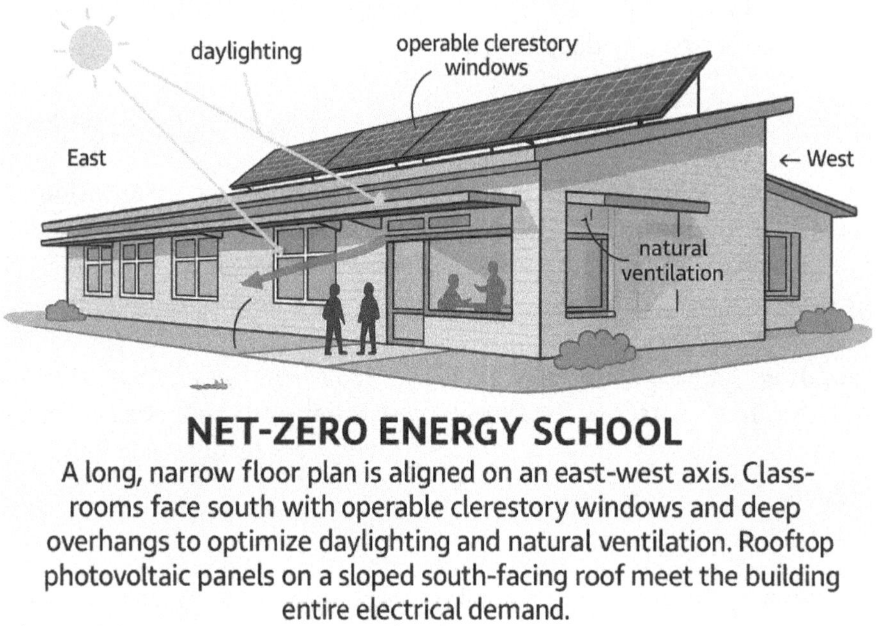

Figure 3.2 A net-zero energy school

2. **Prevailing Winds & Natural Ventilation**

Understanding and responding to prevailing wind patterns is a key strategy in passive environmental design. Proper alignment and detailing of buildings can help harness natural breezes to improve **thermal comfort**, reduce mechanical cooling loads, and increase **energy efficiency**, particularly in moderate and warm climates.

Conversely, site planning should also defend against unwanted winter winds, which can lead to **heat loss, occupant discomfort, and increased energy use**. The strategic use of landscaping, topography, and built forms helps mitigate these effects while preserving ventilation where desired.

Key Strategies:

- **Align Buildings to Take Advantage of Cooling Summer Breezes**
  Orient openings, courtyards, and key living spaces perpendicular to prevailing summer winds to allow breezes to naturally cool interior spaces. This is particularly effective in hot and humid regions, where airflow can improve occupant comfort and reduce dependency on air conditioning.
- **Provide Windbreaks Against Cold Winter Winds Using Vegetation or Site Forms**
  Use dense vegetation (such as evergreen trees), berms, walls, or service structures on the north or northwest side of a building (in the Northern Hemisphere) to block harsh winter winds. This improves envelope performance and reduces heating demand.
- **Position Operable Windows or Breezeways for Cross-Ventilation**
  Locate operable windows or vents on opposite or adjacent walls to encourage cross-ventilation. Breezeways, atriums, or open-air corridors can serve as passive ventilation routes, especially when combined with stack-effect strategies like clerestories or vented skylights. In a hot, humid climate, cross-ventilation is encouraged by aligning openings with prevailing breezes.

Design Implications:

- **Building Layout and Massing:**
  Buildings designed to channel or capture prevailing winds often feature open floor plans, narrow floor plates, or multiple courtyards. Avoid deep floor plates that block airflow.
- **Window Design and Placement:**
  Operable windows, transoms, and vents should be located based on wind direction and use patterns. Low inlets and high outlets (stack effect) can increase airflow efficiency. Casement windows that open outward into the wind are more effective than sliding windows for directing airflow.
- **Landscaping and Wind Management:**
  Trees, shrubs, and hedgerows can serve as natural wind filters. Evergreen trees offer year-round protection, while deciduous trees can shade summer winds without blocking winter sun.
- **Climate-Specific Considerations:**
  In desert climates, wind can carry dust and sand, so passive ventilation strategies should be combined with filters or courtyards that settle airborne particulates. In coastal or humid climates, attention must be paid to preventing condensation and mold by allowing air to flow and removing excess moisture. In a **cold climate**, the main building entrance is recessed and located on the **leeward side** of the structure, avoiding exposure to the prevailing northwestern winter winds. In a **tropical resort**, guest units are oriented perpendicular to the coastline to allow prevailing ocean breezes to pass through, eliminating the need for air conditioning.

Examples:

- A **residential project in Southern California** is designed with living areas facing southwest to capture cool coastal breezes. Large operable windows and sliding glass doors facilitate cross-ventilation throughout the house, significantly reducing HVAC use.

- A **mountain retreat in Colorado** uses a dense line of conifer trees on its northern edge as a windbreak against harsh winter winds. This vegetative buffer improves thermal comfort and reduces the building's heating load.
- In a **tropical school campus in Florida**, classrooms are arranged in a U-shape around an open courtyard. The plan captures prevailing trade winds and directs them through operable louvered windows on both sides of each classroom, promoting passive cooling and indoor air quality.
- In a **hot and humid climate** like the Gulf Coast, a school building is designed with **large operable windows** on the east and west sides of classrooms to align with the **prevailing southeast winds**. This layout facilitates **cross-ventilation**, drawing cooler air through the space and reducing the need for mechanical cooling. Outdoor courtyards are located on the leeward side of the building, protected from the stronger afternoon winds and providing comfortable shaded gathering areas.

Tools for Analysis:

- **Wind rose diagrams** (available through NOAA or local weather data) can help designers understand dominant wind patterns. (Figure 3.3)
- **Site-specific studies** or even on-site testing with flags or smoke during different times of year may be necessary for complex conditions.

Note: Wind roses *are graphical charts that characterize the speed and direction of winds at a location. Presented in a circular format, the length of each "spoke" around the circle indicates the amount of time that the wind blows **from** a particular direction. Colors along the spokes indicate categories of wind speed.*

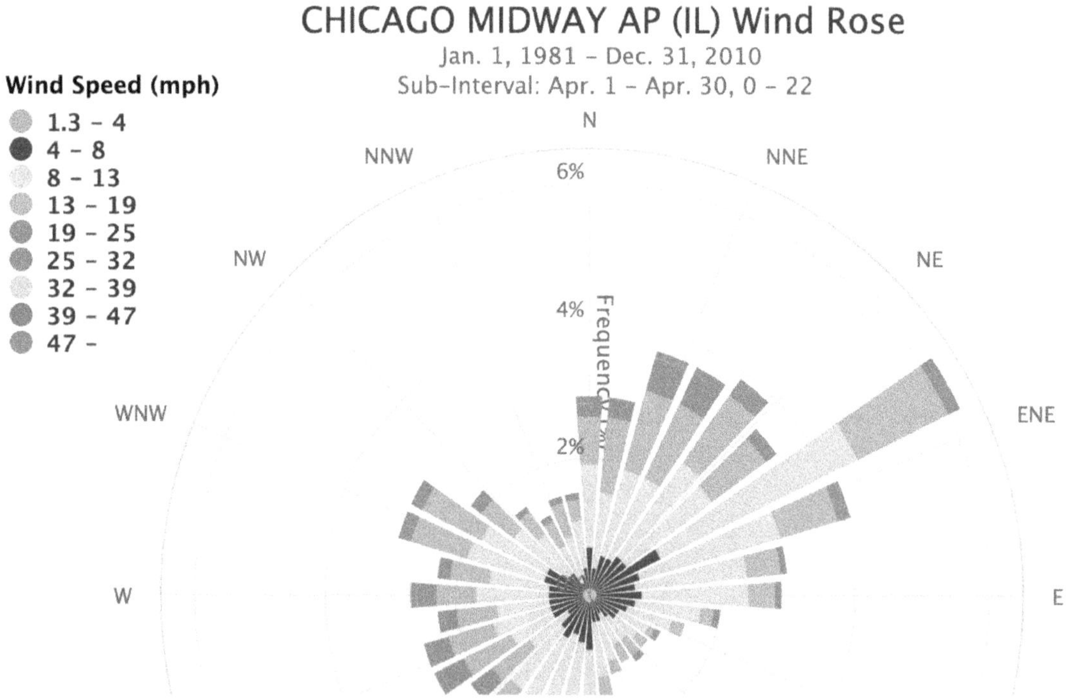

Figure 3.3 Chicago Midway April Wind Roses Chart (Source: Climate.gov)

3. **Topography**

Topography refers to the natural surface features of a site, including elevation changes, slopes, ridges, valleys, and other landforms. Proper integration of a building with its topography helps reduce environmental impact, construction costs, and long-term site maintenance needs.

Design Considerations:

- **Respect natural contours** by designing with the slope instead of against it. Designing to follow the natural contours to minimize grading and environmental disruption.
- Minimize **cut-and-fill** operations to preserve site integrity, prevent erosion, and reduce costs.
- Consider **drainage flow**, especially in sloped areas, to avoid water intrusion and ensure proper site runoff.
- Use the terrain to create **multi-level spaces**, natural daylighting opportunities, or viewsheds.
- In steeper sites, include **retaining walls**, terraces, or stepped forms to manage level changes safely and attractively.

Example:

A hillside site may support a stepped foundation that works with the slope rather than requiring costly cut-and-fill operations.

On a **steep hillside site** in Northern California, a custom home is designed with a **stepped foundation**, where each level of the house follows the existing slope of the land. Instead of leveling the entire site with extensive grading (cut and fill), the architect utilizes the topography to divide the house into three staggered floor levels. This reduces **site disturbance**, lowers excavation costs, and offers panoramic views from each level. Additionally, **native vegetation** is preserved, reducing erosion and maintaining the site's ecological health.

Additional Applications:

- In a **rural park project**, walking trails are designed to follow the contours of the land, minimizing the need for retaining structures and blending naturally with the environment.
- A **museum** located on a sloped urban lot uses the site's incline to create a **below-grade entrance** that opens into a daylight-filled atrium, maintaining a modest street presence while offering dramatic spatial experiences inside.

Benefits of Topography-Based Design:

- Reduces need for structural soil retaining measures
- Enhances passive strategies (e.g., thermal insulation through earth sheltering)
- Improves site drainage and reduces risk of stormwater-related issues
- Offers aesthetic opportunities such as multi-level views and terraced outdoor spaces

4. **Views**

Views refer to the visual connection between the occupants of a building and the surrounding environment. This can include **natural landscapes**, **urban skylines**, **bodies of water**, or even **internal courtyards**. Thoughtful orientation and placement of buildings and spaces can

maximize desirable views while minimizing exposure to **undesirable elements** such as highways, industrial zones, or service areas.

Design Considerations:

- Placing the building or orienting rooms to take advantage of scenic or urban views while blocking undesirable ones.
- Orient frequently occupied or high-value spaces (e.g., living rooms, hotel rooms, offices) to face **scenic or significant views** such as mountains, coastlines, parks, or city landmarks.
- Place less critical or support spaces (e.g., storage rooms, restrooms, back-of-house areas) on the sides of the building facing **undesirable views** or noisy elements.
- Use **window placement**, **balconies**, and **roof terraces** to frame and enhance access to attractive views.
- Incorporate **landscaping** or **screens** to buffer unattractive views and provide visual privacy where needed.
- Consider **seasonal variations** and **sun angles** when designing views—glare and overheating can impact comfort.

  Example:
  A **hotel** located near a mountain range and a busy freeway is carefully sited to take full advantage of its context. Guest rooms and public spaces like the **lobby and restaurant** are oriented toward the **mountains**, with large panoramic windows and balconies to emphasize the view. Meanwhile, **service areas**, such as laundry facilities and mechanical rooms, are positioned on the **back side** of the building, which faces the freeway. This not only enhances the guest experience but also helps buffer traffic noise.

Additional Applications:

- In an **urban high-rise apartment**, upper-floor units are rotated slightly to capture views of a nearby riverfront, while lower floors use planted terraces to screen adjacent rooftops and alleyways.
- A **university library** is designed with large reading rooms facing a tranquil campus quad, while stacks and storage areas are positioned along the rear, facing a parking lot.
- In a **museum**, a café is placed along a glazed wall overlooking a public sculpture garden, turning a casual space into a signature destination.

Benefits of View-Oriented Design:

- Enhances user satisfaction, productivity, and well-being (particularly in schools, offices, and healthcare environments)
- Increases real estate value and desirability of spaces
- Strengthens the connection between the built environment and the natural or urban context
- Improves passive surveillance and safety in public areas when views extend to active or visible zones

Design Tools and Strategies:

- **View analysis diagrams** during site planning phases
- **3D modeling** and sun studies to understand visual impacts over time
- **Strategic massing and elevation studies** to determine optimal sightlines

5. **Site Adjacencies**
   Site adjacencies refer to the careful planning and positioning of new buildings in relation to existing site elements to ensure logical, efficient, and functional relationships. This includes considering proximity to existing infrastructure such as roads, utilities, pedestrian pathways, public spaces, and other buildings.

   Proper attention to site adjacencies enhances connectivity, supports operational efficiency, promotes safety, and improves the overall user experience. The placement of a new structure should complement existing site functions and anticipate future growth or changes in use.

   Key considerations include:
   - **Access and Circulation:** Ensuring that vehicular and pedestrian access to the new building integrates smoothly with existing traffic patterns and walkways. Ensuring functional relationships between the new building and existing elements such as roads, utilities, buildings, and pedestrian paths.
   - **Utilities and Infrastructure:** Positioning the building in a way that allows for cost-effective connections to water, electricity, sewer, and communication lines.
   - **Programmatic Relationships:** Aligning the function of the new building with nearby facilities to enhance convenience and reduce redundancy.
   - **Environmental Context:** Considering solar orientation, wind patterns, noise sources, and views in relation to nearby structures.

   Examples:
   - A new **campus dining hall** is sited adjacent to a **student housing complex** to allow for easy and convenient access for residents, reducing the need for transportation and encouraging community interaction.
   - A **research laboratory** is located near an **academic building** housing related departments to foster collaboration between faculty and students.
   - A **parking structure** is placed at the edge of campus near a **main access road**, minimizing traffic congestion within pedestrian-heavy zones.
   - A **public library** is built adjacent to a **community park**, creating opportunities for shared programming and offering patrons a pleasant environment for study and relaxation.

   By thoughtfully analyzing and planning site adjacencies, designers and planners can create integrated, user-centered environments that maximize the utility and accessibility of both new and existing facilities.

6. **Environmental Risks**
   Environmental risk assessment is a critical component of site planning and development, aimed at ensuring the long-term safety, sustainability, and resilience of a project. This involves identifying and avoiding areas susceptible to natural hazards such as flooding, wildfires, landslides, earthquakes, or unstable soils.

   Addressing environmental risks early in the planning process helps reduce potential damage to buildings, ensures occupant safety, lowers long-term maintenance and insurance costs, and supports regulatory compliance. In many cases, site mitigation strategies must also be incorporated when avoidance is not possible.

Key considerations include:

- **Flood Zones:** Avoid building in designated floodplains when possible. If unavoidable, implement mitigation measures such as elevating structures, using permeable surfaces, or designing water diversion systems.
- **Wildfire-Prone Areas:** Maintain defensible space by clearing vegetation around structures, using fire-resistant materials, and providing emergency access routes.
- **Unstable or Erodible Soils:** Conduct geotechnical studies to assess soil conditions and implement appropriate foundation designs or soil stabilization techniques.
- **Seismic Zones:** In earthquake-prone regions, buildings should be designed with structural reinforcements and flexible foundations that meet or exceed seismic codes.
- **Climate Adaptability:** Consider future environmental risks associated with climate change, such as sea level rise, increased storm intensity, or prolonged drought.

Examples:

- A **coastal development** elevates its buildings on **piers** or **stilts** to minimize damage from **storm surge** and **rising sea levels**, as seen in many hurricane-prone areas like the Gulf Coast or Southeast Asia.
- A **mountain resort** is sited away from known **avalanche paths** and includes **rockfall barriers** and **retaining walls** to manage steep terrain hazards.
- A **residential community** in a wildfire-prone region like **Southern California** uses **non-combustible exterior materials**, installs **sprinkler systems**, and ensures **firebreaks** are in place.
- A **hospital in an earthquake zone** such as **Northern California** is designed with **base isolation technology** to remain operational after seismic events.
- A **commercial complex** near a floodplain incorporates **green infrastructure**, such as **bioswales and rain gardens**, to manage stormwater runoff and reduce flood impact.

By thoroughly evaluating environmental risks and incorporating adaptive strategies, project teams can create safer, more resilient developments that protect both people and investments over the long term.

7. **Planning & Zoning Regulations**

Planning and zoning regulations are legal frameworks established by local jurisdictions to guide land use, building form, and development density. These rules ensure orderly growth, protect property values, and promote public health, safety, and welfare. When designing a new project, adherence to these regulations is essential to gain approvals, avoid costly redesigns, and maintain compliance throughout the development process.

Zoning ordinances typically define allowable land uses (e.g., residential, commercial, industrial), as well as specific dimensional and spatial requirements. Planners, architects, and developers must analyze these constraints early to ensure the proposed project fits within the legal boundaries of the site.

Key considerations include:

- **Setbacks:** Required distances between a structure and the property lines. These ensure adequate light, air, privacy, and emergency access.
- **Height Limits:** Maximum allowable building heights based on zone type or context (e.g., preserving neighborhood character or protecting viewsheds).
- **Floor Area Ratio (FAR):** The ratio of a building's total floor area to the size of its lot, which regulates building mass and density.
- **Easements:** Legal rights allowing utilities, pedestrians, or other access through part of the property—often restricting construction in those areas.
- **Lot Coverage and Open Space Requirements:** Restrictions on how much of a site can be covered by structures, encouraging greenery and stormwater management.
- **Use Restrictions:** Dictating what types of activities or occupancies are allowed on the site.

Examples:

- A **commercial building** on an **urban corner lot** must align with the **street wall** to maintain the visual rhythm of the block and meet a required **10-foot rear setback** for service access and deliveries.
- A **mid-rise apartment complex** is designed within the allowable **FAR of 3.0**, which governs the total square footage across multiple stories based on lot size.
- A **warehouse** in an industrial zone is restricted to a **maximum height of 50 feet** to avoid obstructing nearby airport flight paths.
- A **single-family home** in a suburban neighborhood must observe a **15-foot front setback**, **5-foot side setbacks**, and preserve a **utility easement** along the rear of the property for underground cables.
- A **mixed-use development** is only approved after demonstrating compliance with a **special overlay district**, which may include stricter façade requirements, signage guidelines, or pedestrian-friendly ground-floor uses.

Failing to respect planning and zoning regulations can lead to permit denials, legal disputes, fines, or even the need to demolish or reconfigure completed structures. By integrating these constraints into the early phases of design, project teams can streamline approvals, reduce risk, and contribute to a more cohesive urban fabric.

8. **Access and Circulation**

Access and circulation refer to the systems and design strategies that enable people and vehicles to move efficiently, safely, and intuitively through a site. A well-planned circulation network enhances usability, improves safety, supports operational efficiency, and creates a positive experience for users—whether they are pedestrians, drivers, cyclists, or service personnel.

Designing for effective access and circulation requires careful consideration of entry points, internal pathways, hierarchy of movement, and separation of different types of traffic (e.g., public vs. service). It should also account for accessibility standards (e.g., ADA compliance), emergency access, and wayfinding clarity.

Key considerations include:

- **Pedestrian Circulation:** Walkways, crossings, ramps, and plazas should be direct, accessible, and connected to public transit stops, parking areas, and building entrances.
- **Vehicular Access:** Driveways, drop-offs, and parking lots must be clearly defined and designed to minimize conflict with pedestrian movement.
- **Emergency and Service Access:** Provide unobstructed, dedicated routes for fire trucks, ambulances, trash removal, and deliveries without disrupting regular user flow.
- **Wayfinding and Signage:** Entrances, exits, and key destinations should be intuitive and clearly marked to support ease of navigation.
- **Bicycle Infrastructure:** Consideration for bike lanes, storage, and safe crossings encourages multimodal transportation and reduces reliance on cars.

Examples:

- A **hospital** design places the **emergency entrance** directly accessible from the **main road**, ensuring rapid response, while a **separate service road** leads to back-of-house areas for **deliveries and waste removal**, minimizing disruption to patients and visitors.
- A **university campus** uses a network of **pedestrian-only pathways**, connected by **bicycle racks**, and located away from vehicular roads to prioritize student safety and promote walking and cycling.
- A **shopping center** features a **central drop-off plaza** near the entrance with **clearly marked pedestrian crossings**, and **parking areas** located at the periphery to reduce vehicular congestion near retail areas.
- A **hotel** separates **guest vehicle access** from **staff and service vehicle routes**, ensuring a smooth guest experience while keeping operational functions discreet and efficient.
- A **multifamily residential complex** includes **ADA-accessible ramps and elevators**, as well as **covered walkways**, ensuring year-round comfort and inclusivity.

Incorporating thoughtful access and circulation strategies from the outset of a project leads to safer, more intuitive environments that serve both users and operators. It also reduces long-term operational challenges and contributes to a building's overall functionality and success.

9. **Programmatic Needs**

Programmatic needs refer to aligning the site layout and building placement with the intended functions, activities, and user groups of a project. A successful design supports the day-to-day operations of the facility, enhances the user experience, and optimizes spatial relationships between different functional areas.

Understanding how people will use a building—and how various spaces relate to one another—guides decisions about location, orientation, access, and proximity to other program elements. This ensures the building not only works well on paper but also functions smoothly in real-world use.

Key considerations include:

- **Functional Flow:** Align building placement with intended use and functional flow. Arrange spaces to support logical movement patterns between related uses (e.g., from entry to reception to workspaces).

- **User-Specific Needs:** Consider the needs of specific user groups such as children, elderly individuals, staff, or visitors, and tailor locations accordingly.
- **Noise and Activity Separation:** Place quiet or sensitive areas away from high-traffic or noisy functions.
- **Operational Efficiency:** Optimize adjacencies between spaces with frequent interaction (e.g., storage near delivery zones, labs near classrooms).
- **Security and Supervision:** Design for controlled access and visibility where safety or supervision is critical (e.g., in schools or healthcare settings).

Examples:

- A **school building** places **classrooms for young children** close to the **parent drop-off area** to ease drop-off/pick-up routines and minimize walking distance, while **gymnasiums and athletic fields** are located farther away to reduce noise near learning areas.
- A **corporate campus** arranges **meeting rooms and collaborative spaces** adjacent to **workstations**, while placing **executive offices** in a quieter wing with restricted access.
- In a **hospital**, the **emergency department** is placed near the main road for rapid access, while **inpatient rooms** are located in a quieter rear wing with views and access to green spaces.
- A **university science building** positions **laboratories**, **classrooms**, and **faculty offices** in close proximity to encourage collaboration, while isolating **chemical storage areas** for safety and ventilation.
- A **retail development** situates **anchor stores** at the ends of the site to drive foot traffic past smaller shops, improving visibility and commercial flow.

By prioritizing programmatic needs, designers can create environments that support how people actually use the space—enhancing comfort, efficiency, and long-term success of the facility.

10. Sustainable Design Principles and Material Selection

Sustainable design goes beyond energy efficiency and site orientation—it also involves responsible **material selection** and **resource management** throughout the life cycle of a project. Key principles include choosing materials with **recycled content**, reducing construction and operational **waste**, and specifying products that promote **durability, resilience, and environmental stewardship**.

When selecting building materials, architects and designers should:

- **Prioritize Recycled and Recyclable Materials**
  Use products with high post-consumer or post-industrial recycled content—such as steel, concrete with fly ash or slag, recycled wood products, and reclaimed brick. These reduce demand for virgin materials and lower embodied carbon.
- **Minimize Waste Through Smart Design and Planning**
  Employ modular construction, standard dimensions, and efficient material layouts to reduce offcuts and site waste. Designing for **material efficiency**—such as aligning framing with material sizes—can significantly cut waste.

- **Select Locally Sourced Materials**
  Choose products manufactured or harvested near the project site to reduce transportation emissions and support local economies.
- **Specify Low-Impact, Durable Materials**
  Select materials that are long-lasting, low-maintenance, and non-toxic. Materials with low VOC (volatile organic compound) emissions, formaldehyde-free components, and cradle-to-cradle certifications promote indoor air quality and lifecycle sustainability.
- **Design for Disassembly and Reuse**
  Where possible, incorporate reversible detailing and non-permanent fasteners so materials and components can be reused or recycled at the end of the building's life.
- **Adapt Materials for Resilience**
  In climate-sensitive or disaster-prone areas, materials should be chosen for **resistance to moisture, fire, and temperature extremes**. For example, concrete with supplemental cementitious materials may improve durability and reduce emissions.
- **Plan for Construction Waste Management**
  Incorporate a Construction Waste Management Plan (CWMP) that includes recycling goals, separation protocols, and diversion from landfills. Projects seeking LEED or other green certifications may be required to meet minimum diversion percentages.

Examples:

- A school project uses **recycled denim insulation**, **low-VOC paints**, and **flooring made from reclaimed wood**, supporting both sustainability and healthy indoor air quality.
- A commercial building design includes **precast concrete panels with fly ash** as a supplementary cementitious material, reducing embodied carbon while enhancing resilience.
- A modular housing project employs **standardized panel dimensions** and an optimized framing layout, significantly reducing on-site cutting and material waste.
- A coastal community center uses **fiber-cement siding** and **elevated stainless-steel connections** to ensure long-term durability in a high-humidity, salt-air environment.

C. Determine impact of neighborhood context on the project design (U/A) (1-5%)

NCARB Requirements

You will need to consider the character of a site's neighborhood, including proximity to various modes of transit, nearby amenities and public services, utilities, noise pollution, the scale of adjacent buildings, surrounding façade materials, historic precedent, and historic preservation requirements. Understanding how these impact a project is necessary as you determine an appropriate design response for a given architectural context.

Understanding and responding to the **neighborhood context** is essential for creating a design that is both respectful of its surroundings and functionally appropriate. Below are key contextual factors and how they may influence architectural decisions:

1. Proximity to Transit Modes

Access to reliable and varied transportation options is a critical contextual factor in urban and suburban site planning. Proximity to transit—such as **subway or train stations**, **bus lines**, **bike lanes**, and **pedestrian corridors**—can significantly influence the design of a building and its site in terms of **orientation, building entrances, circulation, density, accessibility**, and **parking requirements**.

Design strategies that respond to transit proximity can improve sustainability, reduce car dependency, and support walkable, vibrant communities. They also align with zoning incentives in many jurisdictions, such as **transit-oriented development (TOD)** bonuses, which allow greater height or density in return for reduced parking or mixed-use programming.

Key design implications:

- **Reduced parking requirements** due to increased reliance on public or alternative transportation.
- **Main entrances and access points** oriented toward the nearest transit stop for pedestrian ease.
- **Enhanced pedestrian infrastructure** such as wider sidewalks, lighting, and covered walkways.
- **Increased density or mixed-use programming** in areas well served by transit.
- **Inclusion of bike storage, repair stations, or showers** to encourage active transportation.

Example 1:
A **multifamily residential project** located within 500 feet of a **subway station** reduces the number of on-site parking spaces, increases bike storage, provides a **secure bike room** for residents, and positions the **main lobby entrance** to face the sidewalk leading directly to the transit stop.

Example 2:
A **public library** adjacent to a **bus rapid transit (BRT) line** includes **seating and shelter** near its entrance for waiting passengers and uses **transparent facades** to enhance visibility and safety around the bus stop zone.

Example 3:
A **mixed-use development** near a **light rail station** incorporates a **shared mobility hub** that includes e-bike rentals, scooter docks, and car-sharing spaces, encouraging diverse, car-free transportation options.

By carefully evaluating and integrating nearby transit infrastructure into the design, projects can promote accessibility, reduce environmental impact, and align with broader urban mobility goals.

2. **Nearby Amenities and Public Services**
   The presence of nearby amenities—such as **schools, parks, libraries, healthcare facilities, grocery stores, and retail centers**—can significantly influence the form, function, and community role of a building. These surrounding resources shape user expectations, impact site design, and may even alter the **programmatic mix**, circulation strategy, or community outreach goals of a project.

Proximity to public services and amenities often encourages **shared infrastructure**, increases **foot traffic**, and presents opportunities for **partnerships or joint-use facilities**. Designs that respond thoughtfully to nearby amenities not only enhance the user experience but can also foster stronger ties to the neighborhood and improve long-term viability.

Key design implications:

- **Site layout** may prioritize connections to adjacent public spaces or institutions.
- **Circulation paths and entrances** can align with neighboring facilities to encourage seamless movement.
- **Program elements** may expand or shift to serve broader community needs (e.g., classrooms used for after-hours adult education near a school).
- **Outdoor or transitional spaces** (e.g., plazas, courtyards) may be designed to accommodate spillover from nearby public amenities.
- **Security and visibility** considerations may be heightened near sensitive facilities like schools or clinics.

Example 1:
A **civic center** located next to a **public park** integrates **outdoor seating**, a **flexible plaza for community events**, and **landscaped walkways** that physically and visually connect the building to the park, enhancing both public engagement and daily usability.

Example 2:
A **community health clinic** near an **elementary school** incorporates a **pediatric wing**, **school outreach office**, and **safe pedestrian crossings** to facilitate access for students and families during and after school hours.

Example 3:
A **mixed-use development** adjacent to a **public library and transit stop** includes a **covered walkway**, **shared parking**, and a **public café** that serves both residents and library patrons, creating a vibrant, multifunctional civic hub.

By evaluating and responding to the availability of nearby amenities and services, architects can design projects that are more **integrated, responsive, and valuable** to the community they serve.

3. **Availability of Utilities**
The presence—or absence—of **essential utility infrastructure** such as **water supply**, **sanitary sewer**, **stormwater drainage**, **electricity**, **natural gas**, and **telecommunications** can limit or facilitate both site planning and building design. Understanding existing utility access is critical during the early phases of a project, as it impacts **budget, timelines, infrastructure coordination**, and **sustainability strategies**.

Where utility access is limited or unavailable, alternative systems must be explored, which can influence site grading, environmental impact, and long-term maintenance considerations.

Conversely, access to robust infrastructure may open opportunities for **denser development**, **renewable energy integration**, or **smart building systems**.

Key design implications:

- **Sewer access** affects the type of waste treatment systems (e.g., municipal tie-in vs. septic or package treatment). A remote site without sewer service may require the integration of a septic system, affecting site grading and landscape design.
- **Water availability** influences fire protection systems, irrigation planning, and fixture types.
- **Electric grid capacity** may determine the feasibility of high-load equipment or all-electric building strategies.
- **Telecom infrastructure** is essential for modern systems integration, including building automation, remote work, and security systems.
- **Stormwater systems** guide the placement and size of retention ponds, bioswales, or underground detention systems.

Example 1:
A **remote site without sewer service** may require the integration of a **septic system**, influencing **building placement**, **soil testing**, and **landscape design** to accommodate drain fields and ensure safe separation from water sources.

Example 2:
A **mixed-use development** in a growing urban area may need to **upgrade existing electrical infrastructure** to support EV charging stations, high-efficiency HVAC systems, and a rooftop solar array connected to the grid.

Example 3:
A **new school facility** is sited to align with the **existing utility corridor**, reducing trenching distances and minimizing disruption to the surrounding environment and nearby residential areas.

Example 4:
On a **rural healthcare campus**, lack of reliable high-speed internet may require installation of **private telecom infrastructure** or satellite broadband to ensure connectivity for telemedicine and digital health records.

Evaluating utility availability early ensures that the project can be **cost-effective, code-compliant, and future-ready**, while avoiding costly redesigns or delays due to infrastructure limitations.

4. **Noise Pollution**
Noise pollution is a critical environmental factor that directly influences site planning, building orientation, façade design, and interior layout. Common sources of disruptive noise include **highways**, **railways**, **airports**, **industrial facilities**, and **urban nightlife districts**. Ambient noise from highways, industrial zones, or transit corridors can impact orientation,

material choices, and window placement. Without proper mitigation, excessive noise can negatively affect occupant comfort, productivity, sleep quality, and even long-term health.

Effective acoustic design begins with a thorough **site analysis** to identify primary noise sources, intensity levels, prevailing wind directions, and potential sound reflections. Once identified, various architectural and landscape strategies can be employed to reduce interior and exterior noise exposure.

Key design implications:

- **Building orientation** may be adjusted to shield sensitive uses from noise sources.
- **Room placement** (e.g., bedrooms, classrooms, or exam rooms) is planned to be furthest from the noise.
- **Sound-attenuating materials** such as acoustic glazing, dense wall assemblies, and insulated façades help reduce transmission.
- **Landscaping elements** (e.g., berms, trees, and water features) can serve as sound buffers.
- **Mechanical system design** (e.g., quiet HVAC, acoustic duct lining) prevents compounding external noise with internal mechanical noise.

Example 1:
A **hotel** adjacent to a **rail line** uses **triple-glazed windows**, installs **acoustic insulation** in perimeter walls, and locates **guest rooms** on the opposite side of the building, with **corridors and service areas** facing the tracks to act as a sound buffer.

Example 2:
An **urban high school** near a **busy freeway** or **airport** incorporates **green roofs**, **high-mass concrete walls**, and **interior courtyards**, creating a quieter interior learning environment and usable outdoor spaces shielded from noise (Figure 3.4).

Example 3:
A **residential tower** in a nightlife district uses **acoustic-rated doors and windows**, **white noise-generating systems** in hallways, and strategic **floor plan zoning** to ensure that bedrooms do not directly face bars or clubs.

Example 4:
An **office building** near an industrial zone employs **vegetated berms**, **sound walls**, and **orientates workspaces** toward a quieter side of the site, while placing **storage or utility areas** near the noise source.

Proactively addressing noise pollution ensures better user experience and occupant well-being while demonstrating thoughtful integration into the surrounding context.

Figure 3.4 An urban high school near a busy freeway

5. **Scale of Adjacent Buildings**
   Maintaining visual continuity and human scale is key to integrating new construction into existing neighborhoods. Massing, height, and setbacks are adjusted accordingly. The **scale and massing** of surrounding buildings is a critical contextual factor in designing a project that feels **visually cohesive, respectful, and appropriately integrated** within its neighborhood. When a new development is introduced into an existing built environment—especially in established residential or historic areas—careful attention to **building height, bulk, roofline, setbacks, and architectural rhythm** helps prevent visual or social disruption. Designers must often strike a balance between maximizing usable space and maintaining a **human-scale presence**, particularly when transitioning between different zoning types (e.g., commercial to residential).

Key design implications:

- **Stepping down building height** toward smaller adjacent structures to soften the transition.
- **Aligning rooflines, cornices, or datum lines** with neighboring buildings for visual consistency.
- **Adjusting setbacks and façade articulation** to echo the rhythm and scale of nearby structures.
- **Using landscaping, screens, or buffer zones** to reduce visual dominance or overlook issues.
- **Incorporating transitional architectural elements** (e.g., porches, balconies, bays) that relate to residential scales.

Example 1:
An **infill office building** adjacent to **2-story townhouses** steps back its upper floors and adds a **landscaped transition zone** at the rear lot line, preserving sunlight access and reducing perceived mass.

Example 2:
A **new apartment building** next to a **row of single-family homes** uses **a tiered façade**, **punched windows**, and **brick cladding** to visually connect with the finer-grained residential context while gradually increasing height toward a main street.

Example 3:
A **cultural center** in a low-rise historic district uses a **flat roof, horizontal banding**, and **stone materials** that match surrounding buildings, even though its interior volumes are larger, maintaining harmony with the context.

Example 4:
A **mixed-use project** transitions from **retail frontage** on a busy avenue to **residential units** on a quieter side street by adjusting the **building height from six to three stories**, and using different entry styles to reflect the adjacent uses.

Maintaining sensitivity to adjacent building scale not only improves neighborhood acceptance and planning approvals but also contributes to a **coherent, pedestrian-friendly urban fabric** that respects both context and community.

6. **Surrounding Façade Materials and Color Palettes**
Local materiality informs aesthetic decisions that promote visual cohesion. While not always strictly required, aligning with or thoughtfully contrasting surrounding finishes helps the project feel rooted in place.

The **materials and color palettes** used on surrounding buildings strongly influence how a new development is perceived within its context. Whether the goal is to **blend in harmoniously** or to **stand out with respectful contrast**, a thoughtful response to local materiality can reinforce a project's **sense of place**, **cultural identity**, and **architectural continuity**.

While some sites—especially those in historic districts or design overlay zones—require strict material compatibility, others offer more flexibility. Regardless, careful selection of façade materials and finishes contributes to **visual cohesion**, improves **public perception**, and often facilitates smoother review and approval processes with planning authorities and communities.

Key design implications:
- **Referencing nearby materials** (e.g., brick, stone, wood, stucco) helps anchor the project in the local vernacular.
- **Color selection** should be sensitive to the tone, warmth, and reflectivity of adjacent façades.
- **Material transitions** can help signal changes in program or scale while maintaining overall cohesion.
- **Modern materials or detailing** may be introduced to provide contrast while still respecting contextual patterns.
- **Durability, maintenance, and environmental performance** of materials must also be considered alongside aesthetics.

Example 1:
A **cultural center** in a **historic district** uses **locally sourced red brick** to align with adjacent buildings, but introduces **contemporary detailing** like metal window frames, deep recesses, and clean lines to distinguish the new structure while maintaining material continuity.

Example 2:
A **modern office building** in a **neighborhood of stone-clad civic structures** uses **stone veneer at the ground level** and transitions to **glass and metal panels** above, establishing a dialogue between traditional and contemporary expressions.

Example 3:
A **multi-family housing project** adjacent to **mid-century stucco homes** incorporates **warm-toned stucco and natural wood accents**, reflecting the surrounding color palette while using **updated materials** and **vertical articulation** to create a fresh identity.

Example 4:
A **public library renovation** retains the **original limestone façade** of the historic structure and uses **dark metal cladding** on a modern addition, clearly delineating the old from new while achieving a complementary visual balance.

By responding to surrounding materials and color palettes—whether through alignment or contrast—designers can ensure that new projects feel **intentional, contextual, and visually connected** to their built environment.

7. **Historic Precedent and Preservation Guidelines**
Historic buildings or districts may impose formal requirements related to façade design, rooflines, materials, and allowable alterations.

When working within or adjacent to **historic buildings**, **landmark sites**, or **designated preservation districts**, architects must adhere to formal guidelines that ensure new construction or alterations are compatible with the **architectural, cultural, and historical significance** of the surrounding context.

These guidelines—often governed by **local preservation commissions**, **national registries**, or **zoning overlays**—define acceptable approaches to **façade treatment**, **materials**, **roof forms**, **fenestration patterns**, and even **color schemes**. The goal is to preserve the visual integrity of heritage areas while allowing for thoughtful, context-sensitive evolution.

Understanding and respecting these regulations is critical to the **approval process**, the **success of community engagement**, and the long-term stewardship of the built environment.

Key design implications:
- **Façade proportions and detailing** must often align with historical precedents.
- **Original materials and features** (e.g., cornices, windows, columns) may need to be preserved or replicated.

- **Rooflines, parapets, and dormers** must respect the massing and rhythm of adjacent structures.
- **Additions and infill** must be clearly distinguishable from historic fabric while maintaining visual harmony.
- **Signage and lighting** are typically subject to strict placement and style requirements.

Example 1:
A **commercial renovation** in a **designated landmark district** must **retain the original cornice line**, **match the historical window proportions**, and **preserve the storefront geometry**, all of which are conditions for **approval by the local historic preservation commission**.

Example 2:
A **museum addition** next to a **19th-century civic building** uses **glass and metal cladding** to differentiate the new structure, while aligning roof height and window rhythm with the original to show respect without imitation.

Example 3:
A **residential conversion** of a **historic warehouse** includes restoration of the **brick exterior and arched window openings**, while inserting modern balconies in a recessed, steel-framed addition on the rooftop to reduce visual impact from the street.

Example 4:
A **public library** in a historic town center uses **wood siding, double-hung windows, and a gable roof** to match the surrounding vernacular, while introducing **solar panels** and a **modern rear extension** that is minimally visible from the main thoroughfare.

Working within preservation guidelines requires a nuanced balance between **historical integrity** and **contemporary function**. Successful projects are those that **honor the past**, engage with the present, and contribute positively to the future of the neighborhood.

8. Community Character and Cultural Identity
Cultural influences, socioeconomic patterns, and neighborhood identity should inform how a building presents itself to the street, who it serves, and how it engages the public realm.

The **cultural identity** and **social fabric** of a neighborhood play a vital role in shaping responsible, inclusive architectural design. A building should not exist in isolation—it should **reflect, support, and enhance** the character of the community it serves. This includes responding to **cultural traditions**, **local values**, **demographics**, and **economic realities** that define the area's unique identity.

Designing with community character in mind requires more than stylistic alignment—it involves creating places that feel **welcoming, meaningful, and accessible** to a broad spectrum of people. This often includes **programmatic alignment**, **public engagement**, and the **use of symbols, materials, languages, and forms** that resonate with the local population.

Key design implications:

- **Façade articulation and signage** can express neighborhood diversity or cultural significance.
- **Programming and space planning** should reflect how the community gathers, celebrates, learns, or heals.
- **Public art, murals, and installations** can tell local stories and empower residents.
- **Engagement during the design process** ensures that community voices shape the final outcome.
- **Material choices and building form** may draw from cultural vernacular or traditions.

Example 1:
A **community arts center** in a **diverse neighborhood** incorporates **multi-language signage**, **flexible indoor/outdoor event spaces**, and **rotating exhibitions of local artists**, ensuring the center reflects and celebrates the cultural richness of its users.

Example 2:
A **health clinic** in an underserved immigrant community includes **private consultation rooms**, **bilingual staff and signage**, and a **welcoming lobby** designed with culturally familiar colors, textures, and artwork to reduce stigma and improve access to care.

Example 3:
A **neighborhood library** in a historically Black community partners with local historians to design a **permanent exhibit wall**, uses **brick detailing from local architectural traditions**, and offers **community rooms** named after local civil rights leaders.

Example 4:
A **market hall** in a predominantly Latino district uses **colorful tilework, open-air arcades, and culturally relevant wayfinding graphics** to reflect local heritage and create a vibrant, familiar gathering space.

Example 5:
A **housing project** in a working-class neighborhood incorporates **communal courtyards**, **shared kitchens**, and **childcare facilities** to support multigenerational living, reflecting both socioeconomic needs and cultural living patterns.

Design that authentically responds to community character not only enhances user engagement—it helps foster **pride, equity, and belonging**, making architecture a tool for strengthening identity rather than imposing uniformity.

9. **Environmental and Microclimatic Factors**
Local wind patterns, sunlight access, vegetation, and topography affect building orientation, shading, and energy use. These contextual factors can also enhance comfort and sustainability.

Environmental and microclimatic conditions—including **sunlight access**, **prevailing wind patterns**, **topography**, **vegetation**, **humidity**, and **urban heat island effects**—have a profound impact on how a building performs and how comfortable it is for its occupants. These factors must be carefully studied during the site analysis phase to inform decisions around **building orientation, massing, materials, shading strategies, and passive energy design**. By responding intelligently to these natural conditions, designers can significantly improve **thermal comfort**, **daylighting**, **natural ventilation**, and **energy efficiency**, often reducing the building's reliance on mechanical systems. Additionally, engaging with the site's topography and ecological features can enhance **stormwater management**, **biodiversity**, and the overall experience of the space.

Key design implications:

- **Solar orientation** determines the placement of windows, overhangs, and interior spaces to optimize daylight and minimize heat gain or loss.
- **Wind exposure** influences the layout of openings, placement of courtyards, and the use of windbreaks or screens.
- **Topography** affects drainage, accessibility, and building entry points.
- **Vegetation and landscape** can be leveraged for shading, cooling, erosion control, and stormwater absorption.
- **Urban conditions** such as narrow streets or surrounding tall buildings may necessitate design interventions to bring light and air into the site.

Example 1:
On a **dense urban site** with **narrow streets and tall neighboring buildings**, a **south-facing courtyard** is incorporated at the center of the building to draw in **natural light and cross-ventilation** deep within the building footprint, improving indoor air quality and reducing energy use for lighting and cooling.

Example 2:
A **school in a hot, arid climate** is oriented with **minimal east and west exposures**, deep roof **overhangs**, and **courtyards planted with native trees**, providing passive shading and cooling for outdoor learning spaces.

Example 3:
A **mountainside residence** is designed to **follow the natural slope** of the site, minimizing grading and maximizing views while using **thermal mass walls** and **earth-sheltered construction** to regulate internal temperatures.

Example 4:
An **office building near a coastal site** takes advantage of **prevailing ocean breezes** by incorporating **operable windows and breezeways**, reducing reliance on mechanical ventilation and improving user comfort.

Example 5:
A **community center** in a humid environment is designed with **high ceilings**, **ventilation towers**, and **deep porches** that encourage passive cooling and airflow, while minimizing indoor heat buildup.

By designing in harmony with environmental and microclimatic factors, architects not only reduce energy use and operational costs but also create **healthier, more resilient buildings** that are attuned to their place and climate.

Chapter Four

Codes & Regulations (16-22%)

NCARB Requirements
In this section, you'll look at the codes and regulations relevant to the planning phase of a project.

A. Apply zoning and environmental regulations to site and building design (U/A) (3-7%)

NCARB Requirements
It is important to understand the regulations that govern the outside of a building, including standard and accessible parking requirements, allowable building size based on FAR, site coverage, property line setbacks, and requirements for managing stormwater runoff.

1. Floor Area Ratio (FAR) and Building Size Limitations

Floor Area Ratio (FAR) determines the allowable building floor area relative to the total site area, which directly affects the building's massing, height, and density.

FAR is a zoning metric that controls building density by establishing the ratio between the total **gross floor area** of a building and the **total lot area**. FAR is one of the most fundamental tools in zoning, used to regulate **building size**, **bulk**, and **development intensity**, particularly in urban areas.

FAR affects:

- **Overall building volume** and possible number of stories
- **Massing strategy** (horizontal vs. vertical emphasis)
- **Site layout**, including open space, parking, and landscape areas
- **Project feasibility**, especially for developers balancing usable area with zoning compliance

FAR limits must often be balanced with other zoning restrictions like **height limits**, **lot coverage**, and **setbacks**, which can further constrain the buildable envelope.

Example 1:
On a **20,000 sq. ft. lot** with a **FAR of 2.0**, the maximum allowable **gross floor area** is:
20,000 sq. ft. (lot size) × 2.0 (FAR) = **40,000 sq. ft. total floor area**
This can be achieved through different configurations:

- A **4-story building** with a **10,000 sq. ft. footprint**, or
- A **2-story building** with a **20,000 sq. ft. footprint**, depending on **height limits**, **setback requirements**, and **parking needs**

Example 2:
In a **downtown commercial district**, a **FAR of 6.0** may be allowed, enabling high-rise construction. A 10,000 sq. ft. lot could yield **60,000 sq. ft. of floor area**, supporting a tower with multiple levels of office or residential use, depending on zoning overlays and special permits.

Example 3:
In a **low-density residential zone** with a **FAR of 0.5**, a **15,000 sq. ft. lot** is limited to only **7,500 sq. ft.** of floor area. This restriction ensures smaller buildings and more green space, preserving neighborhood character.

Key Considerations:
- FAR can be increased in some jurisdictions through **bonuses** (e.g., affordable housing, public open space, green roofs)
- **Basements, mezzanines, or parking levels** may or may not count toward FAR depending on local code
- **Transfer of Development Rights (TDR)** programs may allow FAR to be transferred from underutilized lots to others

Understanding and applying FAR properly helps ensure the proposed building respects zoning limits while maximizing developable potential.

2. **Setbacks and Property Line Restrictions**
Setbacks are zoning requirements that establish the minimum distance a building or structure must be placed from the **site's property lines**. These spatial buffers help preserve **fire safety**, **access to light and air**, **neighborhood character**, **privacy**, and **emergency access**. Setbacks typically vary based on the **type of use** (residential, commercial, industrial), **building height**, and **adjacent zoning designations**.

There are typically four setback zones:
- **Front setback** – Distance from the front property line (often aligning with the street)
- **Side setbacks** – Distances from each side property line
- **Rear setback** – Distance from the back property line
- **Special or contextual setbacks** – Additional setbacks for corner lots, sloped sites, or buffer zones near incompatible uses (e.g., industrial next to residential)

Example 1:
A **multifamily housing project** on an urban lot may be required to maintain:
- A **15-foot front yard setback** (preserving street character and allowing for landscaping),
- **10-foot side setbacks** (ensuring light, privacy, and access to windows), and
- A **20-foot rear setback** (for services, open space, or emergency access).

These setbacks constrain the **buildable area**, directly influencing **building footprint**, **massing**, and **site layout** (such as parking, patios, or service access).

Example 2:
In a **suburban single-family residential zone**, typical setbacks might include:

- **25 feet front**
- **5 feet sides**
- **20 feet rear**

This enforces low-density development and ensures separation between houses for safety and privacy.

Example 3:
A **commercial building** on a **main street with a "build-to line"** may have a **0-foot front setback** to promote pedestrian activity and urban continuity, while maintaining rear or side setbacks for delivery access or mechanical yards.

Design Implications:
- **Building placement and massing** must conform within the "buildable envelope" defined by setbacks.
- **Setback encroachments** (e.g., porches, canopies, balconies) may be allowed or restricted depending on local code.
- **Infill or irregular lots** may require **variance applications** if strict setbacks prevent reasonable development.
- **Fire separation distance** regulations work in tandem with setbacks to prevent fire spread between structures on adjacent lots.

Understanding and applying setback rules is crucial for **code compliance**, **site efficiency**, and **neighborhood integration**. Early coordination with zoning maps and local ordinances ensures a viable site strategy and avoids costly redesigns or permit delays.

3. **Site Coverage and Impervious Surface Limits**
   Zoning often regulates the **percentage of the site that can be covered by buildings, pavement, or other impervious materials** to control runoff and maintain green space.

   Site coverage and **impervious surface limits** are zoning tools used to regulate how much of a lot can be covered by structures and hard surfaces such as buildings, driveways, walkways, and patios. These limits aim to manage **stormwater runoff**, reduce **flood risk**, preserve **natural infiltration**, and mitigate **urban heat island effects**—especially in environmentally sensitive or high-density areas.

   Two related metrics are commonly used:
   - **Building coverage**: The percentage of the lot covered by the building footprint (including roof area).
   - **Impervious surface coverage**: The percentage of the lot covered by surfaces that prevent water absorption, such as asphalt, concrete, or rooftops.

   Many zoning ordinances require a **minimum percentage of pervious or landscaped area**, which supports **stormwater infiltration**, **tree planting**, and **open space**.

   Example 1:
   A zoning code may limit **total impervious coverage to 60%** on a 10,000 sq. ft. site, allowing

only **6,000 sq. ft.** for buildings, driveways, and walkways. The remaining **4,000 sq. ft.** must be **pervious**, such as turf, landscaped areas, or porous paving, which helps reduce runoff and promote on-site water infiltration.

Example 2:
In a **low-impact development (LID)** zone, a commercial site is restricted to **50% impervious surface**, encouraging the use of **bioswales, rain gardens, and permeable pavers** in parking areas to manage stormwater naturally.

Example 3:
In a **residential neighborhood near a lake**, local zoning limits **building coverage to 35%** and **total impervious coverage to 50%**, to protect water quality. A proposed home must be designed with a compact footprint, minimal paved driveway, and extensive native planting.

Design Implications:
- Architects must carefully balance **programmatic needs**, **parking requirements**, and **circulation paths** with coverage limits.
- Use of **green infrastructure**, such as **green roofs**, **rainwater harvesting**, and **permeable paving systems**, can help projects meet coverage thresholds while enhancing environmental performance.
- Some municipalities offer **coverage bonuses** for sustainability features, such as **tree planting**, **stormwater reuse systems**, or **green roofs**.
- **Calculations** must include **all impervious surfaces**, including sheds, patios, and walkways—not just the main building.

Complying with site coverage and impervious surface regulations is essential for achieving **environmental compliance**, gaining **permitting approval**, and contributing to the long-term **resilience and sustainability** of the built environment.

4. **Parking and Accessible Parking Requirements**
   Parking requirements are governed by a combination of **local zoning ordinances**, the **Americans with Disabilities Act (ADA)**, and in many jurisdictions, the **International Building Code (IBC)**. Together, these standards regulate:
   - **Minimum number of total parking spaces** based on occupancy or use
   - **Required number and design of accessible parking spaces**
   - **Location, size, signage, and path-of-travel requirements**
   - **Site circulation and layout**

   1) Zoning Code Requirements
   Zoning ordinances typically establish:
   - Minimum or maximum number of parking spaces per use type (e.g., residential, commercial, assembly)
   - Requirements for **bicycle**, **EV**, or **compact** parking
   - Landscaping buffers and setbacks for surface lots
   - Parking reductions for transit-rich areas or shared-use developments

Example:
A **10,000 sq. ft. community center** with a zoning requirement of **1 space per 300 sq. ft.**:
10,000 ÷ 300 = **34 parking spaces required**

2) ADA and IBC Requirements for Accessible Parking
The **Americans with Disabilities Act (ADA)** and **International Building Code (IBC)** both require a minimum number of **accessible parking spaces**, determined by the **total parking count**.
IBC Reference:
- IBC Section **1106** (2021 edition) aligns with ADA Standards but may vary slightly by jurisdiction.
- Where IBC is adopted, it supplements ADA by reinforcing **technical specifications**, especially regarding **accessible routes**, **van spaces**, and **parking structure requirements**.

Figure 4.1 Accessible Parking Table (per IBC & ADA):

Total Parking Spaces	Required Accessible Spaces
1–25	1
26–50	2
51–75	3
76–100	4
101–150	5
...	...

- **1 in every 6 accessible spaces must be van-accessible**
- Van-accessible spaces must be at least **132 inches wide** with a **60-inch aisle**, or **96 inches wide** with a **96-inch aisle**
- Signage must include the **International Symbol of Accessibility**, and be mounted **60 inches minimum above ground to centerline**

3) Example Scenarios with Zoning + ADA/IBC Integration
Example 1: Community Center
- **10,000 sq. ft.** @ 1 per 300 sq. ft. = **34 spaces** (per zoning)
- **Accessible spaces required:** 2 (IBC/ADA for 26–50 total spaces)
- **At least 1** must be **van-accessible**
- Spaces located near **main entrance**, with a **continuous accessible route**

Example 2: Medical Facility
- **6,000 sq. ft. clinic** @ 1 per 200 sq. ft. = 30 spaces
- **Medical use requires 10% of parking to be accessible** → 3 spaces
- Van-accessible design and clear signage required per IBC §1106.5

Example 3: Multifamily Housing (24 Units)
- Zoning requires 1.5 spaces per unit → 36 total
- **ADA/IBC accessible spaces** = 2 minimum (26–50 total range)
- Path to residential lobby must be **fully accessible and compliant with IBC Chapter 11**

4) Additional Design Considerations (Zoning + IBC)

- **Accessible Route (IBC 1106.6)**: Must be on the **shortest accessible path** to the accessible entrance and comply with IBC §1104 (Accessible Routes)
- **Slope Requirements**: Max slope of **1:48** in any direction at accessible parking areas
- **Surface Requirements**: Must be **firm, stable, slip-resistant**
- **Parking Structures**: IBC requires the same accessibility standards for garages as surface lots
- **Lighting and Safety**: Some jurisdictions and IBC-adjacent codes require minimum **illumination levels** in accessible areas
- **EV Charging**: If included, EV stalls must **not reduce required accessible spaces**, though some codes require **accessible EV stalls**

Summary

Architects and designers must consider all three regulatory layers when planning parking:

1) **Zoning codes** — define how many total spaces are required and where they can go
2) **ADA Standards** — define how many must be accessible and the minimum standards for use
3) **IBC (Chapter 11)** — reinforces ADA accessibility while adding specific technical requirements for built environments, especially in structured parking or complex facilities

Failure to comply with any of these layers can result in **permit delays**, **legal exposure**, or **barriers to equitable access**.

5. **Stormwater Management Regulations**

Projects must manage **stormwater runoff** to prevent flooding, erosion, and water pollution. This may involve **on-site detention**, **green infrastructure**, or **connection to public systems**.

All development projects—residential, commercial, or industrial—must incorporate robust stormwater management systems to control runoff resulting from rainfall and impervious surfaces. Uncontrolled stormwater can lead to significant environmental and infrastructural issues, including localized flooding, streambank erosion, sedimentation of water bodies, and contamination of groundwater and surface water with pollutants such as oils, heavy metals, and nutrients.

To mitigate these risks, stormwater management regulations typically require the capture, storage, treatment, and gradual release or infiltration of stormwater on-site. These measures ensure that post-development runoff conditions mimic pre-development hydrology to the greatest extent possible. Projects must meet local, state, or federal stormwater codes, often based on performance standards tied to specific storm events (e.g., managing runoff from a 10-, 25-, or 100-year storm event).

Common Stormwater Management Techniques Include:

- **On-site detention and retention systems:** Underground or surface tanks and basins that temporarily store stormwater and slowly release it to prevent surges in downstream flow.

- **Green infrastructure:** Nature-based solutions like green roofs, bioswales, vegetated filter strips, rain gardens, and constructed wetlands that use vegetation and soil to treat and absorb runoff.
- **Low Impact Development (LID) practices:** Techniques such as permeable pavements, infiltration trenches, and rainwater harvesting that promote infiltration and reduce impervious surface area.
- **Pre-treatment devices:** Sediment forebays, oil-water separators, and hydrodynamic separators to remove contaminants from runoff before infiltration or discharge.

Design Implications:
Integrating stormwater management into project design has significant implications for land planning, site layout, architectural detailing, grading, and long-term maintenance. Designers and engineers must consider the following:

- **Site Constraints & Space Allocation:**
  Stormwater facilities require space—bioswales and detention basins occupy surface area, while underground tanks need excavation and structural accommodation. This can affect parking layouts, building footprints, and landscaping plans. For urban infill projects, vertical solutions (e.g., green roofs, cisterns) may be necessary due to space constraints.
- **Grading & Drainage Patterns:**
  Proper grading must direct runoff toward collection points (e.g., swales or drains) without causing ponding near buildings or pedestrian areas. Sites with poor natural drainage may require sub-surface drainage or increased detention volumes.
- **Materials & Pavement Choices:**
  The use of permeable pavers, porous asphalt, or pervious concrete impacts both stormwater infiltration and aesthetic design. These materials may require specific subgrade preparation and maintenance plans to ensure effectiveness over time.
- **Vegetation & Planting Design:**
  Green infrastructure requires careful selection of native or adaptive plants that can tolerate varying moisture conditions. Designers must coordinate landscape design with hydrological performance to ensure long-term functionality and ecological value.
- **Regulatory Coordination & Documentation:**
  Civil engineers and design teams must prepare detailed stormwater management plans—including hydrologic and hydraulic calculations, pollutant removal efficiency, maintenance plans, and erosion control measures—for review and approval by relevant permitting authorities.
- **Maintenance & Lifecycle Considerations:**
  Designers must also account for ongoing maintenance requirements, such as sediment removal from bioswales or periodic cleaning of infiltration systems. These can influence operational budgets and long-term ownership responsibilities.

Examples:

- A **mixed-use development** in a coastal city is designed with green roofs on each building, directing stormwater into terraced rain gardens that overflow into an underground infiltration gallery, reducing runoff to sensitive marine ecosystems.

- A **shopping center** incorporates permeable pavers in its parking lots, with stormwater channeled into bioswales planted with native grasses. An underground storage tank temporarily detains excess water before controlled release to a municipal storm drain.
- A **logistics warehouse** near a floodplain integrates an above-ground detention basin capable of managing a 100-year storm event. The basin doubles as a landscaped feature and includes sediment forebays to enhance pollutant removal.
- A **suburban school campus** uses rain gardens and disconnected downspouts to direct rooftop runoff to planted areas. A rainwater harvesting system collects excess runoff for irrigation, promoting sustainability and educational opportunities.
- A commercial building must include **bioswales**, **permeable pavers**, and an **underground detention tank** to manage a 25-year storm event, meeting local environmental impact requirements.

6. **Height Limits and Bulk Controls**

Zoning codes often impose **maximum building height**, **step-backs**, or **angle-of-light requirements** to maintain scale, preserve views, and reduce shadow impacts on neighboring properties.

Zoning ordinances commonly regulate the height, massing, and overall volume—or "bulk"—of buildings to ensure new development is compatible with its surroundings. These controls are critical for preserving neighborhood character, protecting access to light and air, reducing visual dominance, minimizing shadows on public spaces, and maintaining public view corridors such as toward parks, waterfronts, or historic landmarks.

Typical regulatory tools include:

- **Maximum Building Height:** A fixed vertical limit, often measured from average grade or base flood elevation, that restricts how tall a building can be.
- **Step-backs (or Setbacks at Upper Levels):** Requirements that portions of a building recede as they rise, usually above a certain height (e.g., after the third floor), to reduce the perceived mass from the street level and adjacent properties.
- **Angle-of-Light / Sky Exposure Planes:** Geometric controls (like a sloping plane from the property line inward) that regulate the building envelope to ensure access to daylight and limit overshadowing.
- **Floor Area Ratio (FAR):** A ratio that limits total buildable floor area relative to lot size, which indirectly controls height and bulk.
- **Lot Coverage & Building Footprint:** Restrictions on how much of a site can be occupied by a building, often paired with open space requirements.

These regulations vary by zoning district and may differ for residential, commercial, or mixed-use zones. In urban areas, especially near historic districts or low-rise neighborhoods, height and bulk controls are often more restrictive to preserve context.

Design Implications:

Height and bulk regulations have significant effects on building form, massing strategy, and internal space planning. Designers and architects must carefully interpret and respond to these rules while balancing client objectives, site constraints, and architectural vision.

Key design considerations include:

- **Façade Articulation and Massing Strategy:**
  Step-backs, terraces, and varied rooflines can help break up a building's mass and comply with height-related regulations. These features often serve dual purposes—meeting code while enhancing architectural character and usability (e.g., roof gardens or balconies).
- **Sunlight and Shadow Studies:**
  Projects near sensitive areas such as schools, parks, or low-density housing may require detailed shadow impact analyses to demonstrate compliance and minimize overshadowing. These studies may influence building orientation and massing choices.
- **Maximizing Usable Floor Area within Height Limits:**
  In areas with strict height caps but generous FAR, architects may opt for taller floor-to-ceiling heights or prioritize fewer but more spacious floors. Efficient floor plate design and mechanical layout optimization are essential.
- **Material and Color Choices for Perception Management:**
  Light materials and vertical design elements can reduce perceived bulk, especially in taller buildings with narrow step-backs. These visual strategies may be encouraged or required in design overlay zones.
- **Integration with Adjacent Context:**
  Height transitions at zoning boundaries—e.g., between high-rise and low-rise zones—often necessitate design moves like tapering or partial setbacks to avoid abrupt visual or physical contrasts.

Examples:

- A **mid-rise apartment building** in a transitional zone is limited to a total height of 45 feet. The design must incorporate a step-back of at least 10 feet on upper floors after 30 feet in height to preserve sunlight on the adjacent single-family homes and create a more human-scaled streetscape.
- In a **downtown commercial district**, a mixed-use tower is subject to an angle-of-light regulation requiring that no portion of the building project into a 45-degree plane starting at 60 feet above the sidewalk. As a result, the tower's upper floors are designed as a tapered form to comply while still achieving maximum allowable FAR.
- A **coastal condominium development** must stay under a 35-foot height cap to protect views toward the ocean from a public bluff trail. The design incorporates sloped roofs and upper-level terraces to preserve views while offering premium units with outdoor access.
- A **university building** near a residential zone is limited in height on the side facing the neighborhood. Step-backs and green roofs are used on that elevation, while the opposite façade facing the campus courtyard takes advantage of the full permitted height and bulk.

By accurately interpreting and applying these zoning and environmental requirements, architects ensure that their designs are both **permissible and context-sensitive**, while reducing the risk of costly redesigns, permitting delays, or community opposition.

B. Apply building codes to building design (U/A) (8-12%)

NCARB Requirements
It is also important to assess the building code requirements that impact a project. This includes building height and area limits, maximum occupant load, fire separation, required egress, accessibility, plumbing fixture counts, and energy usage. Architects must be able to compile this information into a code analysis and incorporate it into the design.

On the ARE 5.0 PPD exam—and in professional practice—architects must assess and apply relevant **building codes** that directly affect the layout, massing, performance, and safety of a project. This includes understanding and calculating limits for **building height and area**, determining **occupant loads**, providing **adequate egress**, ensuring **fire protection**, addressing **accessibility**, confirming **fixture counts**, and maintaining **energy code compliance**. Architects must compile this data into a **clear code analysis** and integrate the findings into both the **design and documentation**.

1. Determine Allowable Building Height and Area

The **International Building Code (IBC)** establishes maximum allowable **building height** and **floor area** to ensure fire safety, emergency access, and structural performance. These limits are not fixed—they vary depending on multiple project-specific parameters and can be increased when certain code conditions are met.

☑ **Key Factors That Determine Building Height and Area:**

- **Occupancy Classification**
  Different use groups (e.g., Business, Assembly, Residential, Storage) have different allowable areas and height limits. For example, a school (Group E) has different base allowances than a warehouse (Group S-1).
- **Construction Type**
  The construction type (Types I–V, A or B) reflects the level of fire resistance. More fire-resistant types (e.g., Type I or II) allow greater height and area.
  - *Type I-A*: Non-combustible, high fire resistance → maximum allowances
  - *Type V-B*: Wood-framed, least resistant → limited height and area
- **Presence of Automatic Sprinkler System**
  Fully sprinklered buildings (per NFPA 13) qualify for increases in both height (up to one additional story) and floor area (up to 300%).
- **Number of Stories**
  Limits are also placed on how many floors a building can have above grade. Taller buildings must meet stricter fire protection and structural requirements.
- **Frontage Allowance (Section 506.2)**
  If a building has a significant portion of its exterior walls fronting open space (e.g., a public street or open yard), it may receive an **area increase** because fire department access and smoke dissipation are improved.

☑ **Key Action: Code Analysis Using IBC Chapter 5**

Use **IBC Chapter 5** to identify the **base allowable height and area** per occupancy and construction type. Then apply any **modifiers** based on:

- Section **506.2**: Frontage Increase
- Section **506.3**: Sprinkler Area Increase
- Section **504.2 and 504.3**: Height and Story Increase with Sprinklers

You may also reference **Tables 504.3, 504.4, and 506.2** for determining these thresholds.

📌 Example Scenario 1: Basic Area Increase

A new office building (Occupancy Group B) is designed using **Type IIB construction** and includes:

- 75% of the perimeter fronting open public space
- A fully compliant NFPA 13 sprinkler system

Per IBC Table 506.2:

- **Base allowable area (unsprinklered)** for Type IIB, Group B = 23,000 SF
- **Area increase due to frontage (506.2)**: up to 25% bonus
- **Area increase due to sprinklers (506.3)**: up to 300%

After applying the **area increase formula (IBC Equation 5-1)**:

Aa = At + (At × If) + (At × Is)

Where:

- Aa = Allowable area per story
- At = Tabular area (e.g., 23,000 SF)
- If = Frontage increase factor
- Is = Sprinkler increase factor

The architect determines the building can legally have a **floor plate of 25,000 SF or more**, which accommodates the client's program while remaining compliant.

📌 Example Scenario 2: Height Limitation Due to Construction Type

A developer proposes a 4-story apartment building (Group R-2) using **Type V-A construction.**

- Base max height per Table 504.3: **3 stories / 70 ft**
- Even with sprinklers, **a 4th story is not permitted** in Type V-A unless upgraded to Type III construction.

In this case, to meet the client's program, the architect must:

- Choose a more fire-resistant construction type (e.g., Type III-A),
- Or reduce the number of stories to comply.

☑ Best Practices for ARE 5.0 and Practice

- Always **begin code analysis early** in schematic design. Changing construction type or reducing stories late in design can be costly.
- Show **height and area calculations** clearly in the **code summary sheet** in your drawings.
- Confirm with the **AHJ** if local amendments modify IBC base values.
- Use area increase calculations in Excel or architectural software templates to avoid math errors.

🔍 ARE 5.0 PPD Exam Tips:

- Expect questions requiring identification of allowable area based on given parameters.

- Be ready to apply height and area limits when choosing between different construction types.
- Know when sprinkler systems allow increases—and when they don't.

2. **Classify Occupancy and Construction Type**

Every building must be assigned:

- A **primary occupancy group** (e.g., A, B, E, R, S),
- And a **construction type** (I through V, A or B) based on material combustibility and fire resistance.

☑ **Key Action**: Select the construction type and occupancy early to inform structure, fireproofing, and exit design. Consider if the building is **separated or non-separated mixed use**.

Example:
A mixed-use retail (M) and apartment (R-2) building may be designed as non-separated occupancy, using the most restrictive requirements across the entire structure.

Properly classifying **occupancy** and **construction type** is a critical step in code analysis and a foundational requirement for design development. These classifications inform everything from allowable height and area to fire separation requirements, means of egress, and the level of structural fire protection needed throughout the building.

This step is one of the first in the **building code application process**, as almost every code provision hinges on the accurate identification of:

- The **primary use(s)** of the building (occupancy group)
- The **materials and fire-resistance** characteristics of its structure (construction type)

◇ **Occupancy Classification**

The **occupancy classification** is defined by **IBC Chapter 3**, and reflects the building's function or use. Common occupancy groups include:

Group	Use Example
A	Assembly (theaters, churches)
B	Business (offices, banks)
E	Educational (schools, daycare)
F	Factory/Industrial (manufacturing)
H	High-Hazard (flammable storage)
I	Institutional (hospitals, prisons)
M	Mercantile (retail stores)
R	Residential (apartments, hotels)
S	Storage (warehouses, garages)
U	Utility/Miscellaneous (sheds, barns)

Figure 4.2 Common occupancy groups

🧠 **Note**: A building may contain **more than one occupancy type**. In this case, IBC allows for *separated* or *non-separated mixed-use* configurations.

◇ Construction Type

Construction types are categorized in **IBC Chapter 6** and are designated Types **I through V**, with subtypes **A (protected)** and **B (unprotected)**. The classification is based on:

- Combustibility of structural materials (e.g., steel, concrete, wood)
- Fire-resistance ratings of primary building elements (e.g., walls, floors, roof)

Type	Fire Resistance	Materials	Typical Use Case
Type I	Highest	Non-combustible	High-rises, hospitals, schools
Type II	High	Non-combustible	Commercial offices, parking garages
Type III	Moderate	Non-combustible exterior, combustible interior	Apartments, mid-rises
Type IV	Heavy timber	Combustible (solid wood)	Historic buildings, sustainable projects
Type V	Lowest	Combustible	Residential, small commercial buildings

☑ **Type A** = Fireproofed structural members

☑ **Type B** = No fireproofing (lower fire resistance)

Figure 4.3 Construction types

☑ Key Action: Classify Early in the Design Process

Choosing the correct occupancy and construction type in **schematic design** is essential to:

- Determine allowable **height, number of stories, and area** (IBC Chapter 5)
- Specify **required fire-resistance ratings** for structural and separating elements
- Design appropriate **means of egress** (number and width of exits)
- Coordinate with **consultants** (structural, fire protection, MEP)

Delaying classification can lead to noncompliant layouts, undersized egress paths, or costly redesigns later in the project.

📌 Mixed-Use Buildings: Separated vs. Non-Separated

Mixed-use projects often involve two or more occupancy groups (e.g., residential over retail). IBC allows two approaches:

1) **Separated Occupancy (IBC 508.4)**
   - Occupancies divided by rated fire barriers
   - Each area complies independently with height/area limits
   - Fire ratings depend on occupancy separation table (IBC Table 508.4)
2) **Non-Separated Occupancy (IBC 508.3)**
   - Treated as one combined occupancy
   - Entire building designed to meet most restrictive requirements of all occupancies present

- Simplifies some calculations but may limit design flexibility

🧱 **Example Scenario: Mixed-Use Building**
A proposed **4-story mixed-use development** includes:

- **Retail (Group M)** on the ground floor
- **Apartments (Group R-2)** on the upper floors
- Type **III-A** construction is selected to meet fire-resistance and wood construction goals

👉 **Option A**: Design as **non-separated mixed-use**

- Use most restrictive code provisions between M and R-2
- E.g., fire-rated corridors, increased stair enclosure ratings

👉 **Option B**: Design as **separated occupancy**

- Provide a 2-hour rated floor/ceiling between the retail and residential levels
- Can apply occupancy-specific height/area limits independently

In this case, if the building exceeds the allowable area for Type III-A under non-separated use, the architect may recommend using separated occupancy to increase floor area flexibility.

📝 **ARE 5.0 PPD Exam Tips**

- Expect to classify a building's **occupancy and construction type** from descriptions or plans
- Understand how mixed-use classification affects allowable area and separation requirements
- Know which construction types require **fire-resistive protection of structural elements**

🔍 **Summary**
Occupancy and construction type classification are **fundamental to building code compliance**. These designations determine nearly all fire and life safety requirements and must be applied **accurately and early**. A thorough understanding allows the architect to develop a feasible design, avoid conflicts with code, and set realistic expectations with the client.

3. **Calculate Occupant Load and Egress Requirements**
   Properly calculating the **occupant load** is fundamental to ensuring life safety and code compliance. The occupant load directly informs the **number, size, arrangement, and type of exits**, and influences everything from **door swing direction** to the **requirement for exit signage** and **area of refuge** locations.

   This process is essential early in the design phase to verify that the building can safely accommodate the intended number of people under both normal and emergency conditions.

◇ **What Is Occupant Load?**
Occupant load refers to the number of people a building, space, or room is designed to hold. It is calculated based on:

- **Function of the space**
- **Square footage**
- **Load factor** (occupant load per square foot), specified in the IBC

The calculated occupant load is used to determine:

- The **number of required exits**
- **Minimum egress width** for corridors, doors, stairs, and ramps
- **Exit signage and emergency lighting**
- **Door swing direction and hardware**
- **Areas of refuge** for accessible egress

☑ **Key Action: Use IBC Table 1004.5 and Chapter 10**

- **IBC Table 1004.5** provides the occupant load factor for different uses (e.g., assembly, business, mercantile).
- **IBC Chapter 10** outlines requirements for:
    - Minimum number of exits (Section 1006)
    - Exit width (Section 1005)
    - Travel distance limitations (Section 1017)
    - Door swing direction and panic hardware (Sections 1010.1.2 and 1010.1.10)

◇ **Minimum Number of Exits – IBC Section 1006 Summary**

The **International Building Code (IBC) Section 1006** establishes the **minimum number of exits or exit access doorways** required from **spaces** and **stories** to ensure safe egress during an emergency.

☑ **From Individual Spaces (IBC 1006.2)**

- **1 exit required** if:
    - Occupant load ≤ 49 for most occupancies (≤ 29 for Group I-2 or R)
    - Travel distance to exit ≤ allowable limits (e.g., 75 ft unsprinklered, 100 ft sprinklered for certain occupancies)
- **2 exits required** if:
    - Occupant load > 49 (or > 29 for I-2, R)
    - Travel distance exceeds the limit for one exit
- **3 or more exits** may be required based on:
        - Occupant load thresholds (e.g., > 500 requires 3 exits; > 1,000 requires 4 exits)

☑ **From Stories (IBC 1006.3)**

- **1 exit required** from a story if:
    - Building is sprinklered
    - Occupant load per story:
        - ≤ 49 (Group B, M, etc.)
        - ≤ 10 (Group A, E, I, R)
- **2 exits required** if:
    - Occupant load per story > thresholds listed above
    - Travel distance exceeds maximum for one exit
- **3 or more exits** required for:
    - 500 occupants (3 exits)
    - 1,000 occupants (4 exits)

📌 **Key Design Considerations:**

- **Travel distance** is critical: one exit may be allowed if it's within allowed distance limits
- **Sprinkler systems** allow increased travel distances and sometimes reduce number of required exits
- **Egress paths** must be independent (exits must be remote from one another)

🧠 **Exam Tip (ARE 5.0):**

Know the occupant load and travel distance limits that trigger the need for **an additional exit**. Often tested with a scenario that straddles these thresholds.

📌 **Example Calculation: Classroom Space**

A **classroom** measures **1,000 square feet**.
According to **IBC Table 1004.5**, the load factor for an **educational use** is:

- **20 SF per person**

Occupant Load = 1,000 SF ÷ 20 SF/person = 50 occupants

This occupant load drives several design decisions:

- **Minimum number of exits**: Per IBC Section 1006.2.1, two exits are required for spaces with >49 occupants.
- **Door swing**: Doors must **swing in the direction of egress** travel if occupant load exceeds 50.
- **Exit width**: Minimum clear width of egress doors is **32 inches**; total egress width for 50 occupants (0.2"/occupant for level egress) = **10 inches**. Since doors are typically 36", this is sufficient.
- **Travel distance**: Must comply with **maximum limits** (e.g., 200 ft for fully sprinklered buildings; 150 ft for non-sprinklered).
- **Exit signage**: Required to clearly indicate exits if the path of travel is not immediately obvious.

📋 **Additional Example: Mixed-Use Floor Plan**

A 4,000 SF floor includes:

- 2,000 SF of **retail (Group M)** → Load factor = 60 SF/person → **33 occupants**
- 2,000 SF of **office (Group B)** → Load factor = 150 SF/person → **13 occupants**

Total Occupant Load = 33 + 13 = 46 occupants

- **Number of exits**: One exit is sufficient (IBC 1006.3.1) because occupant load <50 and travel distance is within limits.
- **Egress width**: 46 occupants × 0.2" = 9.2" → One 36" door is adequate.
- **Signage**: May not be required if exit is immediately visible.

🚨 **Common Pitfalls to Avoid**

- Misidentifying occupancy use group (wrong load factor)
- Forgetting to round **up** to the next whole number for occupant load
- Ignoring local amendments or stricter egress width rules
- Failing to update occupant loads after plan changes (e.g., partition changes)

Code Requirement	Triggered When…
2 or more exits required	Occupant load > 49 (Assembly/Business), or travel distance exceeded
Door swing in direction of egress	Occupant load > 50
Panic hardware required	Occupancy = A or E and load > 50
Area of refuge required	Multi-story buildings with accessible floors
Egress stair width ≥ 0.3"/occupant	For stairs (IBC 1005.3.1)
Corridor width ≥ 44"	Occupant load > 50 or serving public assembly

Figure 4.4 Factors That Affect Egress Design

ARE 5.0 PPD Exam Tips

- Be ready to **calculate occupant load** based on given area and function.
- Understand how occupant load drives **egress width**, **number of exits**, and **travel distances**.
- Expect questions on how **exit requirements change** as the occupant load increases (e.g., 49 vs. 50 people).
- Know when **door swing and panic hardware** are required.

Summary

Accurate occupant load calculations ensure proper egress capacity and safe evacuation in emergencies. On the ARE and in real-world practice, architects must integrate these calculations early to avoid noncompliance, redesigns, or costly delays. A solid grasp of IBC Table 1004.5 and Chapter 10 is essential for every code-savvy designer.

4. **Identify and Design Fire Separation Elements**
Fire-resistive construction plays a critical role in safeguarding buildings by **slowing the spread of fire and smoke**, providing time for occupants to evacuate safely, and enabling emergency responders to act. The design and specification of fire-rated assemblies are essential to both code compliance and life safety performance.

Why Fire Separation Matters

Fire-resistive construction is used to:

- **Separate different occupancies** with varying risk levels (e.g., business vs. hazardous storage)
- **Define fire areas** to control and contain fire spread within large buildings
- **Compartmentalize hazards** such as mechanical rooms, elevator shafts, or flammable material storage
- **Protect egress routes**, including exit corridors, stair enclosures, and horizontal exits, to ensure occupants have a safe path during emergencies

These separations are typically achieved using fire barriers, fire partitions, smoke barriers, and horizontal assemblies—all with defined fire-resistance ratings.

☑ **Key Action: Reference IBC Chapter 7 and UL Assemblies**

- Use **IBC Chapter 7** to determine where and what type of fire-resistive assemblies are required.
- Reference **UL (Underwriters Laboratories) assemblies** or other approved testing agencies for rated wall, floor, and ceiling construction details.
- Ensure **fire-rated elements are labeled** on floor plans, sections, and details—including wall types, ratings, continuity, penetrations, and construction notes.

Element Type	Typical Use Case	Fire Rating
Fire Barrier	Occupancy separation, shaft enclosures	1–4 hours
Fire Partition	Corridor walls, tenant separations	1 hour
Smoke Barrier	I-2 occupancies (healthcare), areas of refuge	1 hour
Horizontal Assembly	Separates floors/ceilings between levels	Varies
Fire Wall	Complete separation between buildings or areas	2–4 hours

Figure 4.5 Types of Fire-Resistive Construction Elements

📌 **Example 1: Mixed Occupancy Fire Separation**

A **Group B office** shares a building with an adjacent **Group S-1 storage** area. According to **IBC Table 508.4**, a **2-hour fire barrier** is required between these occupancies when designed as a **separated mixed-use** occupancy.

- This barrier must be continuous from **floor slab to roof deck** (or through the roof if applicable).
- Openings (e.g., doors) within the fire barrier must be **fire-rated assemblies**—in this case, typically 90-minute doors with closers and smoke seals.

📌 **Example 2: Corridor Protection in Group R-2**

An apartment building (Group R-2) includes interior corridors serving dwelling units. Per **IBC Section 1020.1**, these corridors may require **1-hour fire partitions** if they serve more than 10 occupants and do not have direct exits.

- If the building is **sprinklered throughout (NFPA 13)**, fire-resistance requirements for corridors may be reduced.
- However, penetrations in corridor walls (e.g., ducts, piping) still require **fire- and smoke-rated dampers** to maintain wall integrity.

📌 **Example 3: Stair Enclosure Protection**

Stair enclosures that connect multiple floors must be enclosed in **2-hour fire barriers** (in buildings >4 stories) or **1-hour barriers** for shorter buildings, per **IBC Table 1020.1** and Section **1023.2**.

- These stair shafts must be constructed as **continuous fire barriers**, sealed at all penetrations.
- Doors must be **self-closing and fire-rated** (usually 90 minutes for 2-hour walls).

🛠 Additional Considerations

- **Continuity**: Fire barriers must extend **from foundation to roof**, or **floor to underside of floor/roof deck** as required.
- **Penetrations**: Pipes, ducts, conduits, and wires must be protected with **tested firestop systems** (UL listed).
- **Joints**: Construction joints in rated assemblies (e.g., floor-to-wall) require **fire-resistive joint systems**.
- **Openings**: Doors, windows, and glazing in rated assemblies must meet appropriate fire-resistance and **temperature rise** ratings.

📖 ARE 5.0 PPD Exam Tips

- Be familiar with **where fire separations are required** in common scenarios (e.g., mixed-use buildings, egress enclosures).
- Know the difference between a **fire barrier** and a **fire partition**, and what each is used for.
- Understand how fire-rated assemblies are documented on drawings—look for wall tags, UL references, and fire ratings in legends.

🔍 Summary

Fire-resistive construction is an integral part of code-compliant architectural design. By properly identifying where fire-rated separations are needed—and designing them with accurate ratings and tested assemblies—architects help contain fire, protect egress, and ensure safety for all occupants. Use IBC Chapter 7 and UL assembly references to inform your detailing and coordination from the earliest phases of design.

5. **Ensure Accessible Design Compliance**

Accessibility is a foundational requirement in architectural design, ensuring that buildings are usable by all individuals, including those with physical, sensory, and cognitive disabilities. In the United States, accessible design in commercial and public-use buildings is governed by the **Americans with Disabilities Act (ADA)** and the technical provisions of **ANSI A117.1**, as well as **state-specific amendments** like **California Building Code (CBC) Chapter 11B**.

♿ Key Areas of Compliance

To meet ADA and ANSI A117.1 standards, architects must carefully design and coordinate the following elements:

- **Entrances**: At least one accessible entrance must connect to the public way and parking areas. Thresholds must be ≤ ½" high and doors must meet width and force requirements.
- **Restrooms**: Must include accessible stalls with 60" minimum turning radius, grab bars, and properly mounted fixtures (e.g., sinks at 34" max height, mirrors at 40" max to bottom edge).
- **Door Hardware**: Lever or push-type hardware is required; knobs or tight-grasping/pinching hardware are not permitted. Door operation force must be within allowable limits (typically ≤ 5 lbs for interior doors).
- **Vertical Circulation**:
  - **Elevators** must serve all accessible floors.

    - **Ramps** must not exceed 1:12 slope (8.33%) and require landings at top/bottom and every 30 ft horizontally.
    - **Platform lifts** are permitted only in specific situations (e.g., theaters, historic buildings).
- **Clear Floor Space & Turning Radius**:
    - 30" x 48" clear floor space is required at accessible fixtures.
    - 60" turning radius is required in areas like restrooms and accessible kitchens.

☑ Key Action: Dimension and Annotate Accessible Features

On all **plans, sections, and details**, architects must:

- Dimension clearances, mounting heights, and turning spaces
- Call out ADA/ANSI A117.1 references for fixtures and hardware
- Note any applicable **state or local accessibility codes**, such as:
    - **CBC Chapter 11B** (California)
    - **Texas Accessibility Standards (TAS)**
    - **Massachusetts Architectural Access Board (MAAB)**

Local amendments may **exceed** ADA minimums, so always verify with the Authority Having Jurisdiction (AHJ).

📌 Example 1: Door Clearance and Restroom Turning Space

For a **restroom door that swings into the room**:

- Minimum **18" clearance** is required on the **latch side** of the pull side of the door.
- A **5'-0" turning radius** (60" circle) is required inside the restroom for wheelchair maneuverability.
- Doors must not swing into required clearance areas for fixtures (e.g., toilet or sink clearance).

🧠 Tip: The 18" clearance allows a user in a wheelchair to position themselves beside the door for easier opening.

📌 Example 2: Accessible Kitchenette in Office Break Room

An office kitchenette must provide:

- **Forward approach** to the sink with 30" x 48" clear space
- **Counter height ≤ 34"**, with knee space below (27" high, 30" wide, 19" deep minimum)
- **Appliance controls** within accessible reach ranges (typically 15" to 48" above finish floor)

📌 Example 3: Ramp and Stair Coordination

When designing an exterior entry:

- If there are **>½" changes in level**, provide a **ramp with max 1:12 slope**.
- Include **level landings** at the top and bottom of the ramp.
- Provide **handrails** on both sides if ramp rise >6", extending 12" beyond top and bottom.
- Adjacent stairs must have **visual contrast striping** on nosings (required by CBC 11B).

🛠 Best Practices

- Use accessibility symbols consistently on plans (e.g., Ⓐ or ♿ for accessible features).

- Coordinate with consultants (MEP, interiors, civil) to ensure **consistent accessibility** across disciplines.
- Conduct a **thorough plan check** for overlapping clearances, especially in tight spaces like restrooms, exam rooms, or kitchens.
- Confirm **mounting heights** for dispensers, controls, switches, and visual alarms.

📝 ARE 5.0 PPD Exam Tips

- Be prepared to identify **noncompliant elements** from floor plans or elevations.
- Understand the **differences between reach ranges**, clearances, and maneuvering spaces.
- Know how to read and apply **ANSI A117.1 figures**—these are often used in exam questions.
- Expect questions comparing **federal ADA requirements** to **stricter local codes** like CBC.

🔍 Summary

Accessible design is not only a **legal requirement**—it's a critical component of ethical, inclusive architectural practice. Architects must demonstrate a working knowledge of ADA and ANSI A117.1 standards and integrate these principles seamlessly into design documentation. Early coordination and detailed annotations help prevent costly revisions and ensure environments are welcoming to all users.

6. **Verify Plumbing Fixture Counts**

Ensuring that a building has the correct number and type of **plumbing fixtures** is both a code compliance and user satisfaction issue. Undersupplying fixtures can lead to health, safety, and accessibility problems, while oversupplying can add unnecessary cost and space usage. Architects must determine fixture requirements based on **occupant load, occupancy type**, and applicable codes—typically the **International Building Code (IBC) Chapter 29** and/or a **local plumbing code**.

🚻 What Plumbing Fixtures Are Required?

Common required fixtures include:

- **Water closets (toilets)**
- **Lavatories (sinks)**
- **Drinking fountains**
- **Service sinks or janitor closets**
- **Showers** (if applicable by use, e.g., gyms, industrial facilities)

🔍 Key Determinants of Fixture Counts

1) **Occupant Load by Gender**
   - Divide total calculated occupant load **equally between male and female**, unless the use dictates otherwise (e.g., stadiums with known gender skews).
   - Use the **greater calculated fixture requirement** per gender unless unisex fixtures or multi-user rooms are permitted by the code.
2) **Occupancy Type**
   - Fixture quantities vary based on occupancy use group—e.g., office vs. restaurant vs. school.
   - Refer to **IBC Table 2902.1** or local plumbing code equivalent.

3) **Use Group Requirements**
    - Some uses require additional or specialized plumbing elements (e.g., urinals in assembly uses, showers in Group F or H occupancies).
4) **Local Code Modifications**
    - Jurisdictions may amend or replace IBC Chapter 29 with their own plumbing code (e.g., **California Plumbing Code**, **New York City Plumbing Code**).
    - Some local codes have **higher minimums or accessibility-specific provisions**.

☑ **Key Action: Provide Fixture Calculations on Plans**

- Include a **fixture calculation table** in the code analysis sheet or life safety plan.
- Clearly document:
    - Occupant load used for calculation
    - Code reference (e.g., IBC 2021 Table 2902.1)
    - Required vs. provided fixture counts
- Show **actual fixture locations** on floor plans—coordinate with plumbing and interiors consultants.

🧠 **Don't forget**: Plumbing fixtures must also meet **ADA / ANSI A117.1 accessibility requirements**, such as turning clearances, side/rear wall grab bars, accessible heights, and clear floor space.

📌 **Example: Office Occupancy Calculation**

For an office (Group B) with **100 total occupants**:

- Divide occupants evenly:
    - 50 men, 50 women

IBC Table 2902.1 Requirements (Business Occupancy):

- **Toilets**: 1 per **25 occupants** up to 50 →
    - **2 toilets per gender**
- **Lavatories**: 1 per **40 occupants** up to 80 →
    - **2 lavatories per gender**

Resulting Requirements:

Gender	Toilets	Lavatories
Male	2	2
Female	2	2

➡ **Note**: A urinal may substitute for up to **50% of required male toilets**, but at least **1 standard toilet stall** must remain.

📌 **Example: Restaurant with 120 Occupants**

For a **restaurant (Group A-2)** with 120 patrons:

- 60 males, 60 females

Fixture Requirements (IBC Table 2902.1):

- **Toilets (A-2)**:
    - 1 per 75 males → 1 WC
    - 1 per 75 females → 1 WC
- **Lavatories**: 1 per 200 occupants of each gender → 1 each

But if **local health codes** apply (which often do in food service), additional fixtures may be required.

💡 Additional Design Considerations

- **Unisex restrooms** can be used in some occupancies, especially for **small businesses or renovations**.
- **Gender-neutral facilities** may be allowed by local code; when used, fixture counts are often calculated based on **total occupant load**, not split by gender.
- **Multiple occupancy types in one building** may require fixture calculations **by each use group**, particularly when areas are separated (e.g., office + gym + café).
- **Accessibility requirements** include:
  - One accessible toilet and lavatory per restroom
  - Drinking fountains with high and low spouts or bi-level designs

📝 ARE 5.0 PPD Exam Tips

- Expect to calculate fixture counts using occupancy and code tables.
- Understand when **urinals may substitute for toilets** and when they may not.
- Know how fixture requirements change for **separated vs. combined occupancies**.
- Pay attention to **accessibility details**, such as turning space in single-user restrooms.

🔍 Summary

Plumbing fixture counts must be calculated early and verified throughout the design process to ensure compliance, accessibility, and user comfort. By aligning with **IBC Chapter 29**, **ANSI A117.1**, and **local plumbing codes**, architects can provide clear, coordinated documentation and avoid costly plan check delays or redesigns.

Include this on the **Code Analysis Sheet** in your construction documents or as a study reference:

Occupancy Type	**Occupant Load**	**Male**	**Female**	**Required WCs (M/F)**	**Required Lavs (M/F)**	**Other Fixtures**	**Code Reference**
Business (B)	100	50	50	2 / 2	2 / 2	Drinking fountain (1)	IBC Table 2902.1
Assembly (A-2)	120	60	60	1 / 1	1 / 1	Urinal (1 max)	IBC Table 2902.1
Mercantile (M)	200	100	100	2 / 2	1 / 1	Janitor sink	IBC Table 2902.1

🔎 *Adjust this format as needed for different use groups and local amendments.*

Figure 4.6 Fixture Calculation Table Template

📋 Code Summary Graphic Outline

Create a visual cheat sheet with the following sections for your guide:

🚻 IBC Table 2902.1 Fixture Summary (Sample Groups)

Occupancy	Water Closets (WCs)	Lavatories (Sinks)	Drinking Fountains	Service Sinks
Business (B)	1 per 25 (1–50), 1 per 50 (51+)	1 per 40 (1–80), 1 per 80 (81+)	1 per 100 (1 high, 1 low)	1 per floor or tenant space
Mercantile (M)	1 per 500 (M), 1 per 750 (F)	1 per 1,000	1 per 1,000	1 per floor or tenant space
Assembly (A-2)	1 per 75 (M/F)	1 per 200	1 per 500 (hi/lo)	1 required

🧠 Tip: Use these tables to compare quick fixture thresholds across major occupancy groups.

Figure 4.7 IBC Table 2902.1 Fixture Summary (Sample Groups)

♿ Accessibility Highlights (ANSI A117.1 / ADA)

- At least **1 accessible WC and lavatory per restroom**
- Turning radius: **60" min.** inside single-user restrooms
- **Approach clearance** at fixtures: 30" x 48"
- Mirror height: **max. 40"** to bottom edge
- High/low or **bi-level drinking fountains** required

7. **Integrate Energy Code Requirements**

   Energy codes are a critical part of building design and directly influence decisions that affect a building's energy performance, sustainability, and long-term operating costs. During the Project Planning & Design (PPD) phase, architects must understand and integrate relevant energy code requirements into the design documents to ensure compliance and inform system performance goals.

   Three of the most widely adopted energy codes and standards include:

   - **IECC (International Energy Conservation Code)** – Commonly adopted across the U.S., often forming the baseline code.
   - **Title 24 (California Energy Code)** – A rigorous state-specific energy standard with mandatory and performance-based requirements.
   - **ASHRAE 90.1** – A national standard that outlines minimum requirements for energy-efficient design of buildings (excluding low-rise residential).

   These codes and standards govern multiple components of building design, including:

   - **Envelope Performance** – Minimum R-values for insulation and maximum U-values and SHGC for glazing. Proper insulation and high-performance fenestration reduce heating and cooling loads.
   - **HVAC Efficiency** – Sets minimum Seasonal Energy Efficiency Ratio (SEER), Energy Efficiency Ratio (EER), Coefficient of Performance (COP), and Heating Seasonal Performance Factor (HSPF) for heating and cooling systems.

- **Lighting and Controls** – Establishes maximum lighting power densities (LPD) in watts per square foot and mandates control strategies such as occupancy sensors, daylighting zones, and automatic shutoff.
- **Solar Readiness** – For some jurisdictions, buildings must be designed with provisions to support future photovoltaic (PV) systems, including roof space and conduit pathways.
- **Air Leakage** – Requirements for reducing infiltration through continuous air barriers, blower door testing, and sealing of penetrations.

Key Action: Clearly document energy code compliance in construction documents and design narratives. Architects should show:

- **Envelope**: Include R-values for walls, roofs, and floors; U-values and SHGC for windows and doors. These values should meet or exceed minimum code requirements for the applicable climate zone.
- **Mechanical Equipment**: Specify HVAC equipment ratings, including SEER, EER, and COP values, and ensure alignment with the latest code updates.
- **Lighting**: Identify lighting power densities per space type and document lighting control systems.
- **Compliance Path**: Clarify whether the design follows the **prescriptive** path (where each component must meet a specific requirement) or the **performance** path (using whole-building energy modeling to demonstrate compliance or improvement).

Examples:

- **Envelope (IECC Climate Zone 5)**: For residential buildings, windows must not exceed a **U-factor of 0.35** and an **SHGC of 0.40** (per IECC Table R402.1.2) when using the **prescriptive** compliance method.
- **Lighting (ASHRAE 90.1-2019)**: An open office must not exceed **0.79 watts per square foot** in lighting power density. Lighting controls such as occupancy sensors must be included in enclosed spaces like private offices, conference rooms, and storage areas.
- **Roof Insulation (IECC 2021, Climate Zone 3)**: Commercial buildings with low-slope roofs must include **R-30 continuous insulation** (Table C402.1.3).
- **HVAC (ASHRAE 90.1)**: A split-system air conditioner serving a small office (<65,000 Btu/h) must have a **minimum SEER of 14.3**, aligned with DOE 2023 efficiency standards.
- **Solar Readiness (Title 24, Part 6)**: A new low-rise residential project in California must include a **PV-ready zone** on the roof, conduit from the electrical panel to the roof, and appropriate space in the panel for future solar connections.

Design Tip for the ARE 5.0 PPD Exam:

The exam often presents scenarios involving energy performance trade-offs or code compliance paths. Be prepared to:

- Evaluate whether a design meets prescriptive energy code requirements.
- Understand how energy modeling can be used for performance compliance.
- Identify systems or assemblies that may require upgrade to meet code.
- Interpret climate zone data and match it to envelope or mechanical requirements.

By integrating energy code requirements early in the design process, architects can ensure code compliance, reduce future rework, and contribute to high-performance, sustainable buildings.

A strong understanding of these principles will not only help in practice but is also essential for successfully passing the ARE 5.0 PPD exam.

8. Compile a Code Summary and Incorporate into Drawings

A **code summary sheet** is an essential part of a well-coordinated drawing set and plays a crucial role in communicating a building's compliance with the applicable building codes. It presents vital regulatory information in a clear, concise, and organized manner that is easily reviewed by permitting authorities and can be referenced by all members of the design and construction team.

This summary ensures that the design adheres to safety, accessibility, and legal requirements and provides a roadmap for the building's code-related decisions.

📋 **Typical Elements Included in a Code Summary Sheet:**

- **Applicable Codes and Editions**
- List all relevant codes and their adopted editions, such as:
  2021 IBC, 2021 IEBC, 2021 IECC, NFPA 101 (Life Safety Code), ANSI A117.1 (Accessibility)
- **Occupancy Classification(s)**
- Specify the building's use group(s) as defined in the International Building Code (e.g., **Assembly A-3**, **Business B**, **Residential R-2**). Mixed-use buildings should identify all applicable occupancy types and clearly indicate fire separation requirements.
- **Construction Type**
  Indicate the construction type per IBC Chapter 6 (e.g., **Type IIA, Non-combustible**, or **Type VB, Combustible**). This determines fire-resistance ratings and allowable area/height.
- **Allowable vs. Actual Building Area and Height**
  Use tabular data to compare code-allowed and proposed:
  - **Building height (ft and stories)**
  - **Building area (gross or per floor)**
  - **Sprinkler increases** (e.g., IBC Section 506)
  - **Frontage increases** (if applicable)
- **Life Safety Plan Reference**
  Cross-reference sheets showing:
  - Exit access and number of exits
  - Travel distances
  - Exit discharge paths
  - Fire ratings and separation walls
  - Area of refuge (if required)
- **Plumbing Fixture Count Summary**
  Include a table comparing required vs. provided fixtures based on occupancy and occupant load (per IBC Chapter 29 or IPC/UPC). Split fixture counts by sex if applicable and show accessibility compliance (e.g., ADA or CBC).
- **Accessibility Notes**
  Summarize applicable accessibility standards (e.g., **ADA 2010**, **CBC Chapter 11B**) and confirm general site/building features:

  - Accessible route from site entry to building
  - Entry door clearances
  - Elevator access (if multi-story)
  - Toilet room accessibility
  - Tactile signage, turning radii, and reach ranges

☑ **Key Action**:
Place the **code summary sheet near the front of the drawing set**, typically on **Sheet G-001**, so that it is among the first documents reviewed by permitting officials. Ensure that every item in the code summary is **clearly traceable** throughout the rest of the drawing set—for example, occupant loads referenced in the summary should match those used in egress plans and plumbing calculations.

🧠 **Example:**
A 3-story office building (Business Occupancy, Type IIB Construction) has:
- **Code**: 2021 IBC
- **Construction Type**: IIB (Non-rated, non-combustible)
- **Occupancy**: B (Business)
- **Allowable Area** (with sprinkler & frontage increases): 54,000 sf per floor
- **Actual Area**: 42,000 sf per floor
- **Allowable Height**: 75 ft / 3 stories
- **Actual Height**: 59 ft / 3 stories
- **Required Exits per Floor**: 2
- **Provided Exits per Floor**: 2
- **Plumbing Fixtures Required**: 4 male, 4 female
- **Plumbing Fixtures Provided**: 4 male, 4 female
- **Accessibility**: Complies with ADA 2010 and ANSI A117.1, including elevator access and accessible restrooms

This summary would appear in tabular form on G-001 and be supported by architectural plans, elevations, and enlarged accessibility diagrams.

☑ **Summary:**
Understanding and applying building code requirements ensures the **health, safety, and welfare** of occupants and forms the **backbone of legally compliant design**. For the **ARE 5.0 PPD exam**, you will be expected to **calculate**, **reference**, and **document** code-related data—just as you would in real-world practice. This includes verifying allowable areas and heights, calculating occupant loads, determining fixture requirements, and aligning exit strategies with code minimums.

Mastering the integration of code summaries and life safety documentation supports:
- Faster permit approvals
- Reduced plan check comments
- Fewer costly redesigns or delays
- Smoother coordination with engineers, consultants, and code officials

On the exam, you may encounter case studies requiring analysis of code data in a simulated drawing set. You'll need to identify discrepancies, determine compliance, or recommend appropriate code-based design responses. Practice reviewing real code summary sheets and become familiar with how information flows from code to document.

C. Integrate multiple codes to a project design (A/E) (2-6%)

NCARB Requirements

Multiple codes may govern a project, depending on the jurisdiction, funding stream, or other factors. You will need to understand the use of related codes, like energy, mechanical, and universal design; determine how multiple codes are used together; and discern which code governs when a conflict arises.

As outlined by NCARB, candidates must demonstrate an understanding of how multiple codes can apply to a project simultaneously. Projects are often subject to a wide array of building codes and regulations, which may vary based on **jurisdiction**, **project type**, **funding source**, and **occupancy classification**. These can include:

- **International Building Code (IBC)**
- **International Energy Conservation Code (IECC)**
- **ADA (Americans with Disabilities Act) Standards**
- **ANSI A117.1**
- **Local amendments or zoning ordinances**
- **State-specific codes**
- **Federal requirements (for federally funded projects)**

The exam may test your ability to identify, analyze, and resolve conflicts between overlapping code requirements and to correctly apply those codes during the design process.

Key Competencies:

Candidates should be able to:

1. **Recognize and apply various codes that may affect design decisions**

   Architects are responsible for ensuring that their design solutions comply with a range of building codes and standards that may vary by jurisdiction, building type, funding source, and occupancy. Recognizing which codes apply to a particular project is one of the foundational steps in the early phases of project planning and design. Applying those codes effectively requires understanding how they influence space planning, circulation, structural layout, accessibility, energy performance, and more.

   Codes typically fall into several categories:

   - **Model Building Codes** (e.g., International Building Code - IBC)
   - **Energy Codes** (e.g., International Energy Conservation Code - IECC, ASHRAE 90.1)
   - **Accessibility Codes** (e.g., Americans with Disabilities Act - ADA, ANSI A117.1)
   - **Life Safety Codes** (e.g., NFPA 101)
   - **Mechanical, Electrical, and Plumbing Codes** (e.g., IMC, IPC, NEC)
   - **Zoning Ordinances and Municipal Codes**

- **State-specific Codes** (e.g., California Building Code, Massachusetts State Building Code)
- **Special Funding Requirements** (e.g., federal standards for government-funded projects)

Design Implications of Code Application

Codes directly affect multiple design aspects, including but not limited to:

- **Egress and circulation:** IBC governs minimum corridor widths, exit routes, and travel distances.
- **Accessibility:** ADA/ANSI influence door clearances, ramp slopes, restroom layout, and signage.
- **Structural layout:** Seismic codes (e.g., ASCE 7) can impact column spacing and lateral bracing systems.
- **Energy performance:** The IECC or state energy codes may require high-performance glazing, insulation, and efficient HVAC systems.
- **Site and zoning considerations:** Local zoning codes influence building setbacks, height limits, parking, and use restrictions.

Example 1: Apartment Building in a Mixed-Use Zoning District
An architect designing a mid-rise apartment building in a city with a mixed-use zoning overlay must recognize and integrate multiple code layers. The IBC determines occupancy group classifications (e.g., R-2 for residential), fire ratings between units, and egress paths. Meanwhile, the local zoning ordinance may require retail at ground level, limit the building's height to 65 feet, and mandate a minimum number of parking stalls. Simultaneously, ADA requirements ensure all common areas are fully accessible, and state energy codes may require solar readiness or high-efficiency mechanical systems.

Example 2: School Renovation Project
In a public K-12 school renovation, the architect must apply the IBC (for construction and occupancy classifications), NFPA 101 (for life safety and egress), ADA and ANSI A117.1 (for accessibility compliance), and local educational facility guidelines. For example, restrooms must meet ADA standards, egress must account for young children's mobility, and fire separation between classrooms and corridors must comply with fire resistance requirements in IBC Chapter 7. If the project is funded with state or federal grants, additional layers—such as compliance with Uniform Federal Accessibility Standards (UFAS)—may also apply.

Key Takeaways for ARE 5.0 PPD Exam:

- **Read carefully for project context:** The exam may provide details about location, building type, or funding that cue which codes apply.
- **Know the impact of codes on design:** Codes are not abstract—they shape real decisions like corridor width, wall construction, and HVAC strategies.
- **Be aware of overlapping and conflicting codes:** You may need to identify which code takes precedence or how to meet the stricter requirement.
- **Use codes as tools:** Don't think of codes just as constraints—they are also frameworks that protect safety, accessibility, and performance.

2. **Determine applicable codes based on project parameters (location, funding, occupancy, etc.)**
   One of the most critical skills an architect must possess is the ability to identify **which codes apply to a specific project**. This determination is not arbitrary; it depends on a combination of **project parameters** that include:
   - **Project location**
   - **Occupancy type**
   - **Construction type**
   - **Funding source**
   - **Jurisdictional authority (AHJ)**
   - **Project scale and use**
   - **Client type (private, public, federal)**

   These variables will guide which **building codes**, **accessibility standards**, **energy regulations**, **fire and life safety codes**, and **zoning ordinances** must be considered during design.

   a. Location
   Location determines which building code version and amendments apply to a project. While the **International Building Code (IBC)** is widely used across the U.S., **each state or municipality may adopt a modified version** or add its own local amendments.
   - *Example:* A project in **New York City** must comply with the **NYC Building Code**, which is based on the IBC but includes extensive amendments.
   - *Example:* In **California**, projects follow the **California Building Code (CBC)**, which integrates accessibility and seismic requirements that are often stricter than the base IBC.

   🔍 **Tip for the ARE:** Expect questions where you must interpret a project's geographic location and apply the correct version or variant of the code.

   b. Funding Source
   The source of a project's funding can trigger additional code requirements—especially for **public** or **federally funded projects**.
   - **Federal projects** must often comply with:
     - **ABAAS (Architectural Barriers Act Accessibility Standards)**
     - **UFAS (Uniform Federal Accessibility Standards)**
     - **GSA Facilities Standards (P100)** for federal buildings
   - **State-funded schools** may be required to meet educational facility design standards and energy benchmarks beyond what is required by local codes.

   - *Example:* A **federally funded housing project** must comply with both **HUD guidelines** and **ADA accessibility standards**, which may differ from local requirements.

   - *Example:* A **Department of Veterans Affairs (VA) hospital** will require compliance with VA-specific design standards, in addition to the applicable building code.

c. Occupancy Classification

The use or function of a building determines the **occupancy group** (e.g., Assembly, Business, Educational, Residential), which drives decisions about:

- **Egress width and number of exits**
- **Fire protection systems (e.g., sprinklers)**
- **Structural live load design**
- **Accessibility features**
- **Separation requirements between occupancies**

- *Example:* A **theater (Assembly Group A-1)** has stricter requirements for egress and fire protection compared to a **small office (Business Group B)**.

- *Example:* A **daycare facility** for more than 5 children under the age of 2.5 may be classified as **Institutional Group I-4**, triggering special safety provisions such as smoke compartments or direct exits.

d. Jurisdiction and AHJ

Each jurisdiction may adopt and enforce codes differently. The **Authority Having Jurisdiction (AHJ)** interprets and enforces applicable codes and standards. It's critical to consult with the AHJ early in the project to verify:

- The **edition and version** of the code in use
- Any **local amendments**
- Interpretation of ambiguous or conflicting requirements
- Required **special inspections** or **permit procedures**

- *Example:* A city may require **green building certifications** (e.g., LEED Silver or local sustainability mandates) not found in state or national codes.

🔍 **Tip for Practice:** Always schedule early meetings with the AHJ to confirm which codes apply and clarify any conflicts or unique local regulations.

Key ARE 5.0 Concepts:

- Be able to **analyze project parameters** in a given scenario and deduce the correct codes to apply.
- Understand how **occupancy, construction type, and funding** affect code decisions.
- Expect to identify **conflicts between federal and local codes**, and choose the most stringent or governing standard.

3. **Understand how to layer codes together to form a coherent design strategy**
   Designing a code-compliant building is not simply about checking off boxes for individual codes—it's about **integrating multiple overlapping code requirements** into a cohesive, functional, and buildable design. This process is called **"code layering"**: the ability to interpret and apply **multiple codes simultaneously**—such as life safety, accessibility, energy, and local zoning—without compromising design integrity.

In the context of the **ARE 5.0 PPD Exam**, candidates must demonstrate how to synthesize these requirements into a unified strategy that addresses:

- **Life safety and egress**
- **Accessibility and universal design**
- **Energy performance**
- **Structural and fire protection systems**
- **Zoning and site regulations**
- **Client requirements and building program**

a. What is Code Layering?

Code layering refers to the process of **superimposing multiple applicable codes and standards**—which may be issued by different authorities—onto the same project and resolving overlaps or conflicts. Each code may address a different domain of the building, and they often interact with each other.

Think of it as assembling a puzzle where the pieces include:

- **IBC (International Building Code)** – the foundation for occupancy, egress, construction type
- **ADA/ANSI A117.1** – accessibility and usability requirements
- **IECC or Title 24** – energy efficiency, insulation, daylighting
- **NFPA 101** – life safety and egress
- **Zoning Ordinances** – setbacks, height restrictions, FAR, use type
- **Local/State Amendments** – unique AHJ modifications or overlays
- **Special requirements** – LEED, WELL, green building codes, funding requirements

b. Real-World Example: Community Health Center

Project parameters:

- Urban infill site
- Federally funded
- Outpatient healthcare services
- Three stories, 30,000 SF

Applicable codes:

- **IBC 2021**: Establishes occupancy type (B for business), requires two stairwells for egress, and mandates a 1-hour rated corridor system.
- **ADA & ABAAS**: Ensures exam rooms, toilets, reception desks, and elevators are accessible to all users.
- **ASHRAE 90.1 & IECC**: Requires building envelope performance measures such as continuous insulation, daylight sensors, and efficient HVAC zoning.
- **Local Zoning**: Limits height to 40 feet and requires 20% of the lot to be landscaped.
- **NFPA 101**: Imposes stricter travel distance limits in healthcare environments, influencing corridor layout and room grouping.
- **GSA P100 Standards**: Applicable due to federal funding; introduces seismic design considerations and material sourcing requirements.

Code layering in action:

- The architect identifies a conflict between IBC and NFPA 101 regarding **travel distance limits**. NFPA is more restrictive, so the design incorporates an additional stair core to remain compliant.
- While ADA allows a ramp slope of 1:12, **local code** allows only 1:15 for public-facing buildings. The design team uses 1:15 in public areas and reserves 1:12 for back-of-house to comply with both.
- Energy codes require south-facing glazing to have a low SHGC (solar heat gain coefficient), which leads to incorporating **external shading devices**—an aesthetic and functional solution that supports both compliance and design goals.

c. Strategies for Effective Code Layering

- **Start early**: Begin code analysis during schematic design to avoid costly redesigns later.
- **Use code overlays**: Create color-coded plans showing different code impacts—fire zones, ADA paths, egress, and daylighting zones.
- **Default to the most stringent code**: When two codes conflict, apply the stricter one—this is often a safe strategy unless clarified otherwise by the AHJ.
- **Coordinate with consultants**: Mechanical, structural, and life safety consultants must align their systems with architectural intent and code requirements.
- **Document your assumptions**: Keep a running list of code interpretations and approvals from the AHJ for consistency throughout design and documentation.

d. Code Layering in ARE 5.0 PPD Exam

You may encounter questions such as:

- *"Which code governs egress when a conflict arises between the IBC and local fire code?"*
- *"Given a project in a flood zone with federal funding, which additional requirements should be considered?"*
- *"Which of the following design strategies best satisfies both energy and accessibility codes?"*

To answer these, you'll need to think **holistically** and consider how one decision affects multiple systems.

Key Takeaways:

- Layering codes is a **systems-thinking approach** to design that integrates compliance into spatial and functional planning.
- Successful layering avoids code conflicts, minimizes rework, and supports efficient permit review.
- In practice—and on the exam—understanding this integration is crucial for delivering both safe and high-performing buildings.

4. **Resolve code conflicts**

In architectural design, projects often must comply with multiple overlapping codes and regulations that may contain conflicting requirements. These conflicts can arise between:

- Federal, state, and local codes,
- Model codes and local amendments,
- Different codes addressing the same issue (e.g., IBC vs. NFPA 101),
- Special funding agency requirements versus municipal codes.

The key principle when resolving these conflicts is that **the most restrictive or stringent code applies in all cases** to ensure the highest level of safety, accessibility, and performance.

a. Understanding Code Conflicts

When two or more codes regulate the same aspect of a project but have differing requirements, architects must analyze the conflict and apply the **most protective and demanding standard**. This approach not only ensures compliance but often exceeds minimum legal requirements for occupant safety and building performance.

b. Examples of Code Conflicts and Resolution

Example 1: Accessibility Standards Conflict

A federally funded healthcare clinic in California must comply with:

- The **Americans with Disabilities Act (ADA) Standards** (federal),
- The **California Building Code (CBC) Chapter 11B** (state).
  Since CBC typically imposes more stringent accessibility standards than ADA, the design must comply with the **CBC requirements**, ensuring the stricter provisions are met.

Example 2: Energy Code vs. Local Zoning

A commercial office building in a city must adhere to:

- State energy code (IECC 2021) requirements for rooftop solar panels,
- Local zoning ordinance limiting rooftop coverage for solar installations.
  In this conflict, the project must follow the **more restrictive requirement**, which is the zoning ordinance's limit on rooftop coverage—unless a variance is granted—while seeking alternative strategies to meet energy efficiency goals.

Example 3: Fire Code Conflict

A university dormitory project is subject to:

- **IBC (International Building Code),**
- **NFPA 101 Life Safety Code.**

NFPA 101 often has stricter egress travel distance limits than IBC for institutional occupancies. The design must comply with the **NFPA 101 requirement** because it is more restrictive and specific to the occupancy type.

c. Strategies for Resolving Code Conflicts on the ARE 5.0 PPD Exam

- Always identify all applicable codes early in the design process.
- When conflicts arise, apply the **most restrictive or stringent code** to guarantee compliance.

- Consider federal, state, and local codes, as well as any project-specific mandates such as funding requirements.
- Review local amendments carefully, as they often include more stringent regulations.
- Document your code interpretations and decisions clearly to ensure consistency throughout the design and permitting process.

d. Exam Tips

- Remember the core principle: **"The most restrictive code applies."**
- Understand that while zoning ordinances govern land use and may not be overridden by building codes, the strictest requirement across all codes must be followed.
- Code compliance involves harmonizing all applicable codes, applying the most demanding provisions when conflicts occur.
- Use logical reasoning and elimination methods to identify the strictest applicable code in exam scenarios.

Summary

Resolving code conflicts by applying the most restrictive code is essential for safe, compliant, and successful architectural design. This principle helps protect occupants, ensures regulatory approval, and minimizes costly redesigns. Mastery of this concept is critical for success on the ARE 5.0 PPD exam and effective professional practice.

5. **Coordinate with consultants (mechanical, electrical, structural) to ensure code compliance across disciplines**

Successful project planning and design require **close collaboration between architects and consulting engineers**—including mechanical, electrical, and structural specialists—to ensure that all aspects of the building meet applicable codes. This coordination is essential because codes often address systems that span multiple disciplines, and compliance depends on their integrated performance.

a. Why Coordination Matters

- **Interdisciplinary code requirements:** Many codes regulate systems that overlap disciplines—for example, mechanical ventilation affects fire and life safety, electrical systems impact emergency egress lighting, and structural design must meet seismic or wind load codes.
- **Preventing conflicts and omissions:** Early and continuous coordination prevents costly clashes, omissions, or rework caused by inconsistent interpretations or overlooked code requirements.
- **Holistic compliance:** Achieving code compliance is a team effort; one discipline's solution can create issues for another if not carefully integrated.
- **Streamlined permitting:** Coordinated documentation and compliance approaches expedite review and approval by the Authority Having Jurisdiction (AHJ).

b. Key Areas Requiring Consultant Coordination

- **Mechanical Systems**
    - Compliance with energy codes (e.g., IECC, ASHRAE 90.1) for HVAC efficiency.
    - Ventilation rates per ASHRAE 62.1 and local health codes.

- Fire and smoke control requirements (e.g., smoke dampers coordinated with fire protection systems).
- Mechanical equipment location and clearances affecting egress and accessibility.

- **Electrical Systems**
  - Emergency power and lighting per NEC (National Electrical Code) and IBC.
  - Life safety system integration, including fire alarm, security, and communication systems.
  - Energy code compliance for lighting controls, occupancy sensors, and daylight harvesting.
  - Coordination of conduit and wiring paths with architectural elements.
- **Structural Systems**
  - Meeting structural codes (IBC, ASCE 7) for loads, seismic design, and wind resistance.
  - Fire resistance rating of structural elements (e.g., rated walls, columns).
  - Integration of lateral bracing and expansion joints with architectural layouts.
  - Foundation design considering soil conditions and site constraints.

c. Examples of Coordination in Practice

- **Fire and smoke control:** The architect and mechanical engineer collaborate to design smoke control systems that comply with fire codes while maintaining occupant comfort. For example, the mechanical engineer specifies smoke dampers in HVAC ducts, coordinated with fire-rated walls designed by the structural engineer.
- **Energy efficiency:** The architect works with the electrical engineer to incorporate daylight sensors and occupancy controls per the energy code, while ensuring these systems do not interfere with emergency lighting required by life safety codes.
- **Structural and architectural integration:** The structural engineer recommends seismic bracing that requires architectural modifications, such as shear walls or braced frames, which the architect must integrate into the floor plan without compromising programmatic needs.

d. Strategies for Effective Coordination on the ARE 5.0 PPD Exam

- Recognize the interconnectedness of building systems and their impact on code compliance.
- Anticipate potential conflicts early and communicate design intent clearly to consultants.
- Review consultant inputs for code compliance before finalizing construction documents.
- Document coordination efforts and agreed-upon code interpretations.

Summary

Coordination with mechanical, electrical, and structural consultants is critical for ensuring **comprehensive code compliance** across all disciplines. This collaboration supports design integrity, occupant safety, energy efficiency, and successful project delivery—key themes emphasized on the ARE 5.0 PPD exam and vital to professional practice.

6. **Integrate life safety, accessibility, energy, and performance codes into architectural documentation**
   Architectural documentation must do more than convey design intent—it must clearly demonstrate **code compliance** across multiple domains, including **life safety**, **accessibility**, **energy efficiency**, and **overall building performance**. Integrating these requirements into drawings and specifications is essential for successful permitting, construction, and building occupancy.

   For the ARE 5.0 PPD exam, candidates must show they understand **how to embed code requirements into architectural documentation** in a clear, coordinated, and legally enforceable way.

   a. Life Safety Code Integration
   Life safety documentation focuses on protecting occupants in emergencies, such as fire or power failure. Key life safety code elements include:
   - **Egress paths and exit access**
   - **Stair and corridor widths**
   - **Exit signage and lighting**
   - **Travel distance limitations**
   - **Fire separation and smoke compartmentalization**
   - **Fire-rated construction (walls, doors, shafts, etc.)**

   🗎 **Documentation Examples**:
   - **Life Safety Plans**: Show clear exit paths, door swing directions, rated walls, occupant loads per room, exit widths, and travel distances.
   - **Door Schedules**: Indicate fire ratings and hardware for exit doors.
   - **Wall Types Legend**: Include references to UL assemblies or specific fire-resistance ratings.

   b. Accessibility Code Integration
   Accessibility codes such as the **ADA Standards**, **ANSI A117.1**, and local amendments ensure equal access for all users, including people with disabilities. These requirements affect circulation, restroom design, signage, and more.

   🗎 **Documentation Examples**:
   - **Floor Plans**: Clearly dimension accessible clearances around doors, plumbing fixtures, seating, and turning radii in restrooms.
   - **Sections/Elevations**: Show mounting heights for grab bars, signage, counters, and light switches.
   - **Details**: Include curb ramps, door thresholds, and tactile warning strips at stairways.

   ☑ **Best Practice**: Add ADA symbols or accessibility callouts on plans and double-check dimensions during QA/QC reviews.

c. Energy Code Integration

Energy codes (such as **IECC**, **ASHRAE 90.1**, or **Title 24**) regulate how efficiently a building uses energy. Documentation must show compliance in envelope performance, lighting, HVAC systems, and controls.

📄 **Documentation Examples**:

- **Wall Sections**: Indicate R-values of insulation, continuous air barriers, vapor retarders, and thermal breaks.
- **Reflected Ceiling Plans**: Show daylight sensors and lighting zones for energy conservation.
- **Window Schedules**: Include U-values and SHGC (Solar Heat Gain Coefficient) for fenestration performance.
- **Mechanical Coordination**: Ensure building envelope and systems match the mechanical engineer's energy model.

d. Building Performance and Other Codes

Codes may also address **sustainability**, **indoor air quality**, **sound transmission**, or **moisture control**, depending on jurisdiction or program requirements (e.g., LEED, WELL, Green Globes).

📄 **Documentation Examples**:

- **Material Schedules**: Indicate low-VOC products or finishes for indoor air quality compliance.
- **Roof and Wall Assemblies**: Show detailing to prevent thermal bridging or condensation.
- **Schedules or Notes**: Reference code-compliant acoustic assemblies (e.g., STC ratings for walls in multi-family housing).

e. ARE 5.0 PPD Exam Considerations

You may encounter questions requiring you to:

- Select where code information should appear (e.g., life safety info on a floor plan vs. site plan),
- Identify missing or incorrect code documentation in a drawing set,
- Coordinate among disciplines to ensure energy, life safety, and accessibility requirements are fully represented.

☑ **Tip**: Think like a reviewer—if you were the code official, what would you need to see clearly documented to approve this design?

Summary

Integrating life safety, accessibility, energy, and performance codes into architectural documentation ensures the building meets **legal requirements**, protects occupants, and performs efficiently. This is a vital responsibility for architects, and a core skill tested on the ARE 5.0 PPD exam. Drawings should not only communicate design intent but also serve as **compliance tools** that reduce risk and support successful project execution.

7. **Apply universal design principles alongside ADA and ANSI standards**
While the **Americans with Disabilities Act (ADA)** and **ANSI A117.1** establish minimum requirements for accessibility, **universal design** goes beyond compliance to create environments that are usable by **all people, to the greatest extent possible**, without the need for adaptation or specialized design.

In both practice and on the ARE 5.0 PPD exam, architects must understand how to integrate **universal design principles** with code-mandated accessibility standards to achieve spaces that are not only compliant—but inclusive, intuitive, and user-friendly.

a. Distinction Between ADA/ANSI and Universal Design

ADA & ANSI	Universal Design
Mandatory legal standards	Voluntary design philosophy
Applies to people with disabilities	Benefits all users, regardless of age or ability
Enforced by law	Guided by best practices
Minimum requirements	Enhanced usability, comfort, and dignity

Figure 4.8 Distinction Between ADA/ANSI and Universal Design

☑ **Key Point:** Universal design includes ADA/ANSI standards but expands on them to remove barriers for a wider range of people—including children, older adults, and those with temporary injuries or diverse cognitive abilities.

b. The Seven Principles of Universal Design

- **Equitable Use**
  Design is useful and marketable to people with diverse abilities.
  - *Example:* A main entrance without steps benefits wheelchair users, parents with strollers, and delivery personnel alike.
- **Flexibility in Use**
  Accommodates a wide range of individual preferences and abilities.
  - *Example:* Adjustable-height counters in public spaces allow for use by standing or seated individuals.
- **Simple and Intuitive Use**
  Easy to understand, regardless of the user's experience or abilities.
  - *Example:* Signage with pictograms and clear contrast helps non-native speakers and users with low vision.
- **Perceptible Information**
  Communicates information effectively regardless of sensory abilities.
  - *Example:* Elevators that provide audible floor announcements and visual floor indicators.
- **Tolerance for Error**
  Minimizes hazards and adverse consequences of unintended actions.
  - *Example:* A shower with an anti-scald valve to protect users from sudden hot water.
- **Low Physical Effort**
  Can be used efficiently and comfortably with minimal fatigue.

- *Example:* Lever-style door handles require less force and dexterity than round knobs.

- **Size and Space for Approach and Use**
  Provides appropriate size and space regardless of body size, posture, or mobility.
  - *Example:* Wide circulation paths that allow two people (one in a wheelchair) to pass easily.

c. Applying Universal Design in Practice
Even when a building meets ADA and ANSI requirements, it may still pose barriers. Universal design fills these gaps by prioritizing **user dignity, comfort, and independence**.

Feature	ADA-Compliant	Universally Designed
Door width	32" minimum clear	36" for easier access with walkers/strollers
Sink height	34" max	32" with open knee space for all users
Signage	Braille required	Braille + raised lettering + high contrast
Fixtures	Within reach range	Hands-free and sensor-operated for ease of use

Figure 4.9 ADA-Compliant vs. Universally Designed Restroom

d. ARE 5.0 PPD Exam Tips
On the exam, you may be asked to:

- Identify a design strategy that exceeds minimum accessibility requirements.
- Choose between code-minimum and universally inclusive solutions.
- Evaluate plan layouts for equitable, intuitive access.

Summary
While ADA and ANSI standards are required by law to ensure accessibility for people with disabilities, **universal design is a best practice** that improves usability, comfort, and dignity for **everyone**. By applying both in tandem, architects create spaces that are not only compliant—but welcoming, inclusive, and future-ready.

8. **Respond to client and authority having jurisdiction (AHJ) requirements while maintaining compliance**
   An essential responsibility of the architect is to balance the needs and goals of the **client** with the regulatory requirements enforced by the **Authority Having Jurisdiction (AHJ)**. This involves **interpreting, negotiating, and integrating client expectations and jurisdictional feedback** into the design—**without compromising code compliance**.

   On the ARE 5.0 PPD exam, candidates must demonstrate their ability to respond appropriately to stakeholder input while maintaining alignment with applicable codes and professional responsibilities.

a. Understanding Client and AHJ Roles

Client

- Defines project goals, program, budget, and design preferences.
- May request features that stretch standard use (e.g., open plans, high glass walls, unconventional materials).
- Sometimes unaware of code limitations or permitting constraints.

Authority Having Jurisdiction (AHJ)

- Enforces applicable codes, standards, and local ordinances.
- Reviews permit documents and issues approvals, comments, or corrections.
- Has final authority in interpreting code within that jurisdiction.

☑ **Key Challenge:** The architect must **advocate for client goals**, while ensuring **the project meets or exceeds all regulatory requirements**.

b. Common Scenarios Requiring Balance

1) Client Wants Full-Height Glass Walls in a Corridor

- **Code Issue:** If the corridor is rated for fire protection (per IBC), glass may require a specific fire-resistance rating.
- **Resolution:** Propose fire-rated glazing assemblies that maintain transparency while meeting code, or revise corridor classification.

2) Client Requests Inaccessible Rooftop Lounge

- **Code Issue:** ADA and ANSI require accessible routes to all public and common-use areas.
- **Resolution:** Educate the client on accessibility laws and propose a ramp or elevator alternative to meet both code and user experience goals.

3) AHJ Requests Additional Fire Alarm Devices

- **Client Concern:** Added cost or impact on ceiling design.
- **Resolution:** Coordinate with MEP consultants to incorporate devices into an integrated design that aligns with aesthetics and budget.

c. Strategies for Responding Effectively

- **Educate the client**: Help them understand why certain features must be adjusted for code or safety reasons. Use diagrams or code excerpts if needed.
- **Engage early with the AHJ**: Schedule pre-application or early design meetings to clarify local expectations, special amendments, or interpretations.
- **Document everything**: Keep records of decisions, approvals, and communications with both clients and code officials.
- **Offer compliant alternatives**: When client preferences contradict code, present equivalent or better-performing options that satisfy both.
- **Anticipate code questions**: Prepare your permit set with clearly marked code information—occupancy, egress, accessibility, energy, fire protection, etc.—to minimize AHJ review comments.

d. ARE 5.0 PPD Exam Tips

- Look for answer choices that **satisfy both the client and the AHJ**, while preserving **code compliance**.
- Avoid solutions that prioritize design or cost over legal safety/accessibility requirements.
- Be ready to choose alternative strategies that maintain compliance while adapting to new project constraints (e.g., program changes, AHJ feedback, or client goals).

☑ **Strategy:** Always prioritize **code compliance**, but support your clients with **design solutions that align with their intent and the law**.

Summary

Architects must skillfully navigate between **client expectations** and **AHJ code enforcement**. Clear communication, proactive planning, and a commitment to creative problem-solving allow architects to deliver designs that are both **code-compliant** and aligned with the **client's vision**. This balanced approach is essential on the ARE 5.0 PPD exam—and in every phase of professional practice.

9. **Prepare documentation that reflects correct code application during plan check or permit review**

A core responsibility of the architect is to prepare a **complete, coordinated, and code-compliant set of documents** for submission to the Authority Having Jurisdiction (AHJ) during **plan check or permit review**. This documentation must **clearly demonstrate how all applicable codes have been integrated** into the design and must be organized in a way that allows for **efficient review and approval**.

On the ARE 5.0 PPD exam, candidates are expected to understand how to document code compliance accurately, anticipate what reviewers will look for, and ensure that submitted documents stand up to scrutiny from code officials.

a. Purpose of Permit Documentation

Permit documents must:

- Demonstrate **compliance with all applicable codes** (building, fire, energy, accessibility, mechanical, etc.),
- Facilitate **timely review and approval** by the AHJ,
- Minimize corrections, delays, and resubmissions,
- Serve as a **legal record** of design intent and code adherence,
- Reduce the architect's liability by showing due diligence.

b. Key Elements of Code-Compliant Permit Documentation

Here are essential components to include in a permit-ready document set:

1) Code Summary Sheet

- Lists all applicable codes (e.g., IBC, ADA, NFPA, IECC, local amendments),
- States project type, use, occupancy group(s), construction type, and height/area limitations,

- Includes building data: number of stories, fire protection systems, allowable vs. proposed areas,
- Shows code analysis for mixed occupancies (if applicable),
- Identifies the applicable edition year of each code.

2) Life Safety Plans
- Clearly marked egress paths, exit access, and exit discharge,
- Occupant loads per space,
- Exit widths, travel distances, common path of egress, and dead-end corridors,
- Fire-rated construction, including wall types and shaft enclosures,
- Fire extinguisher locations, area of refuge, and exit signs.

3) Accessibility Compliance
- Door swings and clearances,
- Turning radii in restrooms and circulation paths,
- Accessible routes (site and interior),
- Mounting heights for elements like signage, fixtures, and hardware,
- Elevation and detail drawings showing ADA compliance.

4) Energy Compliance
- Building envelope R-values, U-values, SHGC for fenestration,
- Lighting power density calculations and controls,
- HVAC efficiency and controls (often in coordination with the mechanical engineer),
- Location of insulation and thermal breaks.

5) Fire Protection
- Fire alarm systems, smoke detectors, and sprinkler system coverage (in coordination with consultants),
- Fire department access and fire lanes (shown on site plan),
- Fire ratings and penetration protection.

c. Documentation Best Practices for Permit Review
- **Label clearly and consistently**: Use standardized symbols, hatching, and legends.
- **Reference specific code sections**: When in doubt, cite the exact code that supports your design solution (e.g., IBC 1015.1 for guardrail height).
- **Coordinate with consultants**: Ensure structural, MEP, and civil drawings are aligned with architectural plans regarding code-sensitive systems.
- **Avoid missing or contradictory information**: Incomplete or inconsistent details often cause permit delays.
- **Use a permit checklist**: Many jurisdictions provide checklists outlining required documentation—review and incorporate all applicable items.

d. ARE 5.0 PPD Exam Tips

On the exam, you may be asked to:
- Identify missing or incorrect code information on drawings,
- Choose the best location or method to show code compliance,

- Match documentation types with the applicable code requirement.

☑ **Strategy**: Think like a plan checker—what information would you need to confirm the building is safe, accessible, energy-efficient, and within legal limits?

Summary

Preparing documentation that accurately reflects code application is critical for a successful permit process. Permit-ready documents must clearly communicate **how the design complies with all relevant codes**, reduce the likelihood of review comments, and support a smooth path to construction. On the ARE 5.0 PPD exam—and in real-world practice—this competency ensures both regulatory compliance and professional credibility.

Sample Code Summary Sheet

(Include on the first or second sheet of the permit drawing set — typically labeled G-001 or A-0.1)

PROJECT INFORMATION

- **Project Name:** Lincoln Elementary School Addition
- **Project Address:** 1234 School Street, Springfield, CA 90000
- **Building Use:** Educational (K–5)
- **Scope of Work:** New classroom wing and renovation of admin building
- **AHJ:** City of Springfield Building Department

Code Name	Edition
California Building Code (CBC)	2022
California Fire Code (CFC)	2022
California Energy Code (Title 24)	2022
California Plumbing Code (CPC)	2022
California Mechanical Code (CMC)	2022
California Green Building Code	2022
ANSI A117.1	2009
Americans with Disabilities Act	2010

Figure 4.10 Applicable Codes

OCCUPANCY & USE

- **Occupancy Classification:** E – Educational
- **Accessory Occupancies:** B – Administrative Offices
- **Mixed Use Separation:** Not Required (non-separated)

CONSTRUCTION TYPE

- **IBC Type:** Type IIB (Unprotected, non-combustible)
- **Sprinkler System:** Fully Sprinklered (NFPA 13)
- **Fire Alarm:** Yes – Manual and automatic

Description	Allowed (Per CBC)	Proposed
Height (ft)	75 ft	24 ft
Number of Stories	3 stories	1 story
Building Area	23,000 SF	18,500 SF

Figure 4.11 Building Height & Area

EGRESS INFORMATION

- **Occupant Load:** 300
- **Required Exits:** 3
- **Maximum Travel Distance:** 200 ft (fully sprinklered)
- **Exit Width Provided:** 96" total (min. required: 72")

ACCESSIBILITY

- **Accessible Entrances:** All public entrances
- **Elevators:** N/A (1-story building)
- **Accessible Parking Spaces:** 2 van-accessible, 4 standard
- **Restroom Compliance:** All restrooms fully ADA compliant

ENERGY COMPLIANCE (Title 24)

- **Envelope:** R-21 walls, R-38 roof
- **Windows:** U-0.32, SHGC 0.25
- **Lighting:** LED throughout; occupancy sensors in classrooms
- **HVAC:** High-efficiency rooftop units, demand-controlled ventilation

🗺 Annotated Life Safety Plan (Outline Description)

Use your floor plan and create an annotated "Life Safety Plan" with clearly marked code compliance features. Below is a breakdown of key items to label:

■ Key Elements to Include

- **Occupant Loads**:
  - Label occupant load in each room per code table (e.g., 20 SF/person for classrooms).
- **Egress Paths**:
  - Clearly indicate direction of travel to exits using arrows.
  - Identify any common path of travel or dead-end corridors.
- **Exits and Exit Doors**:
  - Mark doors used as exits and include door widths.
  - Show door swing direction (must swing in direction of egress if over 50 occupants).
- **Exit Sign Locations**:
  - Indicate where exit signs are placed and their visibility from different points.
- **Travel Distances**:
  - Draw dashed lines showing travel distances from the most remote point to the nearest exit.
  - Label distance in feet (e.g., Travel Distance = 68').
- **Fire-Rated Assemblies**:

  - Use hatch patterns to indicate rated walls, stair enclosures, or corridors.
  - Note wall types (e.g., "1-HR FIRE BARRIER").
- **Area of Refuge (if required)**:
  - Show location with communication equipment symbol (for multi-story or assembly buildings).
- **Sprinkler System Coverage**:
  - Label that building is fully sprinklered and note riser location.
- **Fire Extinguisher Locations**:
  - Include standard FE symbols and confirm spacing per NFPA 10.

📝 **Recommended Drawing Sheet Label**
A-1.1L – Life Safety Plan (Main Level)
Include a legend for:

- Fire ratings
- Exit signage
- Door swing
- Travel distances
- Fire extinguisher locations
- Occupant load per room

Real-World Example:
Project Type: Public Library in California
Scope: New construction, 2-story, 25,000 SF
Funding: Partially federally funded
Jurisdiction: City level with state and federal requirements

Codes to integrate:

- **IBC (International Building Code):** Base code used for fire-resistance, egress, occupancy classification, etc.
- **California Building Code (CBC):** Includes amendments to the IBC specific to California.
- **Title 24 (California Energy Code):** Mandates high performance in insulation, HVAC, and lighting.
- **ADA Standards & ANSI A117.1:** Required for accessible design, particularly for public facilities.
- **NFPA 101 (Life Safety Code):** Often used to support or supplement local fire code enforcement.
- **Local zoning ordinance:** Governs height restrictions, setbacks, and land use.

During design, the architect identifies that **CBC Chapter 11B** (Accessibility) is more stringent than the federal **ADA Standards**. Because the project is located in California and must meet the CBC, the stricter of the two codes takes precedence. The architect also discovers that the local zoning code limits building height to 30 feet, while the IBC allows 35 feet for the same occupancy and construction type. Since zoning is more restrictive in this case, the 30-foot limit must be followed.

Additionally, due to federal funding, **Section 504 of the Rehabilitation Act of 1973** applies, requiring compliance with **UFAS** (Uniform Federal Accessibility Standards), which must be coordinated with the more current ADA/ANSI guidelines.

10. Tips for the ARE 5.0 PPD Exam:

- Expect questions that ask you to **choose the correct code when a conflict arises**.
- Be prepared to **apply codes simultaneously**, such as integrating IBC occupancy limits with local energy performance criteria.
- Know that in real-world practice, **coordination with consultants** is critical. For example, your MEP engineer may have to meet energy code requirements that affect architectural decisions (e.g., skylight placement, HVAC room sizes).

Chapter Five

Building Systems, Materials, & Assemblies (19-25%)

NCARB Requirements
In this section, you'll focus on other disciplines with which an architect must be familiar, along with the various components that make up the building.

A. Determine mechanical, electrical, and plumbing systems (A/E) (5-9%)

NCARB Requirements
You will need to evaluate mechanical, electrical, and plumbing systems, including their impact on building and site design. Based on function, cost, size, availability, programmatic needs, and energy/water usage, you will need to select the appropriate systems for a building.

In the **Project Planning & Design (PPD)** division of the ARE 5.0, candidates are expected to evaluate **mechanical, electrical, and plumbing (MEP)** systems in terms of their appropriateness for a building project. Your design decisions must consider:

- **Functionality**
- **Cost (initial and operational)**
- **Space and sizing needs**
- **Availability and constructability**
- **Programmatic requirements**
- **Sustainability**, including energy and water efficiency

These systems must be coordinated with architectural and structural design decisions, site characteristics, and code compliance requirements.

☑ **Key Action:**
Select, evaluate, and coordinate MEP systems appropriate to the project's program, climate, code constraints, and performance goals. Be prepared to analyze trade-offs and make recommendations based on energy use, life-cycle cost, and spatial impacts.

▧ **Core Content Areas (1–12):**

1. **Mechanical System Types**
Heating, ventilation, and air conditioning (HVAC) systems are among the most critical components of a building's design. They directly influence occupant comfort, energy efficiency, spatial planning, and operational cost. In the **PPD division of the ARE 5.0**, you'll be expected to evaluate and recommend appropriate HVAC systems based on **building type, size, location, program**, and **code or sustainability goals**.

Understanding the various categories of HVAC systems and their applications will help you make informed decisions both on the exam and in practice.

🌀 All-Air Systems

These systems deliver **conditioned air** through ductwork to maintain temperature and ventilation.

Common Types:

- **VAV (Variable Air Volume)** – Provides varying airflow at a constant temperature. Efficient and responsive.
- **CAV (Constant Air Volume)** – Provides constant airflow with temperature variation; best for single-zone applications.
- **Single-Duct or Dual-Duct Systems** – Use one or two ducts to deliver hot/cold air to various zones.

Advantages:

- Centralized control
- Easy to incorporate ventilation
- Works well for open office plans and large-scale buildings

Disadvantages:

- Large duct sizes require significant ceiling space
- Energy loss through duct leakage

Example:

A **corporate office building** with multiple open-plan zones may use a **VAV system** to efficiently control thermal comfort while maintaining air quality.

💧 Air-Water Systems

These hybrid systems use both air and water to provide heating, cooling, and ventilation.

Common Type:

- **Fan Coil Units with Central Air Supply** – Terminal units (fan coils) handle local temperature, while a central AHU supplies ventilation air.

Advantages:

- Smaller ductwork than all-air systems
- Zonal control at the unit level
- Energy efficient

Disadvantages:

- Requires coordination of both piping and ductwork
- Slightly higher maintenance due to decentralized components

Example:

A **hotel** may use air-water systems to allow **individual room control** while centralizing ventilation for energy efficiency.

💦 All-Water Systems

These systems use **water exclusively** for heating and cooling, with no air delivery for ventilation (requires separate system).

Common Types:

- **Radiant heating or cooling panels**
- **Chilled beams**
- **Hydronic baseboard systems**

Advantages:

- Very energy efficient (water is a better heat transfer medium than air)
- Quiet operation
- Minimal space required for ductwork

Disadvantages:

- Ventilation must be handled separately
- Chilled beams require humidity control to prevent condensation

Example:

A **museum** may use radiant floor heating to maintain **silent operation and thermal stability**, ideal for preserving artifacts.

🏢 Packaged Rooftop Units (RTUs) vs. Central Plants

Packaged RTUs:

- Self-contained units located on the roof
- Common in one-story commercial or retail buildings
- Include heating, cooling, and ventilation in one package

Advantages:

- Simple installation
- Lower first cost
- Space-saving (no mechanical room)

Disadvantages:

- Limited efficiency
- Poor scalability for larger buildings
- Noise and maintenance access issues

Central Plants:

- Consist of central chillers, boilers, cooling towers
- Serve multiple air handling units (AHUs) through piping and ductwork

Advantages:

- Highly efficient for large-scale applications
- Better control and zoning
- Supports LEED and high-performance design

Disadvantages:

- High initial cost
- Requires mechanical rooms and complex control systems

Example:
A **big-box retail store** may use multiple **RTUs**, while a **high-rise office tower** would require a **central plant with chilled water and AHUs**.

🔁 Heat Pumps and VRF/VRV Systems

Heat Pumps: Transfer heat from one place to another. Can be **air-source** or **ground-source** (geothermal).

- **Air-Source Heat Pumps**: Efficient in mild climates; provide both heating and cooling.
- **Ground-Source (Geothermal)**: Use stable underground temperatures; high upfront cost but excellent long-term savings.

VRF/VRV (Variable Refrigerant Flow/Volume):

- Sophisticated heat pump system using refrigerant as the heat-transfer medium
- Simultaneous heating and cooling in different zones
- Excellent zoning flexibility

Advantages:

- High efficiency and precise control
- No ductwork; refrigerant lines are compact
- Ideal for renovations or multi-unit buildings

Disadvantages:

- Higher initial cost
- Requires specialized maintenance
- Less effective in extremely cold climates (for air-source)

Example:
A **multi-tenant office or apartment building** might use a **VRF system** to provide **independent climate control** per unit without large ducts or water pipes.

🛠 Design Tip for the ARE PPD Exam:

Be prepared to choose an HVAC system based on a combination of:

- Building size and type
- Zoning and control needs
- Energy performance goals
- Initial vs. life-cycle cost
- Space constraints (e.g., ceiling plenum)
- Climate zone and sustainability targets

◇ Quick Example Recap:

- A **small retail building** may use **packaged rooftop units (RTUs)** due to low cost and ease of installation.

- A **large corporate office** may use a **central chilled water plant with VAV air distribution** to maximize energy efficiency and control.

2. **HVAC Zoning & Distribution**

In building design, HVAC zoning and distribution are crucial components that directly influence a building's energy efficiency, occupant thermal comfort, system performance, and overall cost. A solid understanding of zoning strategies, ductwork design, and distribution planning is essential for both effective mechanical integration and coordination across architectural disciplines.

HVAC Zoning Strategies

Zoning refers to dividing a building into separate areas, or "zones," each of which is served by its own thermostat and HVAC control. Zoning allows for tailored heating, cooling, and ventilation in different areas depending on use, occupancy patterns, and solar exposure. Thoughtful zoning improves both **thermal comfort** and **energy efficiency**, as it prevents over-conditioning underutilized or naturally tempered areas.

Examples:

- In an office building, east- and west-facing zones may be separated to accommodate different solar heat gains throughout the day. This prevents overheating in the afternoon on the west side while reducing unnecessary cooling on the east.
- In a multi-story residential complex, common spaces (e.g., lobbies, hallways) can be zoned separately from private units, allowing more efficient control and maintenance.

Effective zoning considers:

- **Functionality of spaces** (e.g., conference rooms, open workspaces, lobbies, restrooms)
- **Occupancy schedules**
- **Orientation and solar gain**
- **Building envelope performance**
- **Noise and privacy requirements**

Duct Sizing

Proper **duct sizing** is essential to maintain air velocity within recommended limits, reduce noise, and ensure the system delivers the correct amount of conditioned air to each zone. Oversized ducts may result in wasted materials and higher costs, while undersized ducts can restrict airflow and cause pressure imbalances or noisy systems.

Designers must account for:

- **Friction loss** across long runs
- **Velocity limits** to prevent noise (usually 600–900 FPM for supply ducts in office spaces)
- **Static pressure calculations**
- **Branch duct balancing** to ensure even airflow distribution

Example:

- A classroom located at the end of a long corridor may require a larger branch duct or a booster fan to ensure adequate airflow if static pressure drops significantly along the route.

Plenum Locations

A **plenum** is a space (usually above a suspended ceiling or below a raised floor) that facilitates the return or supply of air. Understanding plenum usage is essential for space planning and coordination with lighting, fire protection, and structural systems.

Return plenums are common in commercial buildings and can help reduce the need for extensive return ductwork. However, plenums must meet code requirements (e.g., be constructed with non-combustible materials and remain free of wiring not rated for plenum spaces).

Example:

- In a large open-plan office, using a return air plenum above the ceiling reduces return duct costs and increases flexibility in layout changes.

Vertical Shaft Planning

Vertical shafts carry HVAC ducts, piping, and other utilities between floors. Early coordination of shaft placement is vital to avoid costly revisions during construction. Proper shaft design ensures that ducts have sufficient clearance to reach different zones while maintaining the architectural layout and avoiding conflicts with structural elements or other MEP systems.

Key considerations include:

- **Stacking of mechanical rooms and risers** across floors
- **Floor-to-floor height** to accommodate ductwork
- **Accessibility for maintenance**
- **Fire-rated enclosures** as required by code

Example:

- In a high-rise building, grouping HVAC risers with plumbing and electrical shafts in a centralized core optimizes space and improves serviceability.

3. **Energy Efficiency and Controls**

Energy efficiency in building systems is a critical focus of sustainable design, regulatory compliance, and long-term operational cost savings. Designers must be familiar with how mechanical systems interact with architectural design decisions and how control strategies contribute to a building's energy performance. This section examines relevant energy codes and explores specific energy-saving control systems, including **economizers**, **demand-controlled ventilation (DCV)**, and **building automation technologies**.

Impact of Energy Codes

Two primary standards govern energy efficiency requirements in buildings:

- **IECC (International Energy Conservation Code):** The IECC sets baseline performance standards for residential and commercial buildings, covering HVAC systems, envelope, lighting, and controls. States may adopt the IECC directly or modify it with local amendments.
- **ASHRAE 90.1:** A more technical and detailed standard, often referenced by the IECC, especially for commercial projects. ASHRAE 90.1 outlines minimum efficiency

requirements for mechanical equipment, insulation, lighting, HVAC controls, and renewable integration.

Designers must ensure that HVAC systems are compliant with these standards, which may dictate:

- Minimum equipment efficiencies (e.g., SEER, EER, COP)
- Insulation levels for ducts and piping
- Zoning and control requirements
- Mandates for automatic setback controls or economizers

Example:

- A new office building design in California must comply with Title 24 (based on IECC and ASHRAE 90.1), requiring occupancy sensors, high-efficiency HVAC equipment, and ventilation that responds to real-time usage levels.

Economizers

An **economizer** is a mechanical device integrated into HVAC systems to bring in outside air for free cooling when exterior conditions permit—typically when the outdoor air is cooler and less humid than the indoor air. This "free cooling" reduces the reliance on mechanical refrigeration and lowers energy consumption.

There are two main types:

- **Airside Economizers:** Bring in cool outdoor air directly.
- **Waterside Economizers:** Use a cooling tower and heat exchanger to cool water without running the chiller.

Example:

- In a temperate climate like Seattle, an office building with airside economizers can rely on outdoor air for cooling during many spring and fall days, significantly reducing chiller operation hours.
  Codes such as ASHRAE 90.1 may require economizers above certain cooling capacities, depending on climate zone and building type.

Demand-Controlled Ventilation (DCV)

Demand-controlled ventilation adjusts ventilation rates in response to occupancy levels, often measured via CO_2 sensors. Rather than running at a constant airflow rate, DCV systems modulate fresh air intake to match the actual needs of the space—improving indoor air quality while minimizing over-ventilation and wasted energy.

Key benefits:

- Energy savings, especially in variable-occupancy spaces
- Enhanced air quality management
- Compliance with ASHRAE 62.1 ventilation standards

Example:

- In a college lecture hall, DCV sensors detect when the room is sparsely occupied and reduce ventilation to minimum levels. When the room fills up for a class, the system ramps up airflow to maintain proper CO_2 levels and thermal comfort.

Programmable Thermostats & Building Management Systems (BMS)

Programmable thermostats allow occupants or facility managers to set temperature schedules that align with occupancy patterns. More advanced systems—like **Building Management Systems (BMS)** or **Building Automation Systems (BAS)**—enable centralized control and monitoring of HVAC, lighting, security, and fire systems.

A BMS can:

- Monitor real-time energy usage
- Provide fault detection and diagnostics
- Control systems based on occupancy, time of day, or weather
- Integrate with renewable energy sources

Example:

- In a high-performance hospital, a BMS maintains constant temperature and humidity in surgical suites, automatically adjusts ventilation rates in waiting areas after-hours, and alerts staff when energy thresholds are exceeded.

  By integrating these technologies, architects and engineers can significantly improve building energy performance while supporting occupant comfort and code compliance.

4. **Electrical System Basics**

   A strong grasp of electrical system fundamentals is essential for architects working on integrated building design. While electrical engineering consultants are typically responsible for detailed calculations and circuit sizing, architects must understand how these systems impact spatial planning, coordination, code compliance, and safety. This section covers **power distribution systems**, **transformer and panelboard locations**, **branch circuit layouts**, and the **space requirements for electrical infrastructure**, especially as governed by the **National Electrical Code (NEC)**.

   Power Distribution Systems

   A building's electrical distribution system begins at the utility service entrance and is distributed throughout the facility via a structured network of feeders, panelboards, transformers, and branch circuits. Power typically enters a building at **medium voltage (e.g., 13.2 kV)** and is stepped down to **low voltage (e.g., 480/277V or 208/120V)** for internal use.

   There are typically three levels of power distribution:

   1) **Main service switchgear or switchboard** – receives power from the utility
   2) **Panelboards** – distribute power to various circuits
   3) **Branch circuits** – deliver power to receptacles, lighting, and equipment

Example:

- A commercial office may have 480/277V for lighting and large HVAC equipment, with step-down transformers providing 208/120V for general receptacles and office equipment.

Transformer Locations

Transformers are used to step voltage up or down between distribution levels. Location and placement affect heat dissipation, noise, and space planning.

- **Dry-type transformers** are often used indoors and located in electrical rooms or closets.
- **Oil-filled transformers** are typically found outdoors due to fire risk and environmental considerations.

Key planning considerations:

- Avoid placing transformers near noise-sensitive areas (e.g., offices or classrooms).
- Provide adequate ventilation due to heat generation.
- Allow access for maintenance and future replacement.

Example:

- A transformer serving a retail space is placed in a rear service corridor, with a louvered door to allow passive cooling.

Panelboards and Branch Circuit Layouts

Panelboards (or breaker panels) are the connection point between feeders and branch circuits. Panel locations should be accessible, clearly labeled, and designed with capacity for future load growth if possible.

Branch circuits are distributed from panelboards to supply lighting, receptacles, appliances, and equipment. Architects should work closely with electrical engineers to:

- Coordinate circuit layouts with interior partitions
- Avoid conflicts with plumbing and ductwork
- Ensure outlets and switches meet ADA and code requirements

Example:

- In a school, classroom panelboards are typically located in custodial closets or utility spaces, serving lighting and receptacles in adjacent rooms.

Electrical Room and Equipment Clearances (NEC)

The **NEC (National Electrical Code)** outlines specific spatial requirements for all electrical equipment to ensure safety, accessibility, and maintenance access. These requirements affect the design and size of **electrical rooms**, **riser shafts**, and **equipment closets**.

Key requirements include:

- **Minimum clearance in front of electrical panels:**
    - **3 feet** (per NEC 110.26(A)) for voltages up to 600V
    - Width of working space must equal the width of the equipment or a minimum of 30 inches
    - Headroom clearance must be at least 6’-8”

- **Dedicated space above panels:**
    - No plumbing, ductwork, or unrelated systems allowed above electrical equipment (NEC 110.26(E))
- **Door requirements:**
    - Electrical rooms with equipment over 1200 amps or over 6 feet wide must have doors that open in the direction of egress and include panic hardware (NEC 110.26(C)(3))

Example:

- In a mid-rise office building, an electrical room serving each floor includes vertical riser access and a 3-foot clearance in front of each panel, as required. The room is placed near the core for easy alignment of risers with upper floors.

Vertical Risers

Vertical risers carry electrical feeders and telecommunications between floors. These shafts should be:

- **Stacked vertically** from floor to floor
- Sized to accommodate conduit, future expansion, and potential fire separation needs
- Located near mechanical shafts or elevator cores for efficiency

Example:

- In a hotel, a dedicated riser shaft runs next to the elevator core, serving floor-by-floor panels for lighting and receptacles in guest rooms.

By understanding the spatial, access, and safety requirements for electrical systems—especially as dictated by the NEC—architects can ensure that designs support functional, compliant, and maintainable electrical infrastructure. Coordination during early design phases helps avoid costly changes later and supports a smoother construction process.

5. **Lighting Systems and Controls**

Lighting plays a critical role in building functionality, occupant comfort, energy performance, and code compliance. Effective lighting design involves more than just selecting fixtures—it requires coordination of lighting types, control strategies, and integration with architectural elements like windows, ceilings, and occupancy zones. In this section, we compare major lighting technologies, examine energy-related performance metrics like **Lighting Power Density (LPD)**, and review essential lighting control systems, including **occupancy sensors**, **daylight-responsive dimming**, and **emergency lighting** requirements.

Lighting Types: LED, Fluorescent, and Daylighting

Understanding the strengths and limitations of different lighting technologies is key for designing efficient and high-performing spaces.

LED (Light Emitting Diode):

- Highly energy-efficient and long-lasting (50,000+ hours)
- Dimmable and compatible with advanced controls
- Minimal heat generation
- Available in a variety of color temperatures and form factors

- Higher upfront cost, but lower life-cycle cost

Fluorescent:
- Less efficient and shorter lifespan than LEDs
- May still be used in existing buildings or retrofit applications
- Contains mercury—requires proper disposal
- Flicker and warm-up time can affect user comfort

Daylighting:
- Uses natural light through windows, skylights, or clerestories
- Reduces dependence on electric lighting
- Must be carefully balanced to prevent glare or overheating
- Works best when integrated with automatic lighting controls

Example:
- A K–12 school uses LEDs for classrooms and hallways, paired with daylighting from large north-facing windows and skylights in corridors. Automatic daylight sensors dim the fixtures when sufficient sunlight is available.

Lighting Power Densities (LPD)

Lighting Power Density (LPD) measures the maximum allowable wattage per square foot for lighting in a given space. LPD is regulated by energy codes like **ASHRAE 90.1** and **IECC**, which set limits based on space type and usage.

- Measured in **watts per square foot (W/sf)**
- Two compliance methods:
  - **Space-by-space method**: Assigns LPD limits to individual room types (e.g., 0.90 W/sf for offices, 1.1 W/sf for classrooms)
  - **Building area method**: Assigns an average LPD for an entire building type

  Designers must select efficient fixtures and control strategies to meet or beat these thresholds.

Example:
- A retail space with a 1.2 W/sf LPD limit uses energy-efficient track lighting and LEDs to stay within code, while enhancing product visibility.

Occupancy and Vacancy Sensors

Occupancy sensors turn lights on when a space is occupied and off when vacant. **Vacancy sensors** require manual activation but turn off automatically. These controls are critical for energy savings and are required in many spaces per IECC and ASHRAE 90.1.

Sensor technologies include:
- **Passive infrared (PIR)**: Detects motion using body heat; best for line-of-sight areas
- **Ultrasonic**: Detects sound/motion; useful in enclosed spaces
- **Dual-technology**: Combines both for better accuracy

Example:

- Private offices and restrooms are often equipped with vacancy sensors to prevent lights from turning on unnecessarily and to save energy when unoccupied.

Daylight-Responsive Dimming

Daylight-responsive dimming systems use photosensors to detect natural light levels and dim artificial lighting accordingly. This ensures that lighting remains comfortable while minimizing energy use during daylight hours.

Key design considerations:

- Place sensors near windows or daylight zones
- Coordinate with window treatments and daylight penetration depth
- Fixtures must be dimmable and compatible with control systems

Example:

- In an open office space, daylight sensors near perimeter windows reduce LED fixture output when morning sun is adequate, while interior fixtures remain at full brightness.

Many energy codes mandate daylight harvesting controls in spaces larger than a certain size (e.g., 150 W of general lighting within 15 feet of windows).

Emergency Lighting and Backup Power

Emergency lighting ensures safe egress and operational continuity during power outages. These systems must meet **life safety codes (e.g., NFPA 101)** and **NEC requirements**, and are typically powered by:

- **Battery-backed LED fixtures**
- **Central inverter systems**
- **Emergency circuits connected to a backup generator**

Critical areas requiring emergency lighting include:

- Egress paths, stairways, and exit doors
- Fire command centers
- Electrical and mechanical rooms
- Health care and mission-critical spaces

Example:

- In a hospital, emergency lighting remains operational for operating rooms, corridors, and exit stairwells using both battery backup units and a central generator tied to essential circuits.

Summary

Lighting systems and controls are a key area of architectural coordination and a high-impact element on both energy performance and occupant satisfaction. Effective lighting design:

- Aligns fixture selection with use and code
- Minimizes energy use through LPD compliance
- Incorporates automated controls to adjust output based on occupancy and daylight

- Ensures reliable emergency egress and safety during outages

  Understanding how lighting integrates with mechanical systems, ceiling design, and window placement is crucial for passing the PPD exam and designing functional, code-compliant buildings.

6. **Plumbing Systems and Layouts**
   Plumbing systems are essential to building functionality, health, and sanitation. For architects, understanding the fundamentals of **supply**, **waste**, and **vent piping systems**, as well as **fixture placement**, **stack coordination**, and **hot water delivery** strategies, is vital for efficient planning and interdisciplinary coordination. Poor plumbing layouts can lead to increased costs, design conflicts, and future maintenance challenges—issues the PPD exam is designed to test your ability to avoid.

Supply, Waste, and Vent Piping Systems
Plumbing systems are broadly divided into:

- **Supply (potable water):** Delivers cold and hot water under pressure to fixtures and equipment.
- **Waste (drainage):** Removes wastewater and directs it to the sanitary sewer system via gravity.
- **Vent piping:** Allows air into the drainage system to maintain proper pressure and prevent siphoning of traps.
  Each of these systems has specific spatial, code, and coordination requirements.

Example:

- In a multi-story apartment building, vertical supply and waste lines must be carefully located and aligned to serve bathrooms and kitchens efficiently while meeting venting and cleanout access code requirements.

Stack Locations and Vertical Coordination
Stacks are vertical piping runs for waste, venting, or water supply. Aligning plumbing fixtures vertically from floor to floor minimizes the number of stacks and reduces costs for piping, firestopping, and coordination.

Key architectural considerations:

- **Stacked toilet rooms** (e.g., restrooms above restrooms) are highly efficient
- Avoid locating plumbing over electrical rooms or critical areas like server rooms
- Coordinate chase walls and shafts early in design

Example:
Toilet rooms in an office building are designed to align on all floors near the core. This allows a single soil stack and vent pipe to serve all levels, simplifying venting and reducing horizontal waste runs.

Pipe Sizing and Routing
Pipe sizing depends on:

- Fixture type and count (known as **fixture units**)
- Anticipated flow rates and pressure needs
- Total developed pipe length (to account for pressure loss)

Cold and hot water supply pipes must maintain adequate pressure at all fixtures, especially in high-rise buildings where pressure drops with height. Pressure zones or booster pumps may be required in taller buildings.

Drain pipes must be sized and sloped correctly (typically 1/4" per foot for small horizontal runs) to allow gravity drainage without clogging. Vent pipes must rise vertically above the flood level rim before changing direction.

Example:

- A commercial kitchen may require larger waste lines to accommodate grease traps and multiple sinks, while maintaining venting to avoid trap siphonage.

Hot Water Delivery Systems: Tank vs. Tankless

Hot water systems can be designed using:

- **Tank-type heaters:** Store and heat a fixed volume of water
    - Simpler and lower cost upfront
    - May require larger footprint
- **Tankless (on-demand) systems:** Heat water as it flows through the unit
    - More efficient and compact
    - May struggle to meet peak demand in large buildings

Recirculation loops are often used in large or elongated buildings to maintain hot water at fixtures and reduce wait time and water waste. These loops keep water moving constantly or on demand using timers or temperature sensors.

Example:

- In a hotel, a central boiler system with a recirculation loop ensures that guests receive hot water immediately without long delays, especially at far ends of the building.

Plumbing Code Considerations

Architects must coordinate plumbing layouts with:

- **IBC and IPC requirements**
- ADA fixture spacing and mounting heights
- Minimum fixture counts by occupancy (e.g., office, restaurant, school)
- Accessibility for maintenance (e.g., valve access, cleanouts)
- Fire separation of plumbing chases in multi-family or mixed-use buildings

Example:

- A school building locates plumbing chases between classroom pairs, allowing restrooms to be back-to-back and simplifying code-required separation from corridors.

Summary

A well-planned plumbing layout:

- Minimizes piping runs by stacking fixtures
- Ensures code compliance and access to shutoffs and cleanouts
- Coordinates pipe sizes and routes to maintain flow and pressure
- Optimizes hot water delivery based on building size and occupancy

As you prepare for the PPD exam, focus on identifying how early architectural decisions (like restroom placement, chase wall coordination, and core design) affect plumbing efficiency and performance. Understand the spatial impacts of plumbing systems and how they must be accommodated within the overall building design.

7. **Fire Protection Systems**

Fire protection systems are a critical component of life safety and code compliance in building design. As an architect, you are not expected to engineer these systems, but you must understand their **types**, **spatial implications**, **location requirements**, and **coordination needs** with other systems and building elements. This section focuses on the primary types of sprinkler systems and introduces key concepts such as **standpipe systems** and **sprinkler zoning**, which often appear on the PPD exam.

Wet-Pipe Systems (Most Common)

Wet-pipe sprinkler systems are the standard fire protection system used in most commercial, residential, and institutional buildings. These systems have pipes constantly filled with water under pressure, allowing immediate response when a sprinkler head is activated due to heat.

Characteristics:

- Simple, reliable, and cost-effective
- Quick response time
- Minimal maintenance requirements

Use Cases:

- Offices, schools, apartments, hotels, retail stores

Example:

- In a typical office building, wet-pipe sprinklers are installed above the ceiling grid and activate individually when a fire triggers the thermal element of the sprinkler head.

Dry-Pipe Systems (Unheated Spaces)

Dry-pipe sprinkler systems are filled with pressurized air or nitrogen rather than water. When a sprinkler head is triggered, air pressure is released, opening a valve that allows water to flow into the pipes and discharge through the open sprinkler.

Characteristics:

- Used where pipes could freeze (unheated or partially heated spaces)
- Slightly slower response than wet systems due to delay in water delivery
- Requires air compressors and monitoring

Use Cases:

- Parking garages, loading docks, unheated storage facilities, attics in cold climates

Example:

- A dry-pipe system is installed in a partially enclosed parking garage to prevent freezing during winter months, while the rest of the building uses a wet system.

Pre-Action Systems (Museums, Data Centers)

Pre-action sprinkler systems combine elements of both wet and dry systems and add a **delayed, two-step activation** process. These systems are filled with air or nitrogen and require a separate detection event—such as smoke or heat detection—**before water is released into the pipes**.

Characteristics:

- Reduces risk of accidental discharge
- Used in areas where water damage would be catastrophic
- More complex and costly to install and maintain

Use Cases:

- Art museums, data centers, archives, libraries, clean rooms

Example:

- A pre-action system in a museum only releases water into the pipes after a heat detector confirms a fire, preventing accidental water damage from false alarms or system failures.

Standpipe and Sprinkler Zones

Standpipe systems are vertical piping systems that provide water access to fire hoses inside the building. They are typically required in:

- Buildings over **three stories or 30 feet high**
- High-rise buildings
- Large assembly occupancies

There are three classes of standpipe systems:

- **Class I:** For fire department use only (2½" hose connections)
- **Class II:** Occupant use (1½" hoses)
- **Class III:** Combination of both

Sprinkler zones divide large or tall buildings into manageable areas based on pressure and flow requirements. Each zone:

- Has its own **valve and flow switch**
- Is limited by **floor area or vertical height**
- Helps maintain adequate pressure and simplify maintenance

Example:

- A 12-story residential tower is divided into two sprinkler zones: one serving floors 1–6, and another serving floors 7–12, with a pressure-reducing valve between them to avoid over-pressurizing lower-floor heads.

Architectural Coordination Considerations

Fire protection systems must be coordinated with:

- **Ceiling systems** (e.g., sprinkler heads cannot be obstructed)
- **Wall finishes** (ratings must be maintained at sprinkler penetrations)

- **Fire-rated construction** (e.g., sprinklers often required in rated corridors and stair enclosures)
- **Accessibility and aesthetics** (consider semi-concealed or concealed heads in exposed areas)
  Designers must also ensure compliance with:
- **NFPA 13** (Standard for the Installation of Sprinkler Systems)
- **NFPA 14** (Standard for Standpipe and Hose Systems)
- **IBC Chapter 9** (Fire Protection Systems)

Summary
Understanding the **type and purpose of fire protection systems** helps architects design safer, more efficient buildings and coordinate effectively with fire protection engineers and contractors. Key takeaways for the ARE 5.0 PPD exam include:

- Wet-pipe systems are the default choice for most applications.
- Dry-pipe and pre-action systems are used in sensitive or freeze-prone environments.
- Standpipes are required in taller buildings or specific occupancies.
- Sprinkler zones help manage pressure and flow in large or vertical structures.

8. **Water Efficiency and Conservation**
   Water efficiency is a critical part of sustainable building design, environmental responsibility, and compliance with both energy codes and green building rating systems. Architects play a key role in reducing potable water demand through the selection of fixtures, layout planning, and the integration of alternative water systems. This section reviews key strategies for improving water efficiency, including **low-flow fixtures**, **dual-flush toilets**, **greywater reuse**, and **rainwater harvesting**, all of which may appear on the **ARE 5.0 PPD exam**.

Low-Flow Fixtures
Low-flow fixtures reduce water consumption at the source without sacrificing functionality. These include faucets, showerheads, urinals, and toilets designed to use significantly less water than conventional models.

- **Standard vs. Low-Flow:**
  - Traditional toilets: ~3.5 gallons per flush (gpf)
  - Low-flow toilets: 1.28–1.6 gpf
  - Low-flow showerheads: ≤2.0 gallons per minute (gpm)
  - Faucets: ≤0.5 gpm for public lavatories

Codes & Standards:

- WaterSense labeling by the **EPA**
- Plumbing fixture performance requirements per **UPC/IPC**
  Example:
- A public school installs 0.5 gpm lavatory faucets and 1.28 gpf toilets, meeting both code and LEED water reduction prerequisites.

Dual-Flush Toilets
Dual-flush toilets provide two flushing options—typically 0.8 gpf for liquid waste and 1.6 gpf for solid waste—allowing users to reduce water use based on need.

Benefits:

- Improved user control over water consumption
- Can contribute to indoor water use reduction credits in LEED and CALGreen

  Example:
- An office building installs dual-flush toilets in all restrooms. With high daily use and user awareness signage, the building reduces water use by over 30% compared to baseline.

Greywater Reuse

Greywater refers to gently used water from sinks, showers, and washing machines that can be treated and reused for non-potable purposes such as flushing toilets or landscape irrigation.

Design Considerations:

- Requires a separate plumbing system
- Treatment and storage requirements vary by jurisdiction
- Subject to state and local codes (e.g., California Plumbing Code Chapter 15)

Applications:

- Large residential projects
- Sustainable commercial developments
- Buildings with high water reuse potential (e.g., dormitories, gyms)

Example:

- A multi-family housing project uses greywater from showers to flush toilets. The system reduces potable water demand by nearly 25%, helping the project earn LEED points and meet local water restrictions.

Rainwater Harvesting

Rainwater harvesting systems collect and store rain from roof surfaces for reuse in irrigation, cooling towers, or even toilet flushing (when treated appropriately).

Components:

- Collection surface (typically a roof)
- Gutters and downspouts
- Filtration and storage tank
- Distribution system

Considerations:

- Local rainfall data and seasonal variation
- Water quality and treatment for intended use
- Roof materials (non-toxic surfaces preferred)

Example:

- A civic building in Portland, OR collects rainwater in underground cisterns for summer irrigation. Combined with native landscaping and drip irrigation, the system drastically reduces municipal water use.

Codes, Standards, and Rating Systems
Projects that implement water-saving strategies often receive recognition or credits under:

- **Title 24 (California)** – mandates water efficiency through CALGreen
- **LEED v4** – offers points under the *Water Efficiency* category
- **International Green Construction Code (IgCC)** – supports water conservation measures
  Example:
  LEED requires a minimum 20% reduction in indoor potable water use over baseline (per EPAct 1992), and provides additional points for higher savings. Similarly, California's **CALGreen** requires low-flow fixtures and promotes greywater systems.

Summary
Water conservation in architectural design supports sustainability, lowers operating costs, and helps projects meet regulatory and certification requirements. Architects can make a significant impact by:

- Selecting high-efficiency fixtures
- Aligning plumbing design with reuse strategies
- Integrating rainwater and greywater systems
- Coordinating with engineers and local code officials to ensure compliance

Key exam tip: Know the basic performance metrics of fixtures and understand where reuse systems are appropriate and code-allowed.

9. **MEP Space Planning**
Effective **MEP (Mechanical, Electrical, and Plumbing) space planning** is essential for integrating building systems without compromising architectural function, aesthetics, or code compliance. Architects must anticipate where systems will be routed, how much space they require, and how they affect building performance and layout efficiency. The PPD exam often tests your ability to **coordinate MEP systems within architectural layouts** while supporting serviceability, constructability, and occupant comfort.

Mechanical Shafts and Chases
Vertical shafts and **horizontal chases** house mechanical ductwork, piping, conduit, and fire protection risers. Planning for these spaces early in design avoids conflicts and ensures smooth integration with structural, architectural, and fire-rated assemblies.

Design tips:

- Align shafts vertically through all floors for efficient distribution
- Group MEP systems in shared chases when possible to conserve space
- Coordinate shaft sizes with engineers based on anticipated system loads and equipment types

Example:

- In a mid-rise apartment building, stacked mechanical chases are located in bathrooms and kitchens to serve plumbing and HVAC vertically, simplifying duct and pipe runs and reducing cost.

Equipment Room Locations

Mechanical and electrical equipment rooms must be strategically located based on system type, service zones, structural capacity, and maintenance access. Common locations include:

- **Ground floor** (near loading docks for large air-handling units or boilers)
- **Basement** (centralized plant systems, water storage tanks, pump rooms)
- **Penthouse/roof** (ideal for air handling units and chillers to maximize vertical duct runs and reduce occupied space impact)

Factors to consider:

- Weight and vibration loads
- Noise control and separation from occupied spaces
- Vertical access for ducts, pipes, and electrical risers
- Accessibility for replacement and service

Example:

- A university lab building places its air handling units in a rooftop penthouse to minimize duct length, while electrical switchgear is located on the ground floor near the main service entry.

Access for Maintenance and Replacement

Designs must accommodate ongoing **access for maintenance, inspection, and eventual equipment replacement**. Inaccessible or poorly located systems can lead to code violations, premature failure, or costly retrofits.

Best practices:

- Provide clearances around HVAC, electrical, and plumbing equipment per code and manufacturer specs
- Ensure mechanical rooms have double doors or removable panels for large equipment
- Avoid placing fixed construction (walls, beams) in front of major access panels
- Coordinate ceiling heights to allow access to ductwork, valves, and sensors

Codes & guidelines:

- Refer to **NFPA 70 (NEC)** for electrical clearance requirements
- Check mechanical code for equipment access space
- Comply with **ASHRAE** maintenance access standards

Example:

- A hospital design includes wide corridor access to the mechanical room and 3 feet of clear space around air handlers, in line with ASHRAE recommendations for service access.

Noise and Vibration Isolation

MEP systems can generate **noise and vibration** that negatively impact occupant comfort and building performance. Architects must work with engineers to mitigate these impacts through smart layout and material choices.

Mitigation strategies:

- Locate noisy equipment away from noise-sensitive spaces (e.g., offices, bedrooms, classrooms)

- Use acoustic enclosures, vibration isolators, and resilient mountings
- Route ducts and piping through acoustically treated shafts
- Separate mechanical rooms from occupied zones with buffer spaces (e.g., storage rooms)

Example:

- In a library, chillers and pumps are located in a basement mechanical room with concrete slab separation and acoustic baffling to prevent noise transfer into reading areas.

Summary

Successful MEP space planning involves:

- **Vertical and horizontal coordination** of shafts and chases
- Strategic **equipment room placement** for efficiency and access
- Adequate **clearances and accessibility** for ongoing service and system longevity
- Thoughtful **noise and vibration control** to protect occupant comfort

While engineers are responsible for MEP system sizing and design, architects must make sure the building can accommodate those systems with minimal disruption, cost, or user discomfort—skills directly tested on the ARE 5.0 Project Planning & Design exam.

10. Sustainability and Renewable Integration

Sustainable building design is no longer optional—it's a standard expectation in contemporary architectural practice. Architects must understand how **MEP (Mechanical, Electrical, and Plumbing)** system decisions directly impact energy consumption, carbon emissions, and long-term operational performance. Effective integration of **renewable energy systems**, **efficient HVAC design**, and **emerging technologies** like **battery storage** and **EV charging** helps projects meet green building standards, local energy codes, and carbon-reduction goals.

This section explores how to align MEP strategies with sustainability targets, including **Energy Use Intensity (EUI)**, **photovoltaic (PV) systems**, and **on-site power generation**.

HVAC's Role in Energy Use Intensity (EUI)

Energy Use Intensity (EUI) measures a building's annual energy use per square foot, expressed as **kBtu/sf/yr**. HVAC systems often account for **30–50%** or more of total energy consumption, making them a primary target for improving performance.

Architectural impacts on HVAC and EUI:

- Orientation and passive design reduce heating/cooling loads
- High-performance envelopes allow for smaller HVAC systems
- Efficient zoning and controls reduce unnecessary conditioning
- Natural ventilation and economizers reduce mechanical cooling demand

Design strategies:

- Right-size HVAC equipment to avoid oversizing
- Use variable speed drives, heat recovery, and demand-based controls
- Coordinate with energy modeling early in design

Example:

- A LEED Platinum office building uses a high-efficiency VRF (Variable Refrigerant Flow) HVAC system and zoned controls to reduce HVAC energy demand, resulting in an EUI of 28 kBtu/sf/yr—well below the baseline.

Photovoltaic (PV) Integration: Inverters and Metering

Photovoltaic (solar) systems convert sunlight into electricity, reducing demand on the utility grid. Architects must coordinate roof and site planning to accommodate panels, electrical infrastructure, and future expansion.

Key system components:

- **Solar panels** (typically on roof or canopy structures)
- **Inverters**: Convert DC electricity to AC; may be string or micro-inverters
- **Meters and monitoring**: Track production and consumption
- **Interconnection equipment**: Ties system to the building electrical panel and utility

Design Considerations:

- Minimize rooftop obstructions and shading
- Coordinate with structural engineers for panel loading
- Provide access paths and maintain clearances for maintenance
- Comply with fire and utility interconnection codes (e.g., NEC Article 690)

Example:

- A community center incorporates a 50-kW rooftop solar array with string inverters and a bi-directional net meter, allowing excess electricity to be sent back to the grid for credit under a net metering agreement.

Battery Storage and EV Charging Stations

Battery storage systems provide backup power, reduce peak demand charges, and support renewable integration by storing solar energy for use after sunset or during outages.

Battery storage planning:

- Allocate secure, ventilated space near the electrical room or in a mechanical yard
- Coordinate fire safety and ventilation per **NFPA 855** and **local codes**
- Plan for monitoring, inverter integration, and access

EV Charging Stations:

Electric vehicle charging infrastructure is becoming a standard feature, particularly in commercial, institutional, and multi-family residential developments.

Levels of charging:

- **Level 1** (120V): Slow; residential use
- **Level 2** (240V): Common in workplaces and public lots
- **DC Fast Charging**: Rapid charging; higher infrastructure cost

Design Implications:

- Reserve electrical capacity and conduit routes for current or future stations
- Consider location visibility, security, and accessibility

- Coordinate with utility provider for demand impact

Example:

- A parking garage includes 10 Level 2 chargers and provisions for future expansion. A lithium-ion battery system smooths peak demand from simultaneous charging and solar output fluctuations.

On-Site Generation vs. Utility Grid Interaction

Projects with on-site power generation (solar, wind, cogeneration) must be planned with a clear understanding of how they interact with the **utility grid**, and whether the building operates in **grid-tied**, **islanded**, or **hybrid** mode.

On-Site Generation Options:

- **Photovoltaics** (PV)
- **Wind turbines**
- **Fuel cells**
- **Combined Heat and Power (CHP)** systems

Grid Interaction Modes:

- **Net metering**: Building sends excess power to the grid, receives credit
- **Grid-tied without storage**: Relies on utility when generation drops
- **Islanded/microgrid**: Operates independently with batteries or generators
- **Demand response**: Adjusts energy usage based on grid conditions or utility incentives

Example:

- A university science building includes PV panels, a battery bank, and a cogeneration unit. During normal operation, it runs grid-tied, but can isolate as a microgrid during outages, maintaining essential lab functions.

Summary

Architects must collaborate closely with engineers to align MEP decisions with **sustainability goals**, optimize **energy performance**, and support **resilience and decarbonization**. Key roles for the architect include:

- Planning solar-ready roofs and battery space
- Selecting HVAC strategies that reduce EUI
- Coordinating MEP systems with renewable energy systems
- Integrating EV infrastructure into site and electrical layouts

These concepts are increasingly common in real-world practice—and on the ARE 5.0 PPD exam—where integrated thinking and sustainability literacy are essential to competent design.

11. Code Requirements for MEP

Understanding **MEP code requirements** is essential for architects—not only to ensure life safety and functionality but also to coordinate with engineers, ensure compliance during plan review, and avoid costly redesigns during construction. While engineers typically specify the technical details of mechanical, electrical, and plumbing systems, architects must understand how these systems are **influenced by code** and how they integrate within the architectural layout.

This section summarizes the most relevant MEP codes for architects, focusing on coordination, accessibility, and exam-critical concepts.

Mechanical Code (IMC)
The **International Mechanical Code (IMC)** governs the design and installation of HVAC systems to ensure safety, comfort, indoor air quality, and fire prevention.

Key areas for architects:

- **Ventilation requirements** (e.g., outdoor air intake per occupancy type)
- **Combustion air** for fuel-burning appliances
- **Duct routing and shaft requirements**
- **Equipment clearances and access**
- **Fire and smoke dampers** in rated assemblies

  Example:
- An apartment building requires corridor supply air and individual exhausts from each kitchen and bathroom. The architect must allocate vertical chases and ensure proper fire separation where ducts penetrate rated walls.

National Electrical Code (NEC / NFPA 70)
The **NEC (NFPA 70)** governs electrical systems, including power distribution, lighting, grounding, and safety requirements. While the electrical engineer handles technical compliance, architects must understand layout and spatial requirements.

Architectural considerations:

- **Clearances around electrical panels** (typically 36" clear in front, per NEC 110.26)
- **Room sizing** for electrical equipment
- **Emergency lighting** and exit signage coordination
- **Outlet spacing and GFCI/AFCI locations**

  Example:
- In a commercial kitchen, NEC requires GFCI-protected outlets within 6 feet of any sink. Architects must coordinate outlet location, cabinetry, and backsplash design accordingly.

Plumbing Codes (IPC or UPC)
Plumbing systems are regulated under either the **International Plumbing Code (IPC)** or **Uniform Plumbing Code (UPC)** depending on jurisdiction. These codes address fixture types, water supply, waste and vent systems, and water conservation.

Architectural responsibilities include:

- Providing sufficient space for plumbing chases and equipment
- Ensuring proper fixture counts based on occupancy (IBC/IPC Table 2902.1)
- Coordinating locations for cleanouts and shutoff valves
- Designing ADA-compliant bathrooms

  Example:
- A school must provide separate male and female toilet facilities based on student occupancy. Using IPC fixture count tables, the architect ensures adequate quantities and locations with vertical alignment for efficient piping.

Accessibility and MEP Requirements (ADA & ICC A117.1)
Accessibility codes (primarily **ADA** and **ICC A117.1**) require that mechanical and plumbing controls be reachable and usable by people with disabilities. These requirements apply to public buildings, workplaces, and multi-family residential occupancies.

Common MEP-related accessibility requirements:
- **Thermostats and controls** must be located within ADA reach range (typically 15"–48" above floor depending on approach)
- **Plumbing fixture controls** (faucets, flush valves) must be operable with one hand and not require tight grasping, pinching, or twisting of the wrist
- **Clear floor space** in front of controls, dispensers, and fixtures
- **Insulation of hot water pipes** under accessible lavatories to prevent burns
  Example:
- In an office restroom, the architect specifies lever-handle faucets and positions the soap dispenser and hand dryer within 40" AFF and above a 30"x48" clear floor space, meeting ADA reach and usability guidelines.

Summary
MEP codes are essential to architectural design—not only for compliance, but also for integration, safety, and user accessibility. Architects should be familiar with:
- The **IMC** for ventilation, duct routing, and HVAC safety
- The **NEC (NFPA 70)** for electrical room layout, clearances, and outlet placement
- The **IPC/UPC** for fixture layout, pipe sizing coordination, and venting
- The **ADA and ICC A117.1** for accessibility to thermostats, plumbing controls, and dispensers

Understanding how MEP codes intersect with architectural design allows you to make smart decisions early, streamline interdisciplinary coordination, and ensure project success—both on the job and on the ARE 5.0 PPD exam.

12. Coordination with Site and Structural Systems
Successful building design requires the seamless integration of **MEP (Mechanical, Electrical, and Plumbing)** systems with both **structural** and **site elements**. Architects must anticipate how ducts, pipes, and conduit will pass through or around structural framing, and how MEP systems will connect to site infrastructure such as water, sewer, power, and storm drainage systems.
Poor coordination in this area can lead to **costly redesigns**, **field conflicts**, or even **code violations**. On the ARE 5.0 PPD exam, you are expected to demonstrate a conceptual understanding of how MEP systems are routed and how they interface with the building's structure and site services.

Structural Framing and MEP Routing
MEP systems often share space with the building's structural framework. Proper coordination ensures that mechanical ducts, piping, and conduit do not interfere with structural integrity or violate clearance requirements.

Key considerations:

- Avoid routing large ducts or pipes through **structural beams or joists** without engineered openings
- Use **framing zones** (e.g., furred ceilings, raised floors) to conceal services
- Coordinate large ducts and main piping with **structural bay spacing**
- Reserve specific corridors or shaft locations for vertical risers

Example:

- In a steel-framed office building, the structural engineer and architect coordinate deeper beams in select bays to allow large ducts to run between them without penetrations, maintaining structural performance and service continuity.

Site Utilities and Building Service Entry

All buildings require utility connections to the site's infrastructure, including **water, sanitary sewer, storm sewer, gas, and electrical service**. The location of these entry points and how they connect to building systems must be carefully planned and coordinated with civil, structural, and MEP engineers.

Architectural coordination includes:

- Establishing utility room locations on the perimeter near incoming services
- Ensuring accessible routes for utility maintenance and metering
- Aligning trench and conduit paths with site layout and grading
- Avoiding conflicts with foundations or footings

Common utility entry points:

- **Water line**: Enters near the main mechanical or plumbing room, often in a below-grade vault
- **Sanitary sewer**: Gravity-fed lines must slope away from the building; exit point determined by invert elevation
- **Electrical service**: Enters through underground conduit to the main electrical room; coordinate with utility provider for transformer placement

  Example:

- A commercial retail center locates its electrical room at the rear corner of the building, adjacent to the utility pad-mount transformer and service drive, minimizing trenching and simplifying service access.

Grading Impacts on Buried Mechanical Vaults and Pump Stations

Topography and grading can significantly affect the feasibility and function of below-grade MEP systems such as:

- **Water meter vaults**
- **Grease interceptors**
- **Sanitary or stormwater pump stations**
- **Backflow preventer vaults**

Design implications:

- Vaults should not be placed in low points where they may flood

- Adequate cover must be maintained over pipes for freeze protection and structural load
- Vault access hatches should be accessible from grade and not obstructed by landscaping or sidewalks
- Storm systems must be gravity-drained unless site grading requires a **sump and pump system**

Example:

- On a sloped healthcare campus, a stormwater detention vault is placed at the lowest site point. However, because the main mechanical room is uphill, a storm lift station is added to ensure proper drainage, with maintenance access coordinated near a service road.

Summary

Architects play a central role in coordinating MEP systems with **site constraints** and **structural design**. Key responsibilities include:

- Reserving zones in floors, ceilings, and walls for MEP routing
- Locating utility entrances to minimize trenching and ensure serviceability
- Coordinating below-grade systems with site grading and civil engineering
- Preventing interference between ducts/pipes and structural framing
  This coordination happens early—typically during schematic and design development phases—and is critical for constructability and long-term building performance. On the ARE 5.0 PPD exam, expect questions that test your ability to **identify logical MEP routing**, **place mechanical rooms relative to utilities**, and **avoid coordination pitfalls.**

🎯 PPD Exam Strategy Tips:

The **Project Planning & Design (PPD)** division of the ARE 5.0 exam assesses your ability to make informed design decisions during early-to-mid project phases, especially regarding building systems integration, spatial planning, and environmental response. Unlike the Project Development & Documentation (PDD) division, which emphasizes detail, PPD tests your ability to evaluate **design-level decisions**—balancing performance, code, cost, climate, and constructability.

Use the following exam strategy tips to stay focused and efficient during your PPD preparation and while working through case studies or multiple-choice questions.

☑ Focus on Basic System Understanding—Not Detailed Engineering

You are not expected to size ducts, calculate wire gauges, or determine exact pipe lengths. Instead, the exam tests your ability to:

- Identify **appropriate systems for building type and climate**
- Understand **functional relationships** between MEP components
- Recognize **coordination issues** with structure and architecture
  🧠 **Tip:** Be familiar with system types like VAV vs. VRF, tank vs. tankless water heaters, or 120/240V vs. 277/480V service—**conceptual differences**, not engineering specs.

☑ Be Prepared to Make Comparative Judgments

Many questions require you to compare systems and select the best one for a given context (e.g., building type, size, climate, program priorities).

Common comparisons:

- **Air-source vs. ground-source heat pumps**
  → Air-source is typically lower first cost but less efficient in cold climates. Ground-source has higher upfront cost but greater long-term efficiency.
- **Packaged rooftop units vs. chilled water systems**
  → Rooftop units are simpler and more common for low-rise buildings; chilled water systems offer better control and efficiency for large or multi-story projects.
- **Tank vs. tankless water heaters**
  → Tankless saves space and provides continuous hot water, but may require higher gas/electrical capacity.

🧠 **Tip:** Know the **advantages, disadvantages, and ideal applications** of each system type.

☑ Watch for System Mismatches in Case Studies

PPD case studies often include **drawings, diagrams, and narrative descriptions** with embedded flaws or mismatches that you must spot. These may involve:

- **Oversized rooftop units**: Waste energy, increase upfront cost
- **Poorly located electrical rooms**: Far from utility access, difficult service entry
- **Inaccessible shutoff valves**: Maintenance hazard, code violation
- **Mechanical rooms with insufficient clearance**: Equipment replacement issues

🧠 **Tip:** Look for red flags in **floor plans**, **MEP diagrams**, or **narrative clues** that suggest poor integration, inefficiency, or impractical design decisions.

☑ Review the Impact of Climate Zone on System Selection

Climate has a direct influence on HVAC strategy, envelope performance, and energy efficiency measures.

Examples:

- **Hot-humid climates** (e.g., Miami): Prioritize systems with latent heat control, like DX units with reheat coils or dedicated outdoor air systems (DOAS)
- **Cold climates** (e.g., Minneapolis): Consider systems with energy recovery and strong heating performance, such as hydronic boilers or ground-source heat pumps
- **Mixed-humid** (e.g., Washington, DC): Balance heating and cooling needs with adaptable systems like VRF or packaged gas/electric rooftop units

🧠 **Tip:** Know which systems perform better in which zones. Refer to the **ASHRAE climate zone map** for reference.

☑ Use Energy Performance and Space Planning Criteria to Guide Decisions

System selection should align with:

- **Energy goals** (e.g., low EUI targets, LEED points, passive strategies)
- **Space availability** (e.g., rooftop area, basement mechanical room, ceiling plenum height)
- **Maintenance accessibility**
- **Architectural program requirements** (e.g., acoustics, zoning needs, equipment visibility)
  Example:
- In a museum, avoid noisy rooftop equipment directly above galleries—opt for quiet interior mechanical zones or remote air-handling units.

🧠 **Tip:** Always ask: *Does this system fit the building type, program, and space constraints?*

📌 **Summary Strategy Checklist**
Before the exam, make sure you:
- Understand the **basic function** and **pros/cons** of key HVAC, electrical, and plumbing systems
- Know how to **match systems to building types** and **climate zones**
- Can recognize **design flaws** in case studies or diagrams
- Remember that **coordination** and **suitability** matter more than numbers or detailed specs

B. Determine structural systems (A/E) (4-8%)

NCARB Requirements
You will need to evaluate different structural systems, including their cost, availability, load capacity, and impact on building design and configuration. Based on these factors, as well as building size and function, you will need to select the appropriate structural system and layout for a building.

NCARB Objective Overview:
As part of the **Project Planning & Design (PPD)** division, you are expected to evaluate and recommend **structural systems** appropriate to a project's size, type, location, function, and design intent. Architects must collaborate with structural engineers but also understand how **structural decisions impact spatial planning, cost, and constructability**.
You should be able to:
- Identify common structural systems
- Assess their advantages, limitations, and suitability for different building types
- Understand how structure affects form, massing, floor-to-floor height, and spatial flexibility

1. Identify Major Structural System Types
Architects must understand the **characteristics, advantages, limitations, and appropriate uses** of the main structural systems commonly used in building design. This foundational knowledge is critical not only for the ARE 5.0 PPD exam, but also for interdisciplinary coordination in real-world practice.

Below are the major structural systems and their typical applications:

🏗 **Steel Frame Systems**
(e.g., moment frames, braced frames, open-web steel joists)
- **Best for:** Mid- to high-rise buildings, long spans, commercial and institutional projects
- **Key characteristics:**
  - High strength-to-weight ratio
  - Prefabricated elements allow faster construction
  - Flexible for large open plans (e.g., offices, airports)
  - Requires fireproofing (spray-on or encased)

- Compatible with curtain wall systems

☑ *Pros:* Long spans, fast erection, high design flexibility

⚠ *Cons:* Higher material cost, vulnerable to fire unless protected

🧠 **Example:** For a mid-rise office tower requiring large open floor plates and maximum tenant flexibility, a **steel moment frame with composite metal decking** provides both structural strength and layout adaptability.

🧱 Concrete Systems

(e.g., cast-in-place slab and beam, post-tensioned slabs, precast panels)

- **Best for:** Parking garages, hospitals, multifamily, labs, institutional buildings
- **Key characteristics:**
  - High thermal mass, improving energy performance
  - Inherent fire resistance
  - Good acoustic separation between floors
  - Allows for flat slab construction (minimizing floor depth)
  - Slower onsite construction unless precast is used

☑ *Pros:* Durable, fire-resistant, quiet, thermally efficient

⚠ *Cons:* Heavy, longer construction time (cast-in-place), more formwork required

🧠 **Example:** A hospital requiring low floor vibration and robust fire protection may use a **cast-in-place concrete flat slab system**, which also accommodates MEP penetrations efficiently.

🪵 Wood Framing Systems

(e.g., platform framing, balloon framing, heavy timber, CLT – cross-laminated timber)

- **Best for:** Residential, small commercial, low-rise buildings (typically up to 4–6 stories depending on code)
- **Key characteristics:**
  - Economical and widely available
  - Easy to modify on-site
  - Fast to build with small labor crews
  - CLT offers enhanced fire resistance and spans of 20–30 ft

☑ *Pros:* Low cost, renewable material, fast construction

⚠ *Cons:* Limited spans and height, combustibility (though mass timber chars predictably)

🧠 **Example:** A 4-story multi-family residential building may use **platform wood framing** over a concrete podium for cost efficiency and code compliance with Type III or V construction.

🧱 Masonry Bearing Walls

(e.g., CMU walls with wood or steel joists)

- **Best for:** Small schools, libraries, religious facilities, or additions
- **Key characteristics:**
  - High fire resistance and durability
  - Adds significant mass, which benefits acoustics and thermal lag
  - Low design flexibility (interior changes are difficult)
  - Works well with steel or wood roof/floor framing

☑ *Pros:* Durable, fireproof, low maintenance

⚠ *Cons:* Not ideal for open-plan spaces or future reconfiguration

🧠 **Example:** A single-story elementary school wing may use **masonry bearing walls** with **open-web steel joists** and a metal deck roof for durability and long-term low maintenance.

⚙ Hybrid Structural Systems

(e.g., steel frame with precast concrete floor planks, wood trusses with steel columns)

- **Best for:** Projects needing custom solutions, phased construction, or performance optimization
- **Key characteristics:**
  - Combines benefits of multiple materials
  - Can optimize construction speed, cost, and performance
  - Requires careful coordination among trades

☑ *Pros:* Tailored to specific needs, performance-optimized

⚠ *Cons:* Coordination complexity, multiple subcontractors and systems

🧠 **Example:** A university science building might use a **steel frame** for its large-span lecture halls, while using **precast concrete planks** in classroom wings to reduce construction time and improve fire resistance.

System Type	Best Use	Typical Span	Fire Resistance	Speed of Construction	Cost Efficiency
Steel Frame	Mid/high-rise, open floor plans	30–50+ ft	Needs protection	Fast	Moderate
Concrete (CIP)	Labs, hospitals, multifamily	20–40 ft	Excellent	Moderate–slow	Moderate–high
Wood Framing	Residential, small commercial	12–30 ft	Limited/Code-based	Fast	High
Masonry Bearing Wall	Institutional, small buildings	16–24 ft (joists)	Excellent	Slow	Moderate
Hybrid	Custom or complex buildings	Varies	Varies	Varies	Varies

Figure 5.1 Summary Table of Structural Systems

🎯 ARE 5.0 PPD Tip:

You are not expected to engineer structural systems, but you **must be able to recommend** the most appropriate system for a building based on:

- **Span requirements**
- **Building height and use**
- **Cost and material availability**

- **Integration with architectural design**
- **Lateral load and code requirements**

2. **Evaluate Structural Span Capabilities**
When selecting a structural system, one of the most critical design criteria is **span capability**—how far a material or system can economically and safely span without intermediate supports. Span affects the **functionality, flexibility, aesthetics**, and **cost** of a building. It also has a direct impact on **column spacing**, **floor-to-floor height**, and the integration of **MEP systems**.

For the ARE 5.0 PPD exam, you are expected to understand **approximate span ranges** of major systems and how those spans relate to planning decisions, not perform structural calculations.

System	Approximate Economical Span	Common Use
Steel Open-Web Joists	30–60 ft	Retail, schools, warehouses, low-rise office
Post-Tensioned Concrete Slabs	Up to 40 ft	Parking garages, hospitals, multifamily
CLT Panels	20–30 ft (based on thickness)	Schools, residential, low-rise commercial
Wood I-Joists or Trusses	20–40 ft	Residential, light commercial
Glulam or Steel Trusses	60–100+ ft	Gymnasiums, auditoriums, worship spaces

🧠 **Important:** These are typical economical spans. Longer spans may be achieved, but often at significantly higher cost or with structural trade-offs (e.g., increased depth, reduced headroom, or custom fabrication).

Figure 5.2 Typical Span Ranges by System Type

Note: CLT panels are Cross-Laminated Timber panels.

🧱 **How Span Affects Building Design**

Understanding span capabilities helps shape the building grid, circulation, and service strategies. Key considerations include:

• **Column Spacing and Grid Layout**

- Wider spans reduce the number of interior columns, allowing for more open and flexible spaces.
- Grid spacing should align with typical bay sizes of the chosen structural system for cost efficiency.

🧠 *Example:* Steel joists in a retail store may span 40 ft with columns spaced every 30 ft, creating a 30x40-ft structural grid ideal for merchandise layouts.

• Ceiling Heights and Floor-to-Floor Dimensions

- Longer structural spans often require deeper beams or trusses, which can affect:
  - Plenum space
  - Ceiling heights
  - Envelope height (zoning impact)
- Flat systems like post-tensioned concrete slabs can minimize depth, maximizing floor-to-ceiling space.

🧠 *Example:* In a hotel, post-tensioned flat slabs reduce total building height while maintaining generous ceiling clearance in guest rooms.

• Integration with MEP Systems

- The depth and layout of structural members affect how ducts, piping, and conduit are routed.
  - **Deeper trusses** allow more space for MEP but require careful coordination.
  - **Solid slabs** may restrict MEP penetrations without embedded sleeves or blockouts.
- Clear span systems help maintain uninterrupted duct runs and reduce coordination conflicts.

🧠 *Example:* An office building with steel wide-flange beams spaced at 10 ft allows MEP ducts to run parallel and perpendicular through a continuous plenum without excessive offsets.

🧠 Example Scenario (PPD Exam Style):

A community gymnasium requires a column-free basketball court with unobstructed views. What structural system is most appropriate?

☑ **Answer:** A **long-span steel truss** or **glulam beam system** would provide the necessary clear span (>70 ft) while supporting a high roof and allowing for suspended lighting, HVAC, and acoustic baffles.

🧠 Additional ARE 5.0 Tips:

- If a **design scenario calls for flexibility**, choose systems with longer spans and fewer columns.
- For **tight ceiling conditions** or **height restrictions**, avoid deep systems like steel trusses unless concealed in a higher volume.
- Always consider **system compatibility with floor layouts** (e.g., post-tensioned slabs don't mix well with frequent penetrations).

📌 Summary

Structural span capability is a key planning and coordination issue. As an architect, you should understand how different systems:

- Support open spaces or column-free areas
- Influence ceiling height and story height
- Affect the layout and routing of MEP systems

Your role is to **evaluate the design context**—such as space use, floor plan flexibility, and height constraints—and select a structural system that supports those goals efficiently.

3. **Consider Cost and Availability**
When selecting a structural system, architects must evaluate **material cost**, **labor availability**, and **regional supply chains**. Even if a system is technically appropriate, its **economic feasibility** and **access to materials and skilled labor** may determine whether it can be implemented efficiently and on budget.

On the ARE 5.0 PPD exam, expect to be asked which structural system is most appropriate based not only on function, but also **cost-effectiveness** and **practical construction constraints**.

Key Factors That Influence Structural System Cost

• **Material Cost**
- Varies significantly depending on region, market conditions, and demand.
- Commodity prices (e.g., lumber, steel) fluctuate—especially for large-scale projects or in areas with limited suppliers.

• **Labor Availability**
- Some materials (e.g., masonry, cast-in-place concrete) are **labor-intensive**.
- Availability of **specialized trades** (e.g., masons, concrete finishers, steel erectors) can affect cost and schedule.

• **Construction Speed**
- Faster systems reduce labor costs and time on site, which can lower overall project costs (e.g., wood framing, precast concrete).
- Systems that require **long curing or staging times** (like concrete) may be less desirable for fast-track projects.

• **Local Codes and Material Limitations**
- Zoning, fire rating, and construction type requirements may restrict low-cost systems (e.g., Type V wood in high-density areas).

System	Cost	Availability	Speed of Construction	Regional Considerations
Wood Framing	Low	Widely available in North America	Very fast	Best for low-rise, limited by fire ratings
Steel Frame	Moderate–High	Common in most urban centers	Fast (prefabricated)	May need crane access; high scrap recyclability
Concrete (CIP)	Moderate–High	Dependent on access to batch plants	Slow (formwork/curing)	Preferred in warm climates for thermal mass
Masonry	Varies (often labor-driven)	Depends on skilled masons	Slow	Common in historic or institutional projects

Figure 5.3 Comparative Summary by System

Note: CIP means Cast-in-place.

🧠 **Example:**
A rural health clinic is being designed in an area where ready-mix concrete delivery is unreliable and labor costs are high for masonry work. Which structural system is likely to be the most cost-effective and efficient?
☑ **Answer: Light wood framing or structural steel**—depending on fire-resistance and span needs—would be more feasible due to quicker construction, fewer labor demands, and greater material access.

🔍 **Additional Considerations:**
- **Remote or rural areas** may lack access to certain building materials (e.g., no local concrete batch plant, no steel fabricator).
- **Urban infill sites** may favor steel due to space constraints and speed of assembly with prefabricated components.
- **Post-disaster or emergency housing** might rely on **panelized wood** or **modular steel frames** for quick deployment and cost control.

🎯 **ARE 5.0 PPD Tip:**
If given a case study where a project is over budget, in a tight timeline, or located in a region with limited trades, the correct system will likely:
- Minimize specialized labor needs
- Use **readily available, cost-effective materials**
- Be compatible with **code constraints** for occupancy and construction type

📌 **Summary**
Cost and availability are **essential filters** for choosing a structural system. The architect's role is not to engineer the system, but to ensure the selected approach is:
- Economical for the location
- Available from local suppliers and trades
- Appropriate for project scale and schedule

Balancing performance with constructability is a critical skill both on the exam and in real-world practice.

4. **Assess Impact on Building Form and Layout**
Structural systems are more than just a means of supporting loads—they shape a building's **architectural expression**, influence **planning flexibility**, and affect **zoning compliance**, **MEP integration**, and **cost-efficiency**. Selecting an appropriate structural system means understanding how it interacts with the **form, proportions**, and **layout** of the building.
On the ARE 5.0 PPD exam, you'll need to identify how different structural systems constrain or enable certain spatial and formal decisions.

🏢 **Key Areas of Impact**

• Building Height and Massing

- Systems with **deep structural elements** (like trusses or deep steel beams) can increase overall building height due to added floor structure thickness.
- Conversely, **flat systems** (like post-tensioned or flat-plate concrete slabs) can reduce building height—important in projects with **zoning height restrictions** or **floor count limits**.

🧠 *Example:* A high-rise multifamily building in a city with a strict height limit might benefit from a **post-tensioned flat slab system**, which minimizes floor depth and allows more usable stories.

• Floor-to-Floor Height Requirements

- Deeper structural systems mean larger **floor-to-floor** heights are needed to maintain minimum **clear ceiling heights** and accommodate MEP systems in the plenum.
- Systems like **waffle slabs** or **beam-and-slab concrete** may require more depth but can hide ductwork between ribs or beams.

🧠 *Example:* In a commercial office tower, **steel framing with a composite metal deck** may require greater floor-to-floor heights than a flat concrete slab system but offers faster construction and larger spans.

• Façade Flexibility

- The type and location of **structural supports** impact where windows and openings can be placed.
- **Steel and concrete frame systems** allow more flexibility for **curtain walls or punched openings** because exterior walls are often non-load-bearing.
- **Masonry bearing walls** or **tilt-up concrete panels** can limit façade articulation due to their structural role.

🧠 *Example:* An office building with a full-glass curtain wall would likely use a **steel or concrete frame**, where the façade is supported independently of structural loads.

• Column Spacing and Interior Flexibility

- Longer spans mean fewer columns and more open floor plans.
- Closely spaced columns (as in masonry or wood framing) limit layout flexibility and future adaptability.
- Institutional and commercial buildings benefit from **wide column grids** for interior planning freedom.

🧠 *Example:* A **cast-in-place concrete flat slab** allows for **widely spaced columns**, ideal for flexible tenant build-outs in an office or lab setting.

Structural System	Floor Depth	Span Capability	Façade Flexibility	Interior Planning Flexibility
Steel Frame	Moderate	Long (30–60+ ft)	High (non-load-bearing)	High (fewer columns)
Post-Tensioned Concrete	Shallow–moderate	Medium (up to 40 ft)	High	High
Masonry Bearing Walls	Shallow	Short (16–24 ft)	Low	Low
CLT Panels	Moderate	Medium (20–30 ft)	Moderate	Moderate
Wood Framing	Shallow	Short–medium (20–30 ft)	Moderate	Low–moderate

Figure 5.4 Summary Table: Structural System Influence on Layout

🧠 **ARE-Style Example Scenario:**

A 10-story residential tower is being designed on a constrained urban site with a 120 ft zoning height limit. Which structural system would best support maximizing the number of floors while maintaining comfortable ceiling heights?

☑ **Answer:** A **post-tensioned concrete flat slab system**, which minimizes floor depth and avoids dropped beams, allowing more usable floors within the overall height limit.

🎯 **ARE 5.0 PPD Tip:**

When answering questions related to structural system selection:

- Think **vertically**: How does floor system depth impact total building height?
- Think **horizontally**: How does the structural grid affect space planning?
- Think **visually**: Will structural elements restrict window placement or impact the exterior expression?

📌 **Summary**

Structural systems are fundamental to the **spatial, visual, and functional outcome** of a building. Architects must balance:

- Height efficiency
- Floor plan openness
- Exterior flexibility
- Construction feasibility

By understanding the **spatial consequences** of each structural system, you'll make better-informed decisions that align with both **design goals** and **code constraints**—a core skill tested in PPD.

5. Understand Lateral Load Resistance Systems

In addition to supporting vertical loads (dead and live), structural systems must resist **lateral forces** from **wind** and **seismic activity**. This is especially critical in mid- to high-rise buildings and projects located in areas with high seismic risk or exposure to strong wind loads.

On the **ARE 5.0 PPD exam**, you'll need to identify the **appropriate lateral force-resisting system** based on building type, design intent, seismic zone, and space planning needs.

🌪 Primary Lateral Force-Resisting Systems

Each system has its own advantages, limitations, and implications for building design:

1) Moment Frames

- **Description:** Joints between beams and columns are rigidly connected to resist rotation, allowing the frame itself to resist lateral forces.
- **Best for:** Buildings requiring **large, unobstructed openings** (e.g., retail storefronts, open offices).
- **Design Implications:**
  - Offers architectural flexibility (few walls or braces)
  - Less stiff than braced frames or shear walls → greater lateral drift
  - Requires stronger, more expensive connections

🧠 *Example:* A museum with open gallery spaces and no internal bracing may use **moment frames** to preserve visual openness.

2) Braced Frames

- **Description:** Diagonal steel members (or X, K, or chevron braces) form triangulated systems to resist lateral forces.
- **Best for:** Structures needing efficient lateral load resistance without the mass of concrete.
- **Design Implications:**
  - Very efficient structurally
  - Bracing can interfere with window openings or circulation paths
  - Often exposed as an architectural feature in industrial or high-tech designs

🧠 *Example:* A multi-story school building in a non-seismic region might use **braced steel frames** along corridor walls for cost-effective stiffness.

3) Shear Walls

- **Description:** Vertical, solid wall elements (typically concrete or reinforced masonry) that resist lateral loads by acting as deep beams.
- **Best for:** High seismic zones and buildings with centralized cores.
- **Design Implications:**
  - Extremely stiff and effective at reducing lateral drift
  - Typically integrated into elevator cores or stairwells
  - Limit flexibility in layout and fenestration

🧠 *Example:* In a **seismic region like California**, a mid-rise apartment may use **cast-in-place concrete shear walls** at stair cores and party walls for robust lateral stability.

4) Diaphragms

- **Description:** Horizontal structural elements (floors and roofs) that transfer lateral loads to vertical resisting elements.
- **Best for:** All buildings—diaphragms are a required part of a complete lateral system.
- **Design Implications:**
    - Must be continuous and rigid enough to act as a unified surface
    - Can be concrete slabs, metal decking with concrete topping, or wood sheathing
    - Diaphragm discontinuities (e.g., large openings) must be addressed in design

🧠 *Example:* A **parking structure** with open floor plans may use **precast concrete floor slabs** as rigid diaphragms, transferring loads to perimeter shear walls.

🧠 **ARE 5.0 PPD Exam Example:**

A 5-story mixed-use building is being designed in a high seismic zone with residential units above retail. What lateral system is most appropriate to minimize drift and ensure occupant safety?

☑ **Answer: Concrete shear walls** located in stair cores and demising walls provide the necessary stiffness, mass, and fire resistance.

System	**Stiffness**	**Architectural Impact**	**Best Use**
Moment Frames	Low–Medium	Minimal interference, more drift	Open plans, high architectural flexibility
Braced Frames	High	Moderate to high visual impact	Cost-effective lateral resistance
Shear Walls	Very High	Limits openings and reconfiguring	Seismic zones, core areas
Diaphragms	Essential	Integrated in floor/roof design	All buildings (horizontal force transfer)

Figure 5.5 Lateral Load Resistance Systems Comparative Summary

🎯 **ARE 5.0 PPD Tips:**

- In **seismic zones**, prioritize **stiff, ductile** systems (e.g., concrete shear walls, special moment frames).
- In **wind-dominant zones**, focus on **efficient bracing or wall systems** that limit sway.
- Evaluate how the **lateral system affects the floor plan**—systems like braced frames may restrict openings or conflict with program needs.
- Watch for **system mismatches** in case studies (e.g., long open spans without proper lateral resistance).

📌 **Summary**

Lateral load-resisting systems are vital for **life safety, occupant comfort, and structural performance**. As an architect, your role is to:

- Understand how different systems behave under lateral loads
- Coordinate their integration into the architectural layout
- Select solutions that balance safety, stiffness, and design intent

This knowledge is frequently tested on the PPD exam, particularly in **case studies**, where you must evaluate system appropriateness based on **climate, location, and building function**.

6. **Coordinate Structural System with MEP and Architecture**
   An effective architectural design must consider how the **structural system integrates with mechanical, electrical, and plumbing (MEP) systems**, as well as how it aligns with **programmatic and spatial needs**. Poor coordination can lead to **costly conflicts**, inefficient layouts, and reduced occupant comfort.

   On the **ARE 5.0 PPD exam**, you'll often be asked to evaluate the spatial and planning implications of structural systems—especially how they affect MEP routing, floor-to-floor heights, and space planning decisions.

   Key Coordination Areas

   • **Beam Depths and Plenum Space**
   - Deeper beams (e.g., drop beams in concrete or large steel girders) reduce available **plenum height**—the space between the structural floor and ceiling, where **ductwork, piping, and conduit** are typically routed.
   - Shallow systems (e.g., **flat slabs**, **post-tensioned slabs**) reduce beam interference and allow **lower floor-to-floor heights**.

   Example: A **waffle slab** system, despite higher cost, offers recessed areas between ribs, creating built-in space for MEP distribution and preserving ceiling height.

   • **Integration of Ducts and Piping Through Structure**
   - Large ducts must often run **between beams**, or require **structural openings** through beams, joists, or slabs.
   - Steel systems typically offer more MEP flexibility due to regular framing bays and fewer solid obstructions.
   - Concrete systems may need **pre-planned blockouts or sleeves** to avoid post-tensioning tendons or reinforcing bars.

   Example: In a **hospital**, which has high MEP demand, a **steel frame with long-span joists** may be preferred to allow uninterrupted duct runs above corridor ceilings.

   • **Floor Penetrations for Risers and Shafts**
   - Vertical MEP elements (e.g., plumbing stacks, electrical risers, HVAC shafts) require floor penetrations.
   - Coordination is critical to ensure **penetrations avoid structural supports**, such as beams, joists, or post-tensioned cables.
   - In slab systems, particularly **post-tensioned slabs**, penetrations must be carefully placed and often cast during construction—not drilled afterward.

   Example: In a **multi-story residential building**, stacking bathroom layouts vertically allows for shared risers and minimizes penetrations through slabs, simplifying coordination.

• Location of Load-Bearing Walls and Columns Relative to Program Areas

- Structural elements should not interrupt key functional spaces (e.g., operable partitions, open-plan offices, auditorium seating).
- In load-bearing wall systems (e.g., masonry or light wood framing), flexibility is limited—altering layouts requires structural redesign.
- Column grids should align with planning modules such as **parking layouts, room dimensions**, and **circulation paths**.

🧠 *Example:* A **conference center** with large meeting rooms benefits from **column-free spans**, so the structure must allow wide bay spacing—often via steel or long-span trusses.

Design Factor	Structural Concern	Design Strategy
Plenum height	Beam depth and spacing	Choose shallow systems or coordinate MEP zones
Duct routing	Obstructions from structure	Plan duct paths between structural members
Shaft location	Alignment with columns and floor penetrations	Stack vertically and pre-plan blockouts
Program space flexibility	Columns or bearing walls in key areas	Use open-span systems or adjust grid spacing

Figure 5.6 Structural Coordination Impacts Summary

🧠 **ARE 5.0 PPD Example Scenario:**

A 4-story academic building has classrooms, labs, and lecture halls. The mechanical system uses large ducts and high air volumes. What structural strategy best supports coordination and future adaptability?

☑ **Answer:** A **steel frame system** with a regular column grid and open web joists allows for flexible MEP distribution, easier future reconfigurations, and efficient ceiling integration.

🎯 **ARE 5.0 PPD Tips:**

- Always evaluate the **MEP impact** of a structural system in floor plans and sections.
- Look for systems that allow **long spans and minimal obstructions** for program-heavy buildings (e.g., labs, hospitals, schools).
- Consider **floor-to-floor height implications** when selecting between systems like flat slabs and drop-beam concrete.

📌 **Summary**

Proper coordination between **structure, MEP systems**, and **architectural layout** is essential for functional, efficient, and code-compliant buildings. As an architect, your role is to:

- Select structural systems that facilitate, not hinder, building performance
- Align structure with program needs and mechanical distribution
- Avoid costly clashes and redesigns through proactive planning

This knowledge is routinely tested on the PPD exam—particularly in **case studies** where you must assess plan sections and identify the best system for coordinated design.

7. **Consider Construction Timeline and Sequencing**
When selecting a structural system, **construction speed** and **buildability** are major factors—especially for projects with aggressive schedules or phased occupancy. The choice of structural system can significantly influence how quickly a project reaches **enclosure**, **dry-in**, and ultimately **occupancy**.

On the **ARE 5.0 PPD exam**, you may be asked to recommend systems that optimize construction sequencing or accommodate fast-track delivery methods, particularly in **design-build** or **developer-driven** projects.

Structural Systems and Construction Speed
Each structural system presents different opportunities (and limitations) in terms of **speed, sequencing**, and **labor requirements**:

• Steel Framing

- **Fabricated off-site**, allowing **simultaneous foundation and superstructure preparation**.
- **Quick on-site assembly** using bolted or welded connections.
- Enables **early enclosure**—skin and roofing can proceed as floors go up.

Example: A **spec office building** may use **steel framing** to complete the structural shell in weeks, making the interior build-out possible while façade work is ongoing.

• Precast Concrete

- Manufactured off-site and shipped to the site, reducing on-site labor.
- Pieces such as **walls, slabs, and stairs** can be set rapidly using cranes.
- **Ideal for repetitive layouts** like parking structures, dorms, or modular residential units.

Example: A **student housing complex** may use **precast concrete planks** and panels to enclose the building quickly and speed up unit turnover before semester start.

• Wood Framing

- Fast and simple to construct on low-rise buildings.
- **Minimal equipment** needed; framing crews are widely available.
- Prefabricated components (e.g., roof trusses, wall panels) can further accelerate assembly.

Example: A **two-story retail strip center** may be framed with **engineered wood** to reduce construction costs and speed up delivery—provided fire code limits are met.

◇ *Limitation:* Building codes typically cap wood framing at **Type V-A construction** and height limits (often 3–5 stories max with added fire protection).

• Cast-in-Place Concrete

- **Slower overall** due to formwork, rebar placement, and curing times.
- Construction proceeds **floor by floor**, with subsequent levels waiting for concrete to gain strength.
- **Weather-sensitive**, especially in cold or wet climates.

Example: A **hospital** requiring high acoustic separation and fire resistance may still justify **cast-in-place concrete**, despite longer timelines, due to its performance benefits.

System	Speed	Off-Site Fabrication	Weather Dependency	Best Use Cases
Steel Frame	Very Fast	High	Low (once erected)	Office, retail, institutional
Precast Concrete	Very Fast	High	Low	Parking garages, dorms, modular units
Wood Framing	Fast (low-rise)	Moderate (trusses, panels)	Moderate–high	Residential, small commercial
Cast-in-Place Concrete	Slower	Low	High	Hospitals, multifamily, seismic areas

Figure 5.7 Construction Timeline Comparison

⏱ Sequencing Considerations

- **Parallel workflows** (e.g., steel fabrication + foundation prep) reduce overall timeline.
- **Dry-in speed** affects when interior trades can begin (mechanical, electrical, finishes).
- Projects under **design-build** or **CM-at-risk** contracts often favor **systems with predictable timelines and fewer on-site unknowns**.

🧠 ARE 5.0 PPD Example Scenario:

A national retailer is constructing a new store in a high-traffic location and wants to open before the holiday season. Which structural approach best supports their schedule-driven priorities?

☑ **Answer:** A **steel frame with precast concrete floor panels**, allowing rapid erection and early enclosure, is ideal for meeting tight deadlines.

🎯 ARE 5.0 PPD Tips:

- Be prepared to **weigh construction duration** as a factor in system selection.
- Consider **climate and seasonality**—cast-in-place may not be suitable for winter starts.
- Watch for **case study clues** like tight schedules, phased occupancy, or fixed opening dates—they signal a need for **faster systems** like steel or precast.

📌 Summary

Construction sequencing is a critical component of architectural decision-making. As an architect, you must:

- Choose structural systems that support project delivery timelines
- Understand how off-site fabrication, weather, and sequencing affect performance
- Coordinate with contractors and engineers to ensure buildability and speed

This insight is tested on the PPD exam, especially when comparing systems or evaluating case studies with schedule pressures.

8. **Match Structural Systems to Building Use and Code Constraints**
Selecting an appropriate structural system requires a careful balance between **building function**, **occupancy classification**, **fire and life safety codes**, and **performance expectations** like vibration control and acoustic separation. On the ARE 5.0 PPD exam, you'll be expected to evaluate how building use and code constraints affect your choice of structural systems.

Your ability to link **occupancy type**, **construction type**, and **programmatic needs** to the most suitable system is essential for sound design decisions.

🏛 Key Factors in Matching Structure to Building Use and Code

The **International Building Code (IBC)** defines five major construction types, with varying degrees of combustibility and fire resistance. This determines which structural materials are permissible.

- **Type I (A/B)** – *Non-combustible, highly fire-resistant*:
→ Requires **concrete or protected steel** (used in high-rises, hospitals, airports)
→ Fireproofing is typically required for all structural components
- **Type II (A/B)** – *Non-combustible with lesser fire resistance*:
→ Often steel with spray-applied fireproofing or concrete
- **Type III** – *Non-combustible exterior walls, combustible interior framing*
→ Common for mid-rise multi-family projects with **wood framing** over concrete podiums
- **Type IV (Heavy Timber)** – *Exposed wood meeting size and char requirements*
→ Mass timber (e.g., CLT) increasingly used under new code allowances
- **Type V (A/B)** – *Combustible (typically wood)*
→ Limited in height and area; used in small-scale commercial and residential projects

Figure 5.8 Construction Type (IBC Types I–V)

🧠 *Example:* A **five-story apartment building** may use **Type III-A** construction: two levels of **concrete podium (non-combustible)** with three wood-framed floors above.

• Occupancy Type and Use Case
Different uses impose different structural demands:

- **Schools / Institutions**
→ Require **fire-resistance**, **acoustic separation**, and sometimes **blast resistance**
→ CMU or concrete walls, steel joists, and slab systems are common
- **Offices / Commercial Buildings**
→ Prioritize **open spans**, **future flexibility**, and **fast construction**
→ Steel frames or composite decks offer these benefits
- **Laboratories / Hospitals**
→ Need **low vibration**, **high strength**, and **service integration**
→ Cast-in-place concrete is often selected for its stiffness and sound isolation
- **Residential / Multi-Family**
→ Emphasize **cost-effectiveness**, **acoustics**, and **modular layouts**
→ Wood framing, CLT, or precast planks are common (depending on fire code)

🧠 *Example:* A **lab facility** should avoid light wood framing or thin slab systems due to **sensitive equipment** needing low deflection and vibration. Instead, it might use **a two-way concrete slab** on columns.

• Fire Resistance and Structural Protection

The IBC mandates **fire-resistance ratings** for structural elements based on occupancy and construction type:

- **Steel framing** often requires **spray-applied fireproofing** or gypsum encasement
- **Concrete slabs** may inherently meet 2-hour ratings without added protection
- **Wood framing** in Type V-A construction must often achieve **1-hour ratings** for structural walls, floors, and roofs

🧠 *Example:* In a **Type I-B hotel**, steel columns must typically be rated for **2 hours**, requiring encasement or intumescent coatings.

• Acoustic, Vibration, and Deflection Control

Different occupancies demand varying levels of structural performance beyond strength:

Use	Primary Concern	Preferred Structure
Libraries	Live load + vibration control	Concrete or steel with vibration damping
Labs	Low vibration, minimal deflection	Cast-in-place or precast concrete
Offices	Open layouts, HVAC integration	Steel frame with composite deck
Housing	Acoustics between units	Wood, concrete plank, or mass timber

Figure 5.9 Occupancies and Preferred Structure

🧠 *Example:* In **multi-family housing**, **precast concrete planks** are often chosen for their **acoustic separation** and **fire resistance**, especially when long corridors are repeated per floor.

Construction Type	Combustible Materials Allowed?	Typical Structure Types
Type I	❌ No	Concrete, steel (non-combustible)
Type II	❌ No (limited allowances)	Steel frame, lightweight concrete deck
Type III	☑ Yes (interior only)	Hybrid concrete podium with wood above
Type IV	☑ Yes (heavy timber only)	Mass timber (CLT, glulam)
Type V	☑ Yes (fully combustible)	Wood studs, trusses, sheathing

Figure 5.10 Quick Reference: Construction Type vs. Material Allowance

🧠 **ARE 5.0 PPD Example Scenario:**
A two-story K–8 school is being designed in a suburban setting. The client wants durable finishes, good acoustics between classrooms, and low cost. What structural system best aligns with these priorities?
☑ **Answer: CMU load-bearing walls with open web steel joists**, meeting fire-resistance and acoustic needs, while keeping construction costs manageable.

🎯 **ARE 5.0 PPD Tips:**

- Always verify the **maximum building height and area** allowed by construction type.
- Link **fire-resistance ratings** with structural member choices—especially in multi-story designs.
- Don't overlook **acoustic separation and vibration control**—especially for **schools, labs, and housing**.
- Watch for **code clues** in case studies (e.g., "Type I-A construction," "non-combustible," or "5-story limit")—they often guide structural selection.

📌 **Summary**
Matching structural systems to building use and code constraints is about balancing:

- **Code compliance** (fire resistance, material limits)
- **Program performance** (acoustics, deflection, layout)
- **Construction type** (IBC Types I–V)
- **Occupancy classification**

A well-informed structural choice ensures **safety, efficiency, and regulatory compliance**—and is a core skill tested throughout the PPD division of the ARE 5.0.

9. Understand Integration with Foundation Systems

The **foundation system** and **superstructure** must work together as a unified whole. As the architect, you must ensure that the structural loads from the building are transferred efficiently to the ground—**without causing settlement, vibration, or conflict with site or building systems**.

On the **ARE 5.0 PPD exam**, you will need to understand how the building's structure affects foundation selection, and vice versa. Integration decisions impact site work, waterproofing, coordination with mechanical rooms, and vertical circulation.

🏗 **Key Considerations in Structure–Foundation Integration**

• Load Paths and Structural System Compatibility
The type of superstructure—steel frame, concrete shear wall, masonry, or wood—affects how loads are distributed to the foundation.

- **Point loads** from **columns** (e.g., steel or concrete frames) typically require **spread footings, piers**, or **pile caps**
- **Line loads** from **bearing walls** (e.g., masonry or wood-framed) distribute more evenly and often use **continuous footings**
- **Shear walls** and **braced frames** introduce **horizontal and overturning forces**, requiring **larger, reinforced foundations**

🧠 *Example:* A **concrete shear wall core** in a high-rise office building will often sit on a **mat foundation** to resist both vertical and lateral loads across a large footprint.

Foundation Type	Best Use Case	Integration Notes
Shallow Spread Footings	Small to mid-rise buildings on stable soils	Simple and cost-effective for bearing wall or column grids
Mat (Raft) Foundation	Heavy loads or poor soils across large area	Reduces differential settlement; often used under shear walls or cores
Deep Foundations (Piles, Caissons)	Tall buildings or poor soils at surface	Transfer loads to deeper strata; required for seismic zones, waterfronts
Slab-on-Grade	Low-rise buildings without basements	Works well with light wood or steel frame structures

Figure 5.11 Foundation Types and Their Applications

🧠 *Example:* A **warehouse** with open steel framing and no basement may use **slab-on-grade with isolated column footings**, reducing excavation and foundation costs.

• Basement Integration and Below-Grade Design

- Structural systems must align with **basement wall supports, slab elevation**, and **retaining wall systems**
- **Column grid** should correspond with **foundation walls and footings**
- Below-grade spaces (e.g., **parking, mechanical rooms, stormwater storage**) require coordination with:
    - Foundation walls and waterproofing
    - Sump pits, under-slab plumbing, and elevator pits
    - Soil bearing capacity and dewatering systems

🧠 *Example:* A **multifamily building** with two underground parking levels may use **concrete shear walls** as both the lateral system and **basement retaining walls**, reducing redundancy.

• Vertical Continuity and Penetration Planning

- Stairs, elevators, and MEP shafts must penetrate floor slabs and foundations cleanly
- Core walls or column grids must **align vertically** from roof to foundation to avoid transfer structures
- Structural walls often double as **foundation walls** in basement conditions

🧠 *Example:* An **elevator shaft** should sit directly on a **reinforced pit slab** connected to the foundation wall, ensuring alignment, waterproofing, and structural continuity.

• Soil Conditions and Structural Loads
Site geotechnical conditions directly influence the foundation approach and the type of superstructure that can be supported:

Soil Type	Implication	Foundation Response
Firm granular soil	Good bearing capacity	Spread footings feasible
Clay or fill soils	Poor bearing, high shrink/swell	Mat foundation or deep piles
Seismic zones	Requires special detailing for overturning and lateral	Deep piles, shear wall integration

Figure 5.12 Soil Conditions and Structural Loads

🧠 *Example:* In a **coastal site** with poor soils, a **precast pile foundation** may be necessary to support a **steel-framed structure**, with pile caps coordinated with column bases.

📐 **ARE 5.0 PPD Exam Scenario:**
A 4-story office building on poor soils has a steel moment frame with large point loads. What foundation system is most appropriate?
☑ **Answer: Drilled piers (caissons)** or **driven piles** with pile caps aligned to column bases, transferring concentrated loads to stable strata.

🎯 **ARE 5.0 PPD Tips:**
- **Match your structure to the soil conditions**—don't assume shallow footings are always viable.
- **Align vertical loads** from columns and walls through floors to foundations—discontinuities require costly transfer beams or mats.
- Watch for **elevator pits, sump pits, and stairs** that interfere with foundations—coordinate early.
- In **case studies**, examine structural grids, foundation notes, and soil reports for integration clues.

📌 **Summary**
Integrating the structural system with the foundation is not just about engineering—it's about planning for **buildability, durability, and efficient space use** below grade. As an architect, your role is to:
- Understand how structural loads flow to the ground
- Coordinate structural and architectural elements with geotechnical realities
- Select foundation strategies that support both the building's form and its function

This coordination is a recurring theme on the PPD exam, especially in scenario-based case studies where **site, structure, and systems must align**.

C. Determine special systems such as acoustics, communications, lighting, security, conveying, and fire suppression (A/E) (1-5%)

NCARB Requirements
You will need to evaluate specialty systems and select the appropriate systems for a building based on the building type, function, and programmatic requirements. You will also need to assess the impact of these specialty systems on the overall design of a building.

NCARB Objective: A/E – ARE 5.0 PPD
You will need to **evaluate and integrate specialty systems** based on building type, function, and user needs. These systems include **acoustics, communications, security, lighting, fire suppression, and conveying systems**. Each of these can significantly impact the design and performance of a facility.

1. Acoustical Systems

Acoustic design is a **fundamental aspect of architectural planning**, directly influencing occupant comfort, privacy, productivity, and well-being. While often overlooked in early design, poor acoustics can compromise a building's performance—particularly in **educational, healthcare, hospitality, and workplace environments**.

Acoustic goals differ by building type and space function. What is acceptable in a **lively restaurant** may be disruptive in a **hospital recovery room**. Therefore, architects must apply tailored solutions that support the **intended program, occupancy, and use** of each space.

Key Considerations:

- **Sound Transmission Control (STC Ratings):**
  STC (Sound Transmission Class) measures how well partitions, doors, floors, windows, and other elements **block airborne sound** from passing between rooms. Ratings typically range from **STC 25 (minimal isolation)** to **STC 60+ (high performance)**.
  → *Used in spaces requiring privacy and noise isolation such as offices, hotel rooms, classrooms, and dwelling units.*
  Note: Partitions must be designed to **extend slab-to-slab** and include **acoustical sealant** at penetrations and junctions to prevent sound flanking.
- **Sound Absorption (NRC Ratings):**
  NRC (Noise Reduction Coefficient) quantifies how much sound a surface absorbs within a room. High NRC materials reduce reverberation and improve **speech clarity**.
  → *Effective in spaces like classrooms, lecture halls, and open-plan offices.*
  Materials such as **acoustic ceiling tiles, wall panels, carpets, perforated wood, and fabric-wrapped systems** help manage sound reflections and enhance comfort.
- **Room Acoustics and Echo Control:**
  The **volume, shape, and material selection** of a room impact how sound behaves. Controlling **reverberation time (RT60)** is especially important in large or specialized spaces like:
  → *Auditoriums, performance venues, houses of worship, and multipurpose halls.*
  Tools such as **diffusers, absorptive baffles, and sound reflectors** are used to optimize clarity and ensure uniform sound distribution.

- **Vibration Isolation:**
  Equipment like **rooftop units, air handlers, elevators, and pumps** can transmit vibration through structure-borne paths. Without mitigation, this can cause **audible humming, rattling, or low-frequency disturbance** in occupied areas.
  → *Particularly critical in hotels, theaters, residential buildings, or labs located under or near mechanical rooms.*
  Common solutions include **spring isolators, resilient mounts, floating floors, and flexible connectors** on ductwork and piping.

🧠 **Example:**
In a **conference center**, meeting rooms adjacent to one another must allow **simultaneous events without acoustic interference**. Walls should achieve at least **STC 55** and extend full height between slabs. Ceilings may include **high-NRC tiles** to control interior reverberation. Penetrations for HVAC, lighting, and sprinkler pipes must be sealed to **limit flanking paths**.

◇ **Design Tips:**
- Use **acoustic seals, door gaskets, and automatic drop seals** at doors separating noise-sensitive spaces.
- **Avoid rigid mechanical connections** between noisy and quiet zones—use **offset ducts** or lined plenums.
- In mixed-use or performance buildings, consider **floating floor assemblies** or **resiliently mounted ceiling systems** for enhanced sound isolation.
- Review **ASTM E90** and **ANSI S12.60** (Acoustics in Schools) as guiding standards for educational facilities.

2. **Communication and IT Systems**
   Modern buildings rely on **robust, scalable communications infrastructure** to support technology-driven operations, connectivity, and user experience. Architects must coordinate closely with consultants and IT planners to ensure that all **low-voltage and data systems** are appropriately accommodated in the design.

   Core components of communication and IT infrastructure include:
   - **Structured Cabling Systems (Cat 6/6A, Fiber Optics):**
     These form the physical backbone of all communication systems.
     → *Category 6/6A cables are commonly used for Ethernet and data transmission, while fiber optic cables are used for high-speed and long-distance connections.*
     Horizontal cabling should run from **Telecommunications Rooms (TRs)** to workstations, with cable length limited to ~295 feet (90 meters) per TIA/EIA standards. Vertical cabling ("backbone") connects between floors via **main and intermediate distribution frames (MDF/IDF)**.
   - **Telecom Room Requirements:**
     These rooms house servers, patch panels, and network switches.
     → *Design should allow for adequate cooling, ventilation, and space for cable management.*
     Large projects may require dedicated **IT equipment rooms** on each floor.

Walls should be treated to reduce electromagnetic interference (EMI), and access must be secure.

- **Wi-Fi Access Point (WAP) Layout and Rack Space:**
  Wireless access depends on **strategically spaced WAPs** to avoid dead zones and bandwidth bottlenecks.
  Server and network racks require consideration of **cooling loads**, **power supply**, and **physical access** for future upgrades.
  → *In high-performance offices or hospitals, a raised floor system may be used to manage dense cabling and HVAC delivery.*
- **Intercom, Paging, and Audio-Visual (AV) Systems:**
  These systems are essential in **educational, institutional, and healthcare environments**.
  → *Classroom intercoms, overhead paging in hospitals, AV conferencing systems in offices, and auditorium projection/sound systems all require integrated infrastructure.* Space must be planned for speakers, screens, control panels, and wiring conduit.
- **BIM, IoT, and Smart Building Integration with BMS (Building Management Systems):**
  Communication systems increasingly tie into **building automation**—integrating lighting, HVAC, security, and energy monitoring through centralized BMS platforms.
  → *Smart thermostats, occupancy sensors, and environmental monitors generate data that's processed and managed via the IT network.*
  ***Note: IoT** stands for **Internet of Things**, and it plays a critical role in enabling real-time data-driven building operations and automation.*

🧠 **Example:**
A **K–12 school** might require telecom closets spaced no more than 200 ft apart per TIA/EIA guidelines. Each classroom may need **AV readiness** with ceiling-mounted projectors, HDMI ports, wireless screen sharing, and voice amplification. Corridors and outdoor areas may require **Wi-Fi coverage** for devices used in 1:1 learning programs.

🧠 **Example:**
In a **hospital**, nurse call systems, patient monitoring, and paging are integrated into the IT backbone. The telecom room must support **redundant fiber loops** and have battery backup or UPS systems to ensure critical operations continue during outages.

◇ **Design Tip:**
Early coordination is essential. Ensure:

- Telecom spaces are located **near vertical risers**
- They are **not placed below restrooms** or mechanical rooms to avoid water damage
- Sufficient **floor and wall space** is allocated for conduit entry and cable trays
- HVAC is designed to maintain equipment within acceptable **temperature and humidity ranges**

3. **Lighting Systems and Controls (Specialty Applications)**
   Lighting design is not only about illumination—it directly impacts a building's **energy efficiency, occupant well-being, visual comfort, and code compliance**. Specialty lighting solutions must address the **function, mood, and spatial performance** required for different

environments. Architects play a critical role in specifying fixture types, layouts, and control strategies appropriate to the programmatic needs of each space.

Specialty Lighting Considerations:

- **Theatrical and Performance Lighting:**
  Used in **auditoriums, theaters, lecture halls**, and performance venues to support various presentation modes and audience experiences.
  → Systems often include **dimmable spotlights, wash lighting, stage lighting grids**, and programmable control panels (e.g., DMX systems).
  Fixtures are typically mounted on **catwalks or lighting battens**, and must be coordinated with **rigging systems, AV, and ceiling structure**.
- **Task Lighting and Glare Control:**
  Specialized lighting for **offices, laboratories, studios**, and classrooms improves visual performance and reduces eye strain.
  → Includes **under-cabinet lights, adjustable desk lamps, and indirect fixtures**. Glare can be mitigated using **louvers, diffusers, shielding, or indirect lighting techniques**, especially near **computer workstations or lab benches**.
- **Lighting Power Density (LPD):**
  A key metric in energy codes like the **IECC (International Energy Conservation Code)** and **California Title 24**, LPD is measured in **watts per square foot (W/sf)** and defines the **maximum allowable power consumption** for lighting by space type.
  → *For example, office spaces may have an LPD limit of ~0.8–1.0 W/sf.*
  Architects must select efficient fixtures (e.g., LED) and use controls to help stay below baseline allowances while meeting visual requirements.
- **Lighting Controls:**
  Intelligent controls improve **energy efficiency**, support **occupant needs**, and are often required by code. Common strategies include:
  - **Occupancy sensors**: automatically turn lights on/off based on presence
  - **Daylight-responsive dimming**: reduces output near windows
  - **Time-based scheduling**: shuts lights off after hours or based on building use
  - **Personal dimming controls**: allow users to adjust task lights at desks
    These systems may be standalone or integrated into a **Building Management System (BMS)**.

🧠 **Example:**

In a **museum**, lighting design must balance **preservation** with **visual experience**. Use **adjustable track lighting** with high **CRI (Color Rendering Index)** to highlight artwork colors accurately, while incorporating **UV-filtering lenses** and dimming controls to protect sensitive materials like textiles and paintings. Lighting is often zoned per exhibit and integrated with motion sensors to reduce energy use when galleries are unoccupied.

◇ **Design Tip:**

Lighting design significantly affects **ceiling system selection** (e.g., suspended vs. drywall), **plenum coordination**, and **fixture access**.
Ensure early coordination with:

- **Mechanical duct layout**

- **Structural elements like beams and trusses**
- **Electrical conduit and control panel locations**

→ *Example conflict:* Recessed troffers may clash with ductwork in a narrow plenum. Consider **surface-mounted fixtures** or **remote driver locations** in such cases.

⚠ Exam Alert – PPD Focus:

- Know how to compare fixture types (LED vs. fluorescent vs. halogen) in terms of **efficacy, longevity, and heat output**
- Be ready to apply **LPD values** and recommend controls that meet or exceed **energy code baselines**
- Understand the relationship between **lighting quality and occupant needs** in specialized spaces (e.g., labs, museums, classrooms)
- Recognize coordination issues in case studies—such as **misaligned lighting layout with HVAC ducts or fire sprinklers**

4. **Security and Access Control**
Security systems are essential components of building design and must be **tailored to the function, occupancy, and threat level** of the facility. Effective security enhances not only **physical protection** but also **user confidence and operational control**. Architects are responsible for ensuring that infrastructure is in place to support integrated electronic security solutions—ideally planned early in schematic design.

Key Components of Security and Access Control Systems:

- **CCTV Surveillance, Motion Sensors, and Facial Recognition:**
Closed-Circuit Television (CCTV) systems are used to **monitor entry points, high-value zones, and public circulation areas**.
→ *Cameras can be analog or IP-based, fixed or pan-tilt-zoom (PTZ), and may include night vision or motion detection capabilities.*
Facial recognition is typically reserved for **high-security applications** like data centers, airports, or government buildings, where personnel tracking and identity verification are critical.
Motion sensors are used for **interior intrusion detection** and **after-hours perimeter control**.
- **Access Control Systems (ACS):**
These include **card readers, keypads, biometric scanners (e.g., fingerprint or retinal), and mobile device credentials** to control access to restricted areas.
→ *Common in hospitals (for staff-only zones), research labs, corporate offices, and schools.*
Access control panels must be connected to an **electronic locking system**, typically powered and monitored via low-voltage systems.
- **Emergency Communication Systems:**
Critical in **campuses, transportation hubs, and parking structures**, these systems include:
    - **Blue phones or call boxes**: direct link to emergency services
    - **Panic buttons**: used in high-risk locations like front desks, exam rooms, or cashier stations

  - **Mass notification systems (MNS)**: provide building-wide alerts via PA systems, digital signage, or mobile push notifications
- **Intrusion Detection and Lockdown Capabilities:**
  Schools, civic buildings, and institutions may require **lockdown functions** that instantly secure doors and notify authorities in the event of a threat.
  Intrusion detection integrates door contacts, glass-break sensors, motion detectors, and alarms.
  → *Systems must balance* ***security*** *with* ***life safety****, ensuring egress is never blocked in an emergency (per IBC and NFPA 101).*

🧠 **Example:**
A **healthcare facility** might implement **keycard access** for staff-only corridors, pharmacies, and data rooms. Public areas like lobbies and waiting rooms are monitored by **CCTV**, while **loading docks and back-of-house entries** are monitored via motion-activated cameras. Sensitive departments such as **maternity wards or operating rooms** may include **biometric access control** and automatic lockdown functions.

◇ **Design Tip:**
- **Security wiring requires conduit, power supply, and equipment rooms**—plan for wall chases, ceiling space, and dedicated panels.
- Coordinate with **electrical, telecom, and architectural layouts** early to avoid retrofit conflicts or incomplete system coverage.
- **Power redundancy** (UPS or emergency circuits) should be provided for life-safety systems like access control and emergency comms.

⚠ **PPD Exam Strategy Notes:**
- Recognize security needs based on **building type** (e.g., school, hospital, museum)
- Understand the **relationship between access control and life safety**
- Be prepared to **identify system coordination needs** (especially conduit routing, panel access, and electrical service)

5. **Conveying Systems (Elevators, Escalators, Lifts)**
Conveying systems are essential for providing **vertical and horizontal circulation** in multilevel, high-occupancy, or accessible buildings. Whether serving office towers, airports, or historic buildings, these systems must be selected based on **height, load capacity, traffic flow, and code compliance**. Architects must understand the **space planning, power, and coordination requirements** of each system type.

Key Conveying System Types:
- **Elevators:**
  Elevators vary by technology, speed, and building height:
  - **Hydraulic Elevators**:
    → Suitable for low-rise buildings (typically 2–4 stories)
    → Slower, lower initial cost, but require a pit and machine room at the base
    → Limited to about 60 feet of vertical travel

- Traction Elevators (Geared or Gearless):
  - **Traction Elevators (Geared or Gearless):**
    → Used in mid-rise to high-rise applications
    → Faster and more energy-efficient for taller buildings
    → Typically require a machine room above the shaft (MR), but gearless options are common in high-rise
  - **MRL (Machine-Room-Less) Elevators**:
    → Ideal for mid-rise buildings where space is tight
    → Use compact gearless machines mounted in the hoistway
    → Save valuable square footage and reduce construction cost for a separate machine room
- **Freight vs. Passenger Elevators:**
  - **Freight elevators** are designed to carry goods and equipment and are rated for higher capacities. They include **rugged interiors, wide doors**, and **dual-speed operation**.
  - **Passenger elevators** prioritize speed, aesthetics, and user experience.
  - Buildings with both uses may need **dedicated shafts** to avoid code violations related to mixed use.
- **Fire and Emergency Service Requirements:**
  - Elevators in high-rise and certain occupancies must include **Phase I and Phase II Firefighter Operation** per **NFPA 72 and ASME A17.1**.
  - In some occupancies (e.g., hospitals), one elevator must serve as an **emergency evacuation elevator** and be connected to **emergency power**.
  - Elevator lobbies may require **smoke containment systems** or **pressurization**.
- **Escalators and Moving Walks:**
  - Common in high-traffic buildings like **transit centers, malls, and airports**.
  - Allow continuous movement and reduce waiting, but require **larger floor openings**, **balustrades**, and **machine pits**.
  - Moving walks are often found in **airports and convention centers** to handle long horizontal distances.
- **ADA-Compliant Lifts:**
  - In buildings where full elevators are impractical—such as **historic renovations, small libraries, or duplex units—vertical platform lifts or inclined stair lifts** can be used.
  - Must comply with **ADA Standards**, including **platform size, controls, and call buttons**.

🧠 **Example:**
A **5-story office building** will typically require a **traction elevator**, either with a **machine room above** or as a **machine-room-less (MRL)** system. MRL designs are often chosen to **conserve rentable floor area**, reduce construction costs, and provide smoother rides with lower maintenance than hydraulic systems.

◇ **Code Tip:**

- **Per ADA**, at least **one elevator must serve all accessible floors** unless an exception (such as a two-level split-level retail space) applies.
- **IBC and NFPA 99/101** may require:
  - **Stretcher-accessible elevators** in healthcare facilities
  - **Emergency power backup** in high-rises, hospitals, and assisted living buildings

- **Rated elevator lobbies** or **smoke partitions** where required by fire code

⚠ **PPD Exam Strategy Notes:**

- Understand the **differences between elevator types**, including their structural and spatial impacts
- Know when **escalators vs. elevators** are appropriate based on occupancy and circulation needs
- Be ready to **evaluate elevator placement in case studies**, especially in terms of **code compliance, accessibility, and space efficiency**
- Consider **shaft sizing**, **overhead clearances**, and **machine space** in design planning
- Watch for **elevator lobby requirements** and fire-rated enclosures on exams

6. **Fire Suppression and Special Hazard Systems**

Fire suppression systems are critical for protecting life and property, and they are **required by code** in most commercial and institutional buildings. The type of system selected must be based on **occupancy classification, fire risk level, environmental conditions, and content sensitivity**. Architects must work closely with fire protection engineers and code officials to ensure early coordination with ceiling layout, structure, and MEP systems.

Common Types of Fire Suppression Systems:

- **Wet-Pipe Sprinkler Systems (Most Common):**
  - Pipes are filled with water at all times, allowing immediate discharge when a sprinkler head is triggered by heat.
  - Most **reliable and cost-effective** system; used in typical office buildings, schools, retail spaces, and residential occupancies.
  - Requires **freeze protection** in cold climates (cannot be used in unheated spaces like parking garages).
- **Dry-Pipe and Pre-Action Sprinkler Systems:**
  - **Dry-pipe systems** are filled with pressurized air and used in **unheated areas** such as loading docks, parking structures, and warehouses in cold climates.
  - **Pre-action systems** require two triggers: heat or smoke detection, and a secondary signal (e.g., from a fire alarm) before water is released.
  - Ideal for **sensitive environments** such as **museums, libraries, or data centers**, where accidental discharge would cause major damage.
    → Often used with **early warning smoke detectors** and integrated alarm systems.
- **Standpipes and Fire Department Connections (FDCs):**
  - Required in **multi-story, high-rise, or large-area buildings** to allow firefighters to access water supply at multiple floors or zones.
  - Types include:
    - **Class I**: for fire department use only (2½" hose connection)
    - **Class II**: occupant-use (1½" hose)
    - **Class III**: combination of both
  - Standpipes must be located in **stairwells or designated fire-rated corridors** and connected to the building's main water supply or fire pump.

- **Clean Agent Systems (Special Hazard Suppression):**
  - Non-water-based fire suppression systems, using gaseous agents like **FM-200, Inergen, or Novec 1230**.
  - Extinguish fire without damaging sensitive electronic equipment or archival materials.
  - Common in **server rooms, data centers, telecom closets, art storage, and control rooms**.
  - Require **airtight enclosure**, **detection systems**, and **agent release delay** with audible/visual warnings for safe evacuation.

System Type	Key Features	Ideal Use Cases	Pros	Cons	Code Reference
Wet-Pipe Sprinkler	Water-filled pipes; discharges immediately upon head activation	Offices, schools, retail, residential	Simple, low cost, reliable	Can freeze in unheated areas; risk of water damage	**NFPA 13**, IBC
Dry-Pipe Sprinkler	Pressurized air in pipes; water released after head opens	Unheated spaces (garages, loading docks)	No freezing; delayed discharge helps avoid false alarms	Slower response; more maintenance	**NFPA 13**
Pre-Action System	Two-step trigger: detection + heat; water held back until both activate	Data centers, museums, libraries	Reduces false discharge; extra protection	Complex; more expensive; requires detection integration	**NFPA 13, NFPA 72**
Deluge System	All heads open; water released simultaneously upon detection	Aircraft hangars, chemical storage, high-hazard areas	Rapid suppression of fast-spreading fire	High water damage risk; needs robust drainage	**NFPA 13, NFPA 15**
Clean Agent System	Gas (FM-200, Inergen, Novec 1230); suppresses fire without water	Server rooms, telecom rooms, archives, control centers	No residue; safe for electronics	Requires airtight enclosure; high cost	**NFPA 2001**

System Type	Key Features	Ideal Use Cases	Pros	Cons	Code Reference
CO_2 System	Displaces oxygen to extinguish fire	Industrial applications, engine rooms (unoccupied areas)	Effective on flammable liquids and electrical fires	Toxic to humans; not for occupied spaces	**NFPA 12**
Standpipe System	Vertical water piping with hose connections	High-rise, large buildings, stairwells for FD access	Enables FD firefighting at multiple levels	Must be supplemented with sprinkler system	**NFPA 14**, IBC
Manual Fire Extinguishers	Portable; various types (ABC, CO_2, water, foam)	All occupancies; required near exits and hazard areas	Fast response; targeted control	Limited reach; user-dependent effectiveness	**NFPA 10**

Figure 5.13 Fire Suppression System Comparison Table

☑ **Quick PPD Study Tips:**

- **Wet-pipe** = default for most conditioned spaces
- **Dry-pipe** = unheated buildings
- **Pre-action** = critical spaces needing water protection
- **Clean agent** = electronics, archives
- **Standpipe** = multi-story fire access

🧠 **Example:**

A **server room** may be protected by a **double-interlock pre-action system**, which only releases water when both a sprinkler head is triggered by heat **and** a smoke detector confirms a fire. To further minimize risk, a **clean agent system** may also be used for quick suppression without water damage.

◇ **Code Tip:**

- **NFPA 13** governs the **design, installation, and spacing** of sprinkler systems in the U.S.
- Sprinkler heads must maintain **minimum clearance below structure or ceiling tiles**, and pipe routing must be **coordinated with lighting, HVAC ducts, and ceiling features**.
- Always verify **ceiling types and plenum depths** to avoid conflicts in T-bar, gypsum, or exposed structure environments.
- **IBC and local codes** dictate which occupancies require sprinklers—e.g., **Group A assembly occupancies over 12,000 sf** generally require full sprinkler coverage.

⚠ **PPD Exam Strategy Notes:**

- Be prepared to **identify when a dry-pipe or pre-action system is required** due to climate or contents
- Understand **when a clean agent system is preferred** over water-based suppression
- Know where **standpipes and FDCs** are required in multi-story or assembly buildings
- Watch for **coordination issues in case studies**, such as:
  - Sprinkler heads blocked by beams or ceiling baffles
  - Insufficient room height for proper head clearance
  - Omission of sprinkler coverage in concealed spaces

7. **Specialty Systems Integration (Acoustics, AV, Security, BMS)**

As buildings become more technologically advanced, the successful integration of specialty systems is critical to meeting performance, comfort, and security goals. These systems must be **coordinated early in design** to avoid costly conflicts in construction and to meet programmatic and code requirements.

Specialty systems should be considered **not in isolation, but in relation to architectural layout, MEP infrastructure, and structural constraints.** Architects must understand their spatial, operational, and environmental implications, even if engineers or consultants are tasked with detailed design.

Key Specialty Systems and Their Design Impacts

1) Acoustics

- **Integration Concerns**:
  - Acoustical wall assemblies must coordinate with partition types, ceiling systems, and door hardware.
  - Room volume and shape affect speech clarity and reverberation time.
  - Mechanical systems must be isolated to prevent vibration and noise transmission.
- **Example**: In a corporate training facility, a combination of **high-STC partitions**, **NRC-rated ceiling tiles**, and **acoustical underlayments** beneath flooring may be used to ensure confidentiality and clarity across adjacent rooms.
- **Design Tip**: Avoid placing mechanical equipment or elevator shafts adjacent to sensitive areas such as **lecture halls, exam rooms, or recording studios**.

2) Audiovisual (AV) and Communication Systems

- **Integration Concerns**:
  - Wall and ceiling surfaces must allow for AV components like speakers, cameras, projection systems, and screens.
  - Power and data must be co-located with equipment at lecterns, walls, or ceilings.
  - Lighting layout and controls should be compatible with AV needs (e.g., zoned dimming in presentation rooms).
- **Example**: A university lecture hall may include **ceiling-mounted projectors**, **motorized screens**, **ceiling speakers**, and **wired/wireless microphone systems** that must be coordinated with acoustic finishes and lighting layout.

- **Design Tip**: Plan **recessed niches** or **equipment closets** for AV hardware and conduit pathways. Avoid conflicts between speaker locations and diffusers, light fixtures, or sprinkler heads.

3) Security and Access Control

- **Integration Concerns**:
  - Security devices (e.g., card readers, CCTV cameras, motion sensors) require **line of sight**, **data cabling**, and **secure mounting points**.
  - Door frames may require special electrical conduit or strike preparations for **electromagnetic locks** or **automatic closers**.
  - Vestibules, security desks, and access control zones must be planned architecturally.
- **Example**: In a medical clinic, controlled access may be required at **staff-only doors**, with **badge readers and CCTV monitoring** in lobbies and pharmacies.
- **Design Tip**: **Plan space for security panels** and electrical coordination early. Ensure all **exterior doors** and sensitive interior rooms (server rooms, labs) have security pre-wiring and power provisions.

4) Building Management Systems (BMS)

- **Integration Concerns**:
  - BMS integrates control over HVAC, lighting, power metering, fire alarm, and security systems.
  - Sensor placement (temperature, CO_2, occupancy) must be coordinated with architectural finishes.
  - Equipment must be accessible for maintenance and located near data infrastructure.
- **Example**: In a LEED-targeted office building, a BMS may link **occupancy sensors**, **daylight sensors**, and **HVAC zoning** to optimize energy use and maintain indoor air quality.
- **Design Tip**: BMS control panels and data hubs require **dedicated wall or closet space**. Plan ahead for sensor wiring paths, and ensure they do not conflict with aesthetic elements or building finishes.

System	Design Considerations	Coordination Issues
Acoustics	Wall/floor/ceiling assemblies, isolation techniques	HVAC noise, partition continuity, structure-borne vibration
AV/Comms	Mounting locations, conduit paths, lighting compatibility	Lighting conflicts, insufficient data/power access
Security	Visibility, power/data access, integration with doors	Door hardware conflicts, delayed conduit planning
BMS	Sensor locations, wall space, maintenance access	Lack of planning for cabling or sensor integration

Figure 5.14 Coordination Considerations Across Systems

🧠 **Example Case Study Scenario (PPD Exam Context):**

You're given a case study with a K–12 school layout. The **media room** is adjacent to the **mechanical room**, and the plan calls for a projector, AV controls, and occupancy sensors. You are asked to evaluate design issues.

☑ **Correct answer** might involve:

- Recommending **acoustic separation (STC-rated wall)** between rooms
- Ensuring **data and AV conduit paths** are coordinated with ceiling layout
- Relocating mechanical room or adding **vibration isolation** to protect AV integrity

☑ **PPD Exam Strategy Tips**

- Look for specialty system conflicts in **case study diagrams** (e.g., security hardware on glass doors, AV screens blocked by light fixtures).
- Know which systems require **early spatial planning** (e.g., telecom rooms, AV closets).
- Understand **acoustical separations** and their detailing at doors, walls, and ceilings.
- Be familiar with **co-location needs**: AV + lighting control, security + electrical, BMS + IT.

8. **Accessibility Considerations in Systems Design**

Accessibility isn't just about clear floor space and ramp slopes—it also affects the design and placement of building systems such as HVAC controls, lighting switches, plumbing fixtures, elevators, and fire alarm components. As an architect, you must ensure these systems are not only functional and code-compliant but also **usable by all occupants**, regardless of ability.

You are expected to understand where accessibility requirements intersect with **MEP and specialty systems** and how to incorporate those into your planning and coordination efforts.

Key Accessible System Integration Areas

1) Thermostats and Environmental Controls

- **ADA Reach Range Compliance**:
  - Operable parts (like thermostats or control panels) must be located within reach ranges:
    - **Forward reach**: 15"–48" AFF (above finished floor)
    - **Side reach**: 15"–48" AFF (or 54" if unobstructed)
- **Design Tip**: Avoid placing thermostats or lighting controls above casework or deep counters without knee and toe clearance.
- 🧠 **Example**: In a classroom, a wall-mounted HVAC thermostat should be installed no higher than 48" AFF and on a wall free from physical obstructions.

2) Lighting and Electrical Switches

- **Control Location and Operation**:
  - Light switches must be operable with one hand and not require tight grasping or twisting.
  - Must comply with **reach ranges** noted above.
  - Dimmers and automated controls (e.g., occupancy sensors) must be easy to override.
- **Design Tip**: Use rocker switches or motion-sensing lights that reset manually. Avoid toggle switches in critical locations.

- 🧠 **Example**: A private office should have a manual light switch within 48" of the door handle side—clearly visible and reachable from a wheelchair.

3) Plumbing Fixtures and Controls

- **Lavatories and Sinks**:
  - Must provide knee clearance (27" min height) and toe clearance (9" min height)
  - Faucets must be operable with one hand and not require more than 5 lbs of force.
- **Toilets**:
  - Flush valves must be on the open side of the toilet and mounted no higher than 44" AFF.
- **Shower and Bath Controls**:
  - Must be located within accessible reach ranges and allow for easy operation.
- 🧠 **Example**: An accessible sink in a break room must use a lever-handle or touchless faucet, with the control mechanism mounted forward of the backsplash for wheelchair users.

4) Fire Alarm Systems

- **Audible and Visual Notification**:
  - Must include **strobes** and **horns** placed at accessible heights (80" max AFF for strobes, min 15" AFF for some devices).
  - Alarms must be synchronized in public areas and visible in restrooms, corridors, and assembly spaces.
- **Pull Stations**:
  - Must be mounted no higher than 48" AFF and operable with one hand.
- 🧠 **Example**: In an office building, a fire alarm pull station near the elevator lobby must be mounted at 42" AFF and installed with clear floor space for wheelchair access.

5) Elevators and Vertical Circulation

- **Car dimensions** must meet **minimum clear space** requirements (e.g., 51" deep x 80" wide for side-entry elevators).
- Control panels must be within reach range and include tactile and Braille markings.
- Audible floor indicators and visual signals are required for those with visual or hearing impairments.
- **Design Tip**: Ensure elevator call buttons and control panels are both **low enough for a seated user** and **readable for those with low vision**.
- 🧠 **Example**: A multi-story government building must provide an elevator with a **hall call button at 42" AFF**, tactile floor indicators, and audible announcements at each floor.

6) Accessible Technology Integration

- **AV Systems**: Microphones and lecterns in public assembly spaces must be accessible in height and reach.
- **Intercoms & Security Systems**:
  - Must be operable from a wheelchair, with video screens at appropriate heights.
  - Emergency phones must include visual indicators for hearing-impaired users.
- 🧠 **Example**: A campus emergency call station ("blue phone") should include a **strobe light, push-button at 42" AFF**, and be located on an accessible route.

☑ **PPD Exam Strategy Tips**

- **Memorize key reach ranges** (15"–48" AFF for most operable parts)
- Watch for **equipment mounting conflicts** in case study diagrams (e.g., controls above counters)
- Be able to **identify non-compliant system elements**, such as fire alarm pull stations mounted too high or switches behind deep obstructions
- Understand the **intersection between ADA and MEP** (e.g., accessible plumbing valves, light switch types, and thermostat height)

D. Determine materials and assemblies to meet programmatic, budgetary, and regulatory requirements (A/E) (4-8%)

NCARB Requirements

You will need to select a building's envelope system, interior materials, and other assemblies based on cost, availability, program requirements, environmental conditions, sustainability requirements, or other factors. You will also need to evaluate the impact of those decisions on building design.

NCARB Objective:

You must understand how to evaluate and select materials and assemblies based on a variety of criteria, including:

- Code compliance and life safety
- Durability and environmental performance
- Thermal and acoustic behavior
- Budget constraints
- Aesthetic and programmatic goals
- Regional climate and material availability

Your decisions will directly influence **performance, constructability, lifecycle costs**, and how well the architecture meets the client's and users' needs.

1. Understand Building Envelope Systems

The building envelope — including roofs, walls, windows, doors, and foundations — serves as the critical barrier between the interior environment and exterior climate. It must balance several performance goals: durability, thermal insulation, moisture control, air tightness, and aesthetics. An effective envelope improves occupant comfort, energy efficiency, and building longevity.

Types of Wall Systems:

- **Cavity Wall (Masonry Veneer + Backup Wall)**
    - Comprises an exterior masonry veneer (brick, stone) separated by an air cavity from an inner backup wall (concrete block or wood framing).
    - The cavity acts as a drainage plane, preventing moisture intrusion and allowing ventilation.
    - Common in institutional and residential buildings for durability and aesthetic appeal.

- **EIFS (Exterior Insulation and Finish System)**
  - A lightweight, continuous insulation system with a textured finish, often used to improve energy performance and aesthetics.
  - Properly detailed EIFS must include drainage planes to prevent trapped moisture and potential rot.

- **Curtain Wall vs. Punched Openings**
  - *Curtain Wall*: Non-structural, glazed or panelized exterior walls suspended from the building frame, providing maximum daylight and modern aesthetics.
  - *Punched Openings*: Traditional framed openings within a solid wall, often smaller windows, more cost-effective and better for controlling solar gain in certain climates.
- **Metal Panel Rainscreen**
  - A rainscreen assembly separates the metal cladding from the structural wall with an air gap, allowing ventilation and moisture drainage.
  - Often used in commercial and industrial buildings for sleek aesthetics and long-term durability.

Roof Assemblies:
- **Low-Slope Roofs**
  - Typically less than 3:12 pitch, these roofs rely on membrane roofing systems such as:
    - Single-ply membranes (TPO, EPDM) for waterproofing and UV resistance
    - Built-up roofing (BUR) composed of multiple bitumen and felt layers
    - Green roofs with vegetation, offering stormwater management and thermal benefits
- **Steep-Slope Roofs**
  - Roofs with pitches greater than 3:12 using traditional materials:
    - Asphalt or architectural shingles for affordability and ease of installation
    - Standing seam metal roofing for durability and weather resistance
    - Clay or concrete tiles for aesthetic appeal and longevity, common in Mediterranean or southwestern climates

Key Envelope Performance Metrics:
- **R-Values (Thermal Resistance)**
  - Measure insulation effectiveness; higher R-values mean better resistance to heat flow.
  - Important to consider in walls, roofs, and slabs to reduce heating/cooling loads.
- **U-Factors (Thermal Transmittance)**
  - Measure total heat transfer; lower U-factors indicate better insulating performance, especially for windows and doors.
- **Air and Vapor Permeability**
  - The envelope must control airflow (to avoid drafts and heat loss) and moisture vapor to prevent condensation, mold, and material degradation.
  - Vapor retarders or retarders are placed based on climate and assembly type (e.g., vapor barrier inside in cold climates).

- **Solar Heat Gain Coefficient (SHGC) for Glazing**
  - Indicates how much solar radiation passes through a window; lower SHGC reduces cooling loads in hot climates, while higher SHGC can be beneficial in cold climates for passive solar gain.

🧠 **Example:**
In cold climates such as the northern US or Canada, an energy-efficient wall assembly often includes continuous exterior insulation (CI) to reduce thermal bridging and a vapor retarder placed on the warm-in-winter side to prevent condensation within the wall. This approach maintains insulation performance and durability by controlling moisture migration and protecting framing materials.

2. **Evaluate Interior Finish Systems**
Interior finish systems contribute significantly to a building's functional performance, safety, aesthetics, and user comfort. The right finish selections must respond to the building's **occupancy type**, **programmatic use**, **durability needs**, **fire code requirements**, and **maintenance expectations**.

Materials Based on Occupancy and Use:
Different building types and interior zones demand specific finish materials tailored to their daily use, hygiene standards, and performance expectations:

- **Healthcare and Education**
  - Require **durable, hygienic, and easily cleanable** finishes due to high foot traffic and infection control standards.
  - Typical materials: sheet vinyl flooring, impact-resistant gypsum board, epoxy wall paint, plastic wall protection, and acoustical ceilings with washable tiles.
- **Offices, Hotels, and Multifamily Housing**
  - Sound privacy is a key concern. Partitions often require **high STC ratings** (≥50) between guest rooms or units.
  - Finishes may include carpet (for sound absorption), wall coverings, or decorative panels.
- **Restrooms and Kitchens**
  - High moisture areas demand **non-porous, mold-resistant** surfaces.
  - Use tile, FRP (fiberglass-reinforced panels), stainless steel, or water-resistant gypsum board (Type X or green board).

Fire Ratings and Flame Spread (Per IBC Chapter 8):
Interior finishes are subject to fire performance testing for **flame spread** and **smoke development**, particularly in exit paths and areas of assembly:

- **Flame Spread Index (FSI)** and **Smoke Developed Index (SDI)** are measured per **ASTM E84** and classified as:
  - **Class A**: FSI 0–25, SDI 0–450 → Highest performance, typically required in **exit corridors**, **stairways**, and **assembly spaces**
  - **Class B**: FSI 26–75
  - **Class C**: FSI 76–200

- **Ceiling and Wall Materials** must meet the appropriate class rating based on location and occupancy type.
  - Example: Lobbies and corridors in an office building typically require **Class A** or **Class B** finishes for code compliance and life safety.

🧠 **Example:**
In a hospital corridor, select **Class A** wall and ceiling finishes that are **washable, impact-resistant**, and capable of withstanding disinfectants. A common specification might include:
- Abuse-resistant gypsum board with a high-performance epoxy or polyurethane coating
- Vinyl or rubber wall base for cleanability
- Acoustical lay-in ceiling tiles with sealed surfaces (for noise control and hygiene)
- Crash rails or wall protection panels to reduce maintenance from gurney or cart traffic

◇ **Design Tips:**
- Coordinate wall finish thicknesses with door and frame details to avoid flush misalignment.
- Avoid finishes with high VOC emissions—choose **low-VOC** or **GreenGuard-certified** products for improved indoor air quality.
- In mixed-use buildings, verify that fire-rated assemblies (e.g., corridor walls, tenant separations) maintain rating integrity after finish installation, including penetrations and trim.

3. **Assess Cost and Constructability**
Material and assembly decisions should balance **initial cost**, **constructability**, and **long-term performance**. Architects must go beyond upfront pricing and consider labor availability, ease of installation, durability, and replacement frequency. Effective choices help maintain project budgets, reduce change orders, and ensure delivery within tight schedules.

Low-Cost vs. Life-Cycle Value
Material selection often involves a tradeoff between **immediate affordability** and **long-term durability**. While some materials appear cost-effective during design, they may incur higher maintenance or replacement costs over the building's life.
- **Vinyl Composition Tile (VCT)**
  - Inexpensive to purchase and install, but requires **regular stripping, waxing, and replacement** (typically every 10–15 years).
  - Common in schools and offices, but increasing maintenance costs over time may negate initial savings.
- **Terrazzo Flooring**
  - High initial cost but extremely durable, low-maintenance, and can last **50+ years** with minimal upkeep.
  - Often used in public buildings such as airports, hospitals, and schools where durability is critical.
- **Fiber Cement Siding vs. Stucco or Brick**
  - Fiber cement is relatively inexpensive and easy to install, but brick or stucco offer **greater thermal mass**, longevity, and reduced maintenance over time.

Prefabrication Benefits

Prefabricated systems offer advantages in **labor efficiency, speed**, and **quality control**, especially under constrained site conditions or accelerated schedules.

- **Precast Concrete Panels**
    - Manufactured off-site under controlled conditions, ensuring consistent quality and faster installation on-site.
    - Especially useful for façades in civic, educational, or institutional projects with repetitive layouts.
- **Modular Wall or Bathroom Pods**
    - Entire assemblies are built off-site, reducing the need for multiple trades on-site and minimizing construction sequencing delays.
    - Ideal for hospitals, hotels, and dormitories with repetitive floor plans.
- **Panelized or SIP (Structural Insulated Panel) Systems**
    - Prefabricated panels that combine insulation and structural support, speeding up the building enclosure process.

🧠 **Example:**

In a **public K–12 school project** with a strict 10-month construction schedule, using **prefabricated wood-framed wall panels** and **precast concrete façade elements** allows the general contractor to rapidly erect the building structure and shell, reducing the need for extensive scaffolding and multiple wet trades. This not only shortens the project timeline but also improves quality control during winter months when site labor and weather conditions pose risks.

◇ **Design Tips:**

- Evaluate **labor availability and skill level** in the region when specifying materials—some high-performance systems may be impractical in areas lacking skilled trades.
- Coordinate with the **contractor and construction manager early** to determine which prefabricated systems might optimize schedule and budget.
- Consider material availability and lead times, especially for **custom or imported finishes**, which may cause delays if not ordered early.

4. **Balance Aesthetic Goals with Performance**

Material selection is a central driver of architectural identity, and must serve **both visual and functional goals**. Architects must thoughtfully balance a building's appearance with performance criteria such as durability, climate responsiveness, energy efficiency, and code compliance. The best designs integrate materials in a way that enhances both **form and function**.

Architectural Expression

Materials define a building's **character**, communicate its use, and reflect its purpose. Selections should support the building's concept, massing, and detailing, while also standing up to user and environmental demands.

- **Exposed Structural Systems**
    - Using structure as finish can reduce costs and reinforce design clarity.
    - *Examples:*

    - Exposed **glulam beams** in a community center convey warmth and sustainability.
    - **Steel trusses** in a gymnasium or train station express lightness and span, while creating dramatic volumes.
- **Finish Grade Concrete vs. Cladding Systems**
  - **Board-formed concrete** or **polished concrete walls** can serve as both structure and finish.
  - Alternatively, cladding systems (e.g., rainscreens or ventilated facades) allow a variety of skins over insulated backup walls, giving more flexibility in finish selection without affecting structural systems.

- **Natural and High-End Materials**
  - *Stone veneers*, **terracotta panels**, **anodized metal panels**, and **fiber cement boards** offer high durability, refined appearance, and long service life when detailed properly.
  - These materials also allow for creative modulation of texture, scale, and rhythm across the façade.

Regional Design Response

Materials and assemblies should be selected in response to the **regional climate**, **cultural context**, and **local resilience needs**. This reinforces a building's sense of place and ensures long-term performance.

- **Climate-Responsive Material Choices**
  - *Southwest US:* Use **stucco** or **integral color CMU** over high thermal mass walls (e.g., ICF or concrete) to help regulate temperature swings.
  - *Coastal Regions:* Require **corrosion-resistant materials** (e.g., stainless steel, marine-grade aluminum) and **hurricane-resistant glazing**.
  - *Wildfire-Prone Areas:* Use **non-combustible siding**, **metal roofing**, and protected vents to resist ember intrusion (per WUI code).
- **Cultural and Aesthetic Relevance**
  - Materials should reflect local traditions where appropriate—*e.g.*, natural stone in the Northeast, timber cladding in the Pacific Northwest, adobe-inspired forms in the Southwest.

🧠 **Example:**

A **civic building in the American Southwest** may use **integrally colored concrete block (CMU)** for both structural and aesthetic purposes. The thermal mass of CMU improves energy performance in the desert climate, while the block's coloration and texture tie into local architectural traditions. If built over **ICF (Insulated Concrete Forms)**, the wall assembly can achieve even better thermal resistance while allowing a stucco finish for a clean, modern appearance.

◇ **Design Tips:**

- Use **material transitions** and joint details to reinforce architectural hierarchy and articulate massing.
- Select materials that align with the **project's maintenance budget and life expectancy**—not just the desired appearance on opening day.

- When exposing materials (e.g., concrete, steel, wood), ensure detailing addresses **weathering, corrosion, and fire resistance**.
- Leverage **sunlight and shadow** to enhance textures—this is especially effective in stone, masonry, or ribbed panels.

5. **Account for Climate and Environmental Conditions**
   Material selection must be climate-responsive to ensure **durability**, **energy efficiency**, and **occupant comfort**. Different regions present unique environmental challenges—such as extreme temperatures, humidity, wind, seismic activity, or salt-laden air—that influence how materials age, perform, and interact with the building envelope.

   Failure to properly match materials to climate conditions can lead to premature failures, excessive maintenance, and compromised building performance.

   Cold Climates (Snowbelt regions, northern latitudes)
   - **Thermal Bridging & Conductivity:**
     Select materials and assemblies with **continuous insulation (CI)** to reduce thermal bridging at slab edges, balconies, and parapets. Use thermally broken window frames and detailing.
   - **Freeze-Thaw Resistance:**
     Avoid porous materials that can absorb water and crack when freezing. Use **air-entrained concrete**, **masonry rated for severe weathering (SW)**, and **non-corrosive ties**.
   - **Snow and Ice Accumulation:**
     Roof assemblies must handle **snow loads**, and detailing must prevent **ice dams** at eaves. Use ventilation, air barriers, and proper insulation to keep roof surfaces cold and dry.
   - *Example Materials:*
     High-performance glazing (low U-factor), insulated metal panels, air barriers with low air infiltration rates.

   Hot & Humid Climates (Gulf Coast, Southeast U.S., tropical zones)
   - **Vapor Permeability & Moisture Control:**
     Wall assemblies should be **vapor-open to the exterior** to allow drying and prevent mold growth. Use breathable sheathing and avoid impermeable interior vapor barriers.
   - **Roofing & Solar Control:**
     Specify **cool roof membranes** (high solar reflectance index or SRI) to reduce heat gain and urban heat island effects. Include shading devices and deep overhangs.
   - **Corrosion Resistance:**
     Materials exposed to high humidity should resist mold, rot, and rust. Choose **stainless steel**, **fiber-cement**, **ceramic tile**, or treated wood for exterior exposure.
   - *Example Materials:*
     Cement board siding, ceramic tile, aluminum framing with thermal breaks, breathable housewraps.

High-Wind / Seismic Zones (Coastal areas, California, Pacific Rim)

- **Impact Resistance:**
  Use **hurricane-rated windows** and **fiber-cement siding** or masonry in wind-prone regions. Lightweight materials may detach in high winds and become hazardous.
- **Anchorage & Fastening:**
  Specify **reinforced connections**, **tie-downs**, and **load paths** that meet IBC or local wind/seismic codes. Materials should maintain integrity under racking or lateral loads.
- **Flexible Assemblies:**
  In seismic zones, select materials that can **absorb movement** or **accommodate drift** without cracking or separating (e.g., flexible sealants, slip joints in cladding).
- *Example Materials:*
  Laminated safety glass, fiber-cement panels with concealed fasteners, reinforced CMU walls, flexible underlayments.

🧠 **Example:**
In **coastal hurricane zones** like the Gulf Coast or Florida, a residential building might use **fiber-cement lap siding** for its resistance to moisture, impact, and UV degradation. Roofing would feature a **wind-rated standing seam metal roof**, securely anchored to handle uplift forces. **Hurricane impact-resistant windows** with laminated glass and reinforced frames protect against flying debris and pressure changes during storms.

◇ **Design Tips:**

- Review **local climate data** (ASHRAE Climate Zone Map) and match materials to heat, humidity, and precipitation trends.
- Consider **regional code amendments** (e.g., Florida Building Code, California's CBC seismic provisions) when selecting exterior assemblies.
- Always integrate **appropriate flashings, sealants, and breathable membranes** to enhance weather-tightness.
- Incorporate **mock-ups and field testing** (e.g., blower door tests, water infiltration testing) to verify performance under expected conditions.

6. **Consider Fire Resistance and Code Compliance**
   Fire safety is a **non-negotiable design requirement** and a key factor influencing material and assembly selection. Architects must ensure that all building elements comply with the **International Building Code (IBC)** and referenced standards (e.g., ASTM, NFPA), particularly with regard to **fire-resistance ratings**, **combustibility**, and **flame spread**. These requirements impact both life safety and code approval during permitting.

   Fire-Rated Construction
   Different parts of a building must achieve specific fire-resistance ratings depending on their function, occupancy, and location within the structure.

   - **1-hour and 2-hour Assemblies:**
     Required for:
     - Occupancy separations (e.g., residential vs. commercial)
     - Egress corridors and stairwells
     - Floor assemblies in mixed-use or podium buildings

- **Construction Types (IBC Chapter 6):**
  - **Type I & II:** Noncombustible (steel, concrete, masonry); wood is generally prohibited.
  - **Type III, IV, V:** Allow varying degrees of combustible materials like wood framing.
- *Example Assemblies:*
  - UL-listed 2-hour gypsum shaft wall around stairwells
  - Fire-rated CMU demising walls in multifamily buildings
  - Concrete floor slabs with fire-resistive ceiling systems

Interior Finish Requirements

Interior wall and ceiling finishes must comply with **flame spread and smoke development limits**, especially in egress corridors and assembly spaces.

- **ASTM E84 (Steiner Tunnel Test):**
  - Class A: Flame spread 0–25 (best performance); required in exits and vertical shafts
  - Class B: 26–75; often allowed in corridors and general areas
  - Class C: 76–200; generally limited to incidental use
- **Fire-Resistive Glazing and Doors:**
  - Glazing in rated walls must be tested and labeled for fire protection (e.g., 45- or 60-minute rated).
  - Doors in rated walls must include appropriate closers, seals, and labeled hardware (per UL 10C).

Firestopping and Penetration Protection

Any penetration through a rated assembly (for MEP systems, conduits, ducts, etc.) must be **sealed with tested and approved firestop systems**. This is a frequent source of failed inspections if not coordinated early.

- **UL-listed assemblies** must be referenced for exact material and installation details.
- Include **intumescent sealants**, firestop collars, and **sleeved penetrations** as required.
- Coordinate closely with MEP engineers and contractors to minimize and properly manage these penetrations.

🧠 **Example:**

In a **Type I podium building**, where a 3-story residential structure sits above a commercial/parking podium, the IBC requires a **2-hour horizontal fire separation** between the mixed occupancies. This might be achieved using a **concrete podium slab** with a UL-rated assembly, ensuring that penetrations (e.g., plumbing, HVAC) are firestopped per UL system details. Additionally, any egress corridors within the podium must be enclosed in **1-hour fire-rated walls with Class A finishes**.

◇ **Design Tips:**

- **Coordinate early** with structural and MEP engineers to minimize complex penetrations through rated assemblies.
- Always reference **UL or Intertek fire-resistance listings** when detailing assemblies.
- Don't forget fire-rated detailing at **joints**, **edges**, and **transitions** (e.g., slab to wall, floor to curtain wall).
- Pay attention to **fire-resistive continuity**—a rated wall must extend from slab to slab or roof deck without gaps or unprotected openings.

- Be aware that **Type IV (Mass Timber) construction**, while combustible, allows for certain fire-resistant assemblies using CLT and char-rate design methods under IBC 2021+.

7. **Evaluate Acoustic Performance**
Acoustic performance is a critical factor in occupant comfort, especially in residential, educational, healthcare, hospitality, and office settings. Poor sound control can lead to complaints, reduced productivity, and compromised privacy.

Architects must evaluate how **building assemblies and materials** affect both **sound transmission between spaces** and **reverberation within spaces**. Acoustic performance should be considered early in the design process to ensure that solutions are integrated rather than added reactively.

Wall and Floor Assemblies
Acoustic ratings for assemblies are typically measured in terms of:
- **STC (Sound Transmission Class):**
  - Measures how well an assembly blocks **airborne sound** (e.g., voices, TV, music).
  - A higher STC means better separation.
  - **STC 50** is generally the minimum target in multifamily housing for demising walls.
- **IIC (Impact Insulation Class):**
  - Measures resistance to **impact sound** (e.g., footsteps, dropped items) through floor-ceiling assemblies.
  - Assemblies with soft floor finishes (e.g., carpet) tend to have higher IIC ratings.
  - A rating of **IIC 50 or greater** is typical for residential construction, sometimes mandated by building codes or local ordinances.

Material Selection
Certain finishes enhance sound absorption and reduce noise transmission:
- **Soft Surfaces:**
  - Carpets, cork, and acoustic vinyl reduce impact noise and echo.
  - Ideal for upper-floor residential units or classrooms.
- **Ceiling Treatments:**
  - Acoustic ceiling tiles (NRC-rated) absorb sound within spaces, helping reduce reverberation.
- **Wall Panels and Baffles:**
  - Perforated wood, fabric-wrapped panels, or foam systems can help absorb and diffuse sound in public spaces.

Construction Detailing
Beyond the material selection, sound performance is heavily dependent on **construction technique**:
- **Staggered or Double Stud Walls:**
  - Physically separates wall faces to reduce sound bridging.
- **Resilient Channels or Clips:**
  - Provide decoupling between drywall and framing, reducing vibration transfer.

- **Slab Edge Isolation:**
  - Required to prevent flanking paths in floor systems; acoustical breaks between units are key.
- **Sealants and Caulking:**
  - Sound flanking can occur at wall perimeters if not properly sealed. Use acoustical sealants around outlet boxes, penetrations, and baseboards.

🧠 **Example:**
In **multifamily residential buildings**, demising walls between units should be designed for **STC 50 or higher**, achieved using staggered stud walls with batt insulation and multiple layers of gypsum board. Floor-ceiling assemblies should include **resilient underlayment below tile or wood flooring**, and **acoustic insulation** between floor joists to improve IIC ratings and reduce complaints about footstep noise.

◇ **Design Tips:**

- Avoid placing **noisy spaces** (e.g., mechanical rooms, laundry, elevators) adjacent to **quiet zones** like bedrooms.
- In healthcare and education, ensure high STC partitions for **exam rooms**, **counseling offices**, or **private meeting rooms**.
- Coordinate acoustic requirements with **mechanical system design**—uninsulated ducts or poorly placed return grilles can compromise otherwise good assemblies.
- Use **Acoustical Consultants** in performance-critical spaces (e.g., auditoriums, recording studios, open offices).

8. **Prioritize Sustainability and Material Health**
In modern architecture, selecting materials is no longer just about aesthetics and durability—it also involves evaluating **environmental impact**, **life-cycle performance**, and **occupant health**. Sustainable materials reduce a building's ecological footprint, help achieve green certifications (e.g., **LEED**, **WELL**, **Living Building Challenge**), and contribute to healthier indoor environments.

Material choices can significantly influence the building's **embodied carbon**, **indoor air quality (IAQ)**, and even **user well-being**. On the ARE 5.0 PPD exam, you may be asked to choose materials that meet sustainability criteria while balancing performance and code requirements.

Product Transparency and Life Cycle

- **EPDs (Environmental Product Declarations):**
  - Quantify a product's environmental impact across its life cycle (including extraction, manufacturing, transport, use, and end of life).
  - Often required in **LEED** and **WELL Building Standard** credits for material optimization.
- **HPDs (Health Product Declarations):**
  - Disclose the chemical content of building materials and assess health hazards.
  - Helps avoid materials with harmful substances like formaldehyde, phthalates, or halogenated flame retardants.

- **Cradle to Cradle**, **Declare Labels**, and **Living Product Challenge** also provide valuable data for selecting environmentally and socially responsible materials.

Sustainable Material Strategies

- **Recycled Content:**
  - Examples: steel, acoustic ceiling tiles, carpet backing, insulation.
  - Reduces reliance on virgin resources and supports circular economy.
- **Rapidly Renewable Resources:**
  - Includes bamboo, cork, linoleum, wool, and strawboard.
  - Typically regrows within 10 years, reducing long-term environmental impact.
- **Locally Sourced Materials:**
  - Reduces transportation emissions and supports regional economies.
  - Often credited in LEED under the "Regional Materials" or "Environmental Impact" categories.

Indoor Air Quality and Material Health

- **Low-VOC Products:**
  - Adhesives, paints, sealants, and composite wood products should meet low-VOC emission standards (per **SCAQMD** or **CDPH** standards).
  - Improves air quality and reduces occupant exposure to toxic off-gassing.
- **Formaldehyde-Free and No-Added Urea Formaldehyde (NAUF):**
  - Important for composite wood products and millwork in schools, hospitals, and offices.
- **Moisture-Resistant and Mold-Resistant Finishes:**
  - Materials like fiberglass-faced gypsum board, tile backer boards, and breathable wall assemblies prevent mold growth, contributing to long-term health and durability.

🧠 **Example:**

A LEED Gold–targeted school design might include:

- **Linoleum flooring** (made from linseed oil and jute—renewable and low-VOC),
- **FSC-certified wood** for doors and millwork,
- **Batt insulation** with 70% post-consumer recycled content,
- Interior paint and adhesives that meet **Green Seal or GREENGUARD** certification.

◇ **Design Tips:**

- Consider **materials with multiple sustainability benefits**—for example, cork provides both acoustic and thermal insulation, is rapidly renewable, and has low VOC emissions.
- Use the **LEED v4 Material and Resources (MR)** credits as a framework for evaluating material impact.
- Don't overlook **operations and maintenance**—select durable finishes that require fewer toxic cleaners and less frequent replacement.
- For **WELL projects**, focus on material health and occupant well-being, including light reflectivity, acoustics, and non-toxic emissions.

9. **Address Maintenance and Durability**

Durability and ease of maintenance are key factors in long-term building performance, especially in **high-traffic, institutional, or public-use facilities**. Material selections must consider the **anticipated wear**, **environmental exposure**, and **ongoing maintenance requirements** to reduce life-cycle costs and ensure user satisfaction.

This is especially relevant in **healthcare**, **transit**, **educational**, and **commercial** settings, where maintenance budgets, cleanliness, and visual appearance directly impact functionality and operations.

Considerations for High-Wear Environments

- **High-Traffic Interiors:**
  - Materials must withstand continuous use, abrasion, and potential impact.
  - Common selections include:
    - **Terrazzo flooring** (extremely durable, long lifespan, low maintenance)
    - **Rubber or linoleum flooring** (resilient, slip-resistant, and sustainable)
    - **Impact-resistant gypsum board** with wall guards or wainscoting in hallways
    - **Corner guards** and **stainless steel kickplates** in doors and corridors
- **Wall and Surface Protection:**
  - Use **vinyl wall coverings**, **fiber-reinforced panels (FRP)**, or **decorative high-impact panels** in areas prone to damage (e.g., schools, hospitals, airports).
  - **Graffiti-resistant coatings** in public or transit environments ease cleaning and reduce vandalism impact.

Exterior Material Durability

- **Environmental Exposure:**
  - Exterior finishes must resist:
    - **UV degradation** (e.g., faded signage, brittle plastics)
    - **Moisture penetration and freeze-thaw cycles** (e.g., cracking masonry or spalling concrete)
    - **Corrosion or rusting** in marine or industrial zones
  - Preferred materials may include:
    - **Fiber-cement siding**, **coated aluminum panels**, or **stainless steel cladding**
    - **Integral color CMU or masonry veneer** with sealed joints
- **Finish Longevity:**
  - Opt for **factory-finished products** with extended warranties (e.g., powder-coated metals, baked enamel panels) to reduce the frequency of repainting or refinishing.

Cleaning and Sanitation Requirements

- **Healthcare and Food Service Environments:**
  - Surfaces must be:
    - **Non-porous**
    - **Seamless where possible**
    - **Resistant to mold, bacteria, and harsh cleaning agents**
  - Use **solid surface countertops**, **sheet vinyl flooring with heat-welded seams**, and **epoxy paint** in sterile environments.
- **Ease of Upkeep:**
  - Smooth, durable materials like ceramic tile, sealed concrete, or high-performance coatings simplify cleaning routines and support hygiene protocols.

Example:

In a **transit station**, design with **graffiti-resistant phenolic wall panels**, **stainless steel benches and fixtures**, and **poured-in-place concrete or recycled rubber flooring** to withstand extreme use, simplify maintenance, and extend the service life of finishes.

◇ **Design Tips:**

- Always review the **owner's operational capacity**—for example, a high-end natural stone might look great but could demand specialty cleaning and sealing that's not feasible for some facilities.
- For **LEED O+M (Operations and Maintenance)** compliance, choose finishes that minimize harsh cleaning agents and reduce maintenance frequency.
- Consider **removable and replaceable panels** in areas that may be vandalized or heavily used, such as elevator cabs, public restrooms, or school corridors.
- Align material warranties and **life-cycle cost** with the building's anticipated service duration—this can help justify up-front investment in higher-performance assemblies.

10. **Coordinate Assemblies Across Disciplines**

Material and assembly choices do not exist in isolation—they must be carefully coordinated with the structural system, MEP (Mechanical, Electrical, Plumbing), fire protection, and building envelope design. Poor coordination can lead to constructability issues, performance failures, cost overruns, and schedule delays. During schematic and design development phases, architects must ensure that all systems **integrate seamlessly** while preserving thermal, structural, and spatial performance.

Key Coordination Areas

- **Wall Thickness and Chase Space:**
  - Interior and exterior walls must accommodate:
    - **Insulation thickness**
    - **Air and vapor barriers**
    - **Ductwork, conduit, fire sprinkler piping, and risers**
  - Coordination is especially critical in **tight shafts**, **multi-trade corridors**, and **rated walls** where space for insulation and service routing must not compromise code compliance.
  - Architects must ensure adequate cavity depth for:
    - **Horizontal fire blocking**
    - **Sound attenuation batts**
    - **Future service access (e.g., for medical gases or telecom in hospitals)**
- **Structural Load Considerations:**
  - Heavy finishes or systems (e.g., **stone veneer, curtain walls, green roofs**) require structural support:
    - **Shelf angles**, **clip systems**, or **reinforced ledgers** may be required
    - Roof structures may need increased capacity to support **mechanical equipment**, **solar panels**, or **rainwater harvesting tanks**
  - Avoid placing point loads or penetrations near **post-tensioned slab zones** without consulting structural engineers.

- **Envelope Penetrations and Transitions:**
  - Each envelope penetration—whether for MEP services, windows, or structural ties—must be detailed to:
    - **Maintain air and vapor control layers**
    - **Preserve continuous insulation (CI)**
    - **Prevent water intrusion** through proper flashing and sealants
  - Frequent problem areas include:
    - **Transitions between wall types**
    - **Roof-wall junctions**
    - **Window heads and sills**
    - **Parapets and balconies**

🧠 **Example:**
A **ventilated metal panel rainscreen system** must be coordinated across several disciplines:
- **Architect**: Designs aesthetic layout and ensures panel size aligns with openings.
- **Structural Engineer**: Confirms support brackets or girts are anchored to structure.
- **Envelope Consultant**: Ensures vapor barrier and flashing are continuous at attachment points.
- **MEP**: Coordinates penetration locations and avoids clashing with vertical duct shafts.
- **Contractor**: Ensures tolerances and sequencing match installation schedule.

◇ **Design Tips:**
- Begin **early coordination** with consultants in schematic design, especially for:
  - **Wall assembly build-up**
  - **Slab edge conditions**
  - **Mechanical shafts and electrical rooms**
- Use **3D BIM modeling** to detect conflicts between structure, systems, and finishes.
- Detail **mockups** for complex façade systems to validate waterproofing, fastening, and layering.
- Where trades intersect (e.g., plumbing stacks through rated walls), verify **UL-listed firestop details** and maintain acoustical performance.

Chapter Six

Project Integration of Program & Systems (32-38%)

NCARB Requirements
In this section, you'll pull together all the decisions from the previous three sections regarding environmental conditions, code, systems, and assemblies. This is the largest section in the Project Planning & Design division.

This section is a cornerstone of the **Project Planning & Design (PPD)** division and represents the most comprehensive and integrated part of the exam. Candidates are expected to demonstrate the ability to synthesize multiple design decisions—gathered from environmental analysis, code compliance, system selection, and spatial planning—into a cohesive architectural solution.

The core of this section is the **integration** of earlier decisions into a fully coordinated architectural design. You must show that you can make informed design choices that respond not only to technical constraints and programmatic needs but also to broader considerations such as sustainability, constructability, and user experience.

A. Determine building configuration (A/E) (8-12%)

NCARB Requirements
As an architect, you will need to resolve the building's configuration based on program and code requirements, selected MEP and structural systems, site and environmental conditions, historic precedent, sustainability requirements, and principles of design logic.

As an architect, determining the building configuration is not a single decision—it is the resolution of multiple, often competing, factors into a functional and coherent architectural form. The design must balance **programmatic requirements** with **building codes**, **MEP (Mechanical, Electrical, and Plumbing)** and **structural systems**, **site and environmental constraints**, **sustainability goals**, **design aesthetics**, and potentially even **historic preservation standards**.

You are not simply laying out spaces; you are orchestrating relationships between systems, site, and structure into a building that works technically, performs sustainably, and inspires visually.

Here are key factors and strategies to consider when determining a building's configuration:

1. **Programmatic Requirements**
A thorough understanding of the client's program is the foundation of any successful building configuration. The program outlines the **functional needs**, **spatial requirements**, **occupancy**

types, and often **hierarchies of use** that the design must accommodate. As the architect, your role is to translate this abstract information into a spatial layout that not only meets these needs but does so in a way that is efficient, flexible, and responsive to site, code, and system integration.

Key considerations when evaluating programmatic requirements include:

- **Space Adjacencies**: Determine which functions should be near or separated from each other. This includes physical proximity, visual connection, acoustic separation, or secure access.
- **Circulation Patterns**: Evaluate both horizontal and vertical circulation. Think about how people move through the building—including users, staff, visitors, and service personnel—and design logical, intuitive routes that minimize travel distances and avoid bottlenecks.
- **Public vs. Private Zones**: Define access control within the program. Public spaces (e.g., lobbies, waiting areas) should be clearly separated from private or restricted areas (e.g., staff lounges, mechanical rooms, or secure storage).
- **Vertical and Horizontal Spatial Needs**: Some program spaces require specific floor-to-floor heights or footprints. Double-height spaces, labs with extensive overhead systems, or industrial equipment zones must be accounted for early.
- **Future Expansion and Flexibility**: Many clients anticipate future growth or change. The building configuration should allow for logical expansion or reconfiguration without major disruptions to existing operations.

Example 1: Healthcare Facility (Expanded)
In a hospital or medical office building, the program often includes tightly defined adjacencies. For instance:

- **Sterile vs. Non-Sterile Areas**: Operating rooms must be adjacent to sterile storage and scrub areas but separated from public corridors.
- **Emergency Access**: There must be **direct access** from an ambulance drop-off zone to trauma bays and imaging rooms.
- **Discreet Circulation**: Separate corridors for patients, staff, and materials (such as laundry and waste) help maintain cleanliness and privacy.

Design Tip: Use a "racetrack" corridor layout to create efficient circulation loops and avoid dead-end corridors. This is especially effective in outpatient clinics or labs where workflow efficiency is critical.

Example 2: University Building
In an academic facility:

- Classrooms may need to be close to restrooms and student lounges.
- Lecture halls should be placed near main entrances for easy access.
- Faculty offices require a quieter zone, away from student gathering spaces.
- Labs might need to be on the ground floor for ventilation exhaust or equipment loading.

Design Tip: Group classroom wings around a shared courtyard or atrium to create a sense of community while maintaining clear circulation and daylight access.

Example 3: Library

In a public library:

- Children's areas should be located near family restrooms and far from silent reading rooms.
- Staff workrooms should be adjacent to book drop-offs and circulation desks.
- Flexible meeting rooms might be located near the entrance to allow after-hours access without opening the full facility.

Design Tip: Use furniture layout and low partitions to help define program areas within open-plan spaces, allowing for future reconfiguration.

Program Evaluation in Practice

During the early design phases—especially **programming** and **schematic design**—you'll often use diagrams such as:

- **Bubble diagrams** to explore adjacencies.
- **Stacking diagrams** to evaluate vertical relationships.
- **Matrix charts** to track space adjacency requirements.

These tools allow you to abstractly test configurations before committing to actual dimensions or structural grids.

Summary

Understanding and resolving programmatic requirements is a critical first step in developing an effective building configuration. On the PPD exam, you may be asked to interpret a space program, identify problematic adjacencies, or select the most appropriate layout based on client needs and contextual constraints.

Your goal is not only to fit the spaces together but to design **intelligent spatial relationships** that support the building's function, performance, and long-term usability.

2. **Code Compliance**

Building configuration must be developed in strict coordination with **local building codes**, which typically follow or are adapted from the **International Building Code (IBC)**. Code requirements dictate many aspects of a building's form and layout—from **life safety and fire protection** to **occupancy classification** and **construction type**.

As a designer, your ability to interpret and apply code early in the design process is critical. A design that fails to comply can lead to costly redesigns, permit rejections, or even life safety hazards during occupancy.

Key Code Considerations for Building Configuration:

- **Occupancy Classification**: Different spaces (e.g., Assembly, Business, Residential, Educational) have specific requirements for egress width, fire separation, and allowable floor area. Mixed-use buildings require particular attention to separation and access.
- **Construction Type (IA–VB)**: A building's construction type affects its allowable height, number of stories, and floor area. More fire-resistive types (e.g., Type I) allow larger and taller buildings.

- **Fire Separations and Fire-Resistance Ratings**: Occupancy groups with differing hazard levels must often be separated by fire-rated assemblies. Corridor walls, stair enclosures, and shaft walls must meet minimum fire-resistance standards.
- **Means of Egress**: Egress requirements include **number of exits**, **travel distance**, **exit width**, **door swing**, and **accessibility**. The placement of stairs, corridors, and exit doors has a direct impact on floor plans and building massing.
- **Vertical Openings and Shafts**: Openings between floors (like atriums or stairwells) must follow strict rules to limit smoke and fire spread, often requiring enclosures or smoke control systems.
- **Height and Area Limitations**: Code tables in the IBC define the maximum **height**, **number of stories**, and **building area** based on occupancy type and construction type. Sprinklers can provide increases in allowable size or height, influencing the design decision to include them.
- **Accessibility Codes (e.g., ADA, ANSI A117.1)**: Accessible routes, door widths, ramps, elevators, and restrooms all must comply with accessibility standards and can influence circulation paths and floor-to-floor heights.

Example 1: Mixed-Use Building (Expanded)
In a five-story mixed-use building with ground-level retail and upper-level residential:

- The **commercial (Group M or B)** and **residential (Group R-2)** occupancies must be separated by a **minimum 1-hour fire-rated floor assembly**, per IBC Chapter 5.
- The ground floor may have **larger occupancy loads**, requiring **wider egress paths** and **additional exits**.
- The building may qualify for a sprinkler-based **area and height increase**, allowing taller construction without moving to a more expensive fire-resistive construction type.
- Egress paths for residential units must be separated from the public retail space, often with a secure, dedicated entry and vertical circulation.

Design Tip: Consider using **two separate vertical circulation cores** for residential and commercial spaces to simplify code compliance and enhance security.

Example 2: Educational Facility
In a K–12 school:

- **Group E occupancy** requires specific exit access travel distances and a certain number of exits based on student count.
- **Assembly spaces** such as gyms or cafeterias may fall under Group A-3 and could trigger different requirements for **egress capacity** and **sprinkler coverage**.
- Certain classrooms may require **direct egress to the exterior** in ground-floor configurations.

Design Tip: Use fire-rated corridor walls with properly rated door assemblies to compartmentalize classroom wings and facilitate phased evacuation.

Example 3: Atrium in an Office Building
If an open atrium is desired in a multi-story office (Group B):

- The atrium must be separated from adjacent floors with a **2-hour fire barrier**, or the design must include an **engineered smoke control system**.

- Exit access paths through the atrium space must comply with **IBC Section 404**, including travel distance and visibility requirements.

Design Tip: Open staircases within atriums can be stunning features, but they **cannot** serve as required egress unless they meet specific enclosure or smoke control criteria.

Code Tools You Should Know:

- **IBC Tables 503 / 504 / 506** – For height, story, and area limitations
- **Chapter 3** – Use and Occupancy Classification
- **Chapter 5** – General Building Heights and Areas
- **Chapter 6** – Types of Construction
- **Chapter 7** – Fire and Smoke Protection Features
- **Chapter 10** – Means of Egress
- **ANSI A117.1 / ADA** – Accessibility standards

Summary

Code compliance is **non-negotiable** and must be embedded in the design process from day one. While creativity drives design, codes set the framework within which that creativity must operate. During the PPD exam, expect to interpret diagrams, identify code violations, or choose code-compliant design solutions.

Smart architects use **code as a design tool**, finding opportunities within its constraints to shape buildings that are not just legal—but safe, functional, and elegant.

3. **Structural System Compatibility**

Selecting an appropriate structural system is a critical part of shaping a building's configuration. The structural system must not only **support the building's loads** but also **respond to the program, spatial layout, and architectural intent**. Early coordination between architectural and structural decisions helps ensure efficient construction, cost control, and long-term performance.

When structure and layout are not compatible, the result may be spatial inefficiencies, compromised flexibility, or expensive structural solutions. Therefore, architects must have a foundational understanding of structural systems to make informed design decisions—especially regarding **spans**, **column placement**, **floor-to-floor heights**, and **material behavior**.

Key Considerations:

- **Column Grid Alignment**: Align column grids with walls, partitions, and circulation paths wherever possible. This ensures that structural elements don't interrupt critical program areas or create unusable floor space.
- **Span Requirements**: Different programmatic uses demand different spans. For example, assembly spaces (like theaters or gyms) and open office layouts require **longer, unobstructed spans**, whereas hotel or apartment buildings typically consist of **repetitive, short-span modules**.
- **System Depth and Floor-to-Floor Height**: Deeper structural systems (e.g., open-web steel joists or deep concrete beams) require more floor-to-floor height, which impacts

building height and volume. Coordination with MEP systems in the ceiling plenum is essential.

- **Integration with Mechanical Systems**: Structural elements must allow for the placement of ducts, piping, and conduit. Systems like **flat slabs** or **beamless concrete** may provide more flexibility for MEP distribution.
- **Vibration and Acoustics**: Structural choices influence floor vibration and acoustic transmission—especially important in performance venues, healthcare facilities, or residential buildings.
- **Material Selection and Construction Speed**: Steel, concrete, timber, and hybrid systems each offer different construction timelines, tolerances, and environmental impacts. The choice of material often relates to availability, climate, budget, and program.

Example 1: Open-Plan Office (Expanded)
In a contemporary office building that prioritizes open workspaces:

- The design may call for **column-free zones** spanning 30 to 50 feet to allow layout flexibility and visibility.
- **Steel wide-flange beams with open web joists**, or **post-tensioned concrete slabs**, are well-suited for this condition.
- Deep beams may be used to reduce the number of structural supports but will require additional ceiling space for MEP routing.

Design Tip: Coordinate column grid spacing with furniture systems and ceiling lighting modules to reduce clashes during the fit-out phase.

Example 2: Parking Garage
In a parking structure, structural efficiency and vehicle maneuverability are key:

- A **cast-in-place concrete system** with regular bays of 18 to 20 feet allows for clear driving lanes and column-free parking stalls.
- **Sloping slabs** or **ramp systems** may be integrated into the structure itself, often using **double-tee precast elements** for speed and efficiency.

Design Tip: Ensure structural bay spacing aligns with standard parking dimensions (typically 9' wide stalls, 24' deep bays).

Example 3: Multi-Family Residential
For apartments or hotels:

- A **repetitive structural grid** supports stacked units with shared plumbing walls and shear cores.
- **Concrete flat plates** or **light gauge steel framing** can reduce structural depth and simplify service routing.
- Columns and shear walls are typically hidden within unit demising walls to maximize usable space.

Design Tip: Use modular design principles to align plumbing chases, mechanical shafts, and structural bays from floor to floor.

Example 4: Cultural or Civic Center
In a performance hall or museum:

- Large spans over 60 feet may be needed for main assembly or exhibit areas.

- **Long-span steel trusses** or **glulam beams** can achieve dramatic open spaces while integrating lighting, mechanical systems, or skylights within the depth of the structure.
- Sound isolation may require **floating floor assemblies** and additional dead load, which impacts structural sizing.

Design Tip: Work closely with acoustic and structural consultants early in the design to avoid costly retrofits or system changes later.

Structural Coordination in Practice
As part of the design team, your role includes:

- Establishing a **preliminary structural grid** in the early schematic phase.
- Verifying that **column locations and beam depths** do not conflict with critical program areas.
- Consulting with structural engineers to determine the **optimal balance** between system performance, cost, and design intent.
- Documenting **typical bay layouts** and ensuring consistency across floor plans.

Summary
Architectural design and structural systems must be developed **in tandem**, not in isolation. On the PPD exam, expect to evaluate whether a structural system supports the proposed design or identify misalignments in floor plans or sections.

A successful building configuration is one in which the structure not only holds up the building—but supports the **function, aesthetics, and flexibility** of the architecture.

4. **MEP System Integration**
Mechanical, Electrical, and Plumbing (MEP) systems are essential to a building's functionality and occupant comfort—but they can also significantly influence the building's configuration, aesthetics, and construction cost. Effective MEP integration requires early coordination during the design process to avoid spatial conflicts, reduce inefficiencies, and ensure that systems work harmoniously with both architectural intent and structural design.

Poor MEP integration can lead to issues such as dropped ceilings, oversized shafts that intrude into usable space, uncoordinated equipment placement, or insufficient service access. On the ARE and in practice, demonstrating foresight in planning for these systems is critical.

Key Considerations for MEP System Integration:

- **Ceiling Heights and Plenum Space**: HVAC ducts, electrical conduit, and plumbing lines often occupy ceiling space above finished areas. Inadequate ceiling clearance can force design compromises, like reducing floor-to-ceiling height or raising the overall building height—both of which have cost and code implications.
- **Shaft Placement and Vertical Alignment**: Plumbing stacks, ductwork, and risers must run vertically through the building and should align from floor to floor. Misaligned shafts can lead to inefficient routing, additional floor penetrations, and potential fire-stopping challenges.
- **Mechanical Room Location**: Mechanical rooms should be located for operational efficiency and minimal distribution distances. This helps reduce energy loss, duct pressure

drops, and pipe runs. In multi-story buildings, mechanical rooms are often placed on the **roof**, **basement**, or **intermediate mechanical floors**.

- **Equipment Access and Maintenance**: All MEP equipment must be accessible for servicing. This includes clearances around air handling units (AHUs), electrical panels, water heaters, and pumps. Design must ensure access doors, ladders, catwalks, and clearance zones are provided.
- **Noise and Vibration Control**: Equipment like compressors, fans, or pumps can generate noise and vibration, requiring strategic placement and isolation—especially in sensitive areas like residential units, theaters, or libraries.
- **Energy Efficiency and Sustainability**: Efficient MEP layouts reduce wasted energy and improve building performance. Consider system zoning, equipment sizing, and passive design strategies to minimize mechanical loads.

Example 1: Multi-Story Residential Building (Expanded)

In a high-rise apartment building:

- Plumbing chases must align vertically between units, typically stacked in wet walls shared between kitchens and bathrooms.
- HVAC systems (like fan coil units or PTACs) must have access to fresh air and exhaust. Vertical duct shafts often need to run alongside plumbing risers.
- Roof space must accommodate **HVAC condensers**, **ventilation exhausts**, and **access hatches**, all while maintaining zoning height limits.

Design Tip: Use a vertical "service core" strategy, where plumbing, electrical, and mechanical systems are grouped together in a central shaft to simplify layout and improve future maintainability.

Example 2: Office Building

In a commercial office building with open-plan layouts:

- HVAC systems often use **variable air volume (VAV)** boxes or **underfloor air distribution** (UFAD). These systems impact floor-to-ceiling heights and require raised floors or significant plenum space.
- Large ductwork must be routed around structural beams, which may require deeper beams or stepped ceilings.
- Electrical systems may include **floor boxes**, **suspended light fixtures**, and **data cable trays**, all of which must be coordinated with architectural and structural components.

Design Tip: Early use of **3D coordination models** (e.g., BIM) helps detect clashes between MEP, structure, and architecture before construction begins.

Example 3: Educational Facility

In a K–12 school or university lab:

- Labs and science classrooms often require **specialized exhaust systems**, **fume hoods**, **dedicated gas and water lines**, and **emergency shut-offs**.
- Mechanical equipment must be separated from instructional areas to avoid noise disruptions.
- Classroom HVAC zoning is often needed to allow for different temperature control in individual rooms.

Design Tip: Group mechanical zones based on solar exposure (e.g., north vs. south wings) to improve HVAC efficiency and user comfort.

Example 4: Historic Renovation Project
In adaptive reuse or historic preservation:

- Existing floor-to-floor heights may be too shallow to accommodate modern ductwork or sprinkler systems.
- Creative solutions like **exposed MEP systems**, **split ductless units**, or **slim-profile equipment** may be necessary to preserve character-defining features while upgrading performance.

Design Tip: Coordinate with preservation consultants to conceal or minimize the visual impact of modern systems in sensitive areas.

Common MEP Systems to Coordinate:

- **HVAC**: Air handlers, ducts, VAV boxes, condensers, heat pumps
- **Plumbing**: Water supply and return, drainage, vent stacks, water heaters
- **Electrical**: Lighting, emergency power, switchgear, conduit routing, control systems
- **Fire Protection**: Sprinkler systems, risers, fire pumps, standpipes

Summary
MEP systems are integral to building function, occupant comfort, and code compliance—but they require **careful planning and coordination** with architectural and structural systems. On the PPD exam, expect to identify appropriate shaft locations, analyze ceiling heights based on MEP needs, or select layouts that allow for efficient system distribution.

Good MEP integration ensures a building that is not only well-designed—but also **functional, efficient, and maintainable** for years to come.

5. **Site Constraints**

The physical, environmental, and contextual characteristics of a site profoundly influence how a building should be configured. **Site constraints** are not obstacles to overcome—they are design opportunities to create a building that is efficient, sustainable, and deeply rooted in its context.

A well-integrated design **responds to the site** by working with natural features rather than against them. This leads to more harmonious forms, improved environmental performance, and often reduced construction costs.

Key Site Constraints That Affect Building Configuration:

- **Topography (Slopes and Elevation Changes)**
  Topography impacts foundation design, floor levels, site drainage, and accessibility. Steeply sloped sites may require stepped buildings, retaining walls, or special grading strategies.

Design Tip: Use split-level or terraced designs to minimize cut-and-fill operations and reduce the need for deep retaining walls.

- **Solar Orientation and Daylighting**
  The building's position on the site affects how sunlight enters interior spaces throughout the day and across seasons. Orientation also influences heat gain, glare, and the potential for passive solar heating or daylighting strategies.

Design Tip: Orient major glazing toward the south (in the northern hemisphere) with horizontal shading devices to balance winter heat gain and summer shading.

- **Prevailing Wind and Microclimate**
  Local wind patterns affect building ventilation, comfort, and mechanical system sizing. In hot climates, passive cooling strategies may leverage prevailing breezes; in cold or windy areas, minimizing exposure can reduce heat loss.

Design Tip: Use landscape features, berms, or building wings to shield courtyards or outdoor spaces from harsh winds.

- **Access Points (Vehicular and Pedestrian)**
  Entrances, service zones, and parking must align with adjacent roadways, zoning setbacks, and circulation requirements. Poorly placed access can lead to circulation conflicts or inefficient layouts.

Design Tip: Separate service and public entries to reduce conflicts and improve site functionality. Consider ADA-compliant paths from parking to main entries, particularly on sloped sites.

- **Noise Sources and Environmental Nuisances**
  Proximity to highways, railroads, industrial sites, or airports introduces acoustic challenges. Configuration must buffer sensitive program spaces (like bedrooms, classrooms, or offices) from these impacts.

Design Tip: Use unoccupied buffer zones (e.g., stairwells, storage, corridors) along the noisy edge, or use courtyard configurations to shelter interiors.

- **Views and Visual Connections**
  Capturing views of water bodies, mountains, city skylines, or open landscapes can enhance the user experience. Likewise, avoiding unsightly or distracting views may shape the building's orientation and fenestration strategy.

Design Tip: Use glazing and room placement to frame desirable views—e.g., placing living or common spaces toward the best outlook, and using solid or service spaces to screen poor ones.

- **Soil Conditions and Subsurface Constraints**
  Soil type affects foundation design, structural system selection, and construction cost. Expansive soils, high water tables, or contaminated ground may restrict excavation or require special solutions.

Design Tip: Engage a geotechnical consultant early to determine if conventional spread footings are feasible or if piles, caissons, or mat foundations are required.

- **Zoning and Legal Boundaries**
  Local zoning codes dictate setbacks, height restrictions, floor-area ratios (FAR),

easements, and use limitations. These constraints can shape the footprint, height, and even massing of the building.

Design Tip: Use zoning envelopes to test various building configurations during early design phases, ensuring the concept complies from the start.

Example 1: Sloped Site (Expanded)

On a steeply sloped site:

- A **stepped foundation** or **split-level design** can allow the building to follow the natural terrain, minimizing earthwork and retaining walls.
- Driveway and ADA access must be carefully coordinated to avoid excessive slopes.
- Opportunities may exist to create lower-level daylight basements or tuck under parking.

Case in Point: A hillside residence could place bedrooms on the upper level and living spaces with patios on the lower, all with access to outdoor views and natural light.

Example 2: Urban Infill Site

In a dense urban environment:

- Zero-lot-line construction may be required, with fire-rated walls on property edges.
- Limited frontage may constrain entry options and service access.
- Solar access could be blocked by neighboring buildings, necessitating strategic use of courtyards, light wells, or skylights.

Design Tip: Use vertical stacking and light wells to bring daylight into deep building plans. Consider rooftop amenities to maximize limited open space.

Example 3: Coastal or Floodplain Site

Near water or in a flood zone:

- The design may need to elevate the structure above the base flood elevation (BFE) using piers or stilts.
- Wind exposure requires lateral load resistance and may influence building orientation and materials.
- Salt corrosion may limit structural and finish material choices.

Design Tip: Use the elevated lower level for parking or unconditioned storage to meet code while protecting living areas from flood risks.

Summary

Understanding site constraints is essential for informed building configuration. Rather than treating these factors as limitations, successful architects use them as **design drivers**—leveraging the site's assets and mitigating its challenges.

On the PPD exam, you'll be asked to evaluate site diagrams, analyze building orientation options, and make strategic design choices that respond effectively to site conditions.

A responsive design is not just more sustainable—it's smarter, more buildable, and often more beautiful.

6. Environmental and Climate Considerations

A well-designed building doesn't just sit on its site—it interacts with its **climate and environment**. Early decisions about **building configuration**, **orientation**, **massing**, and **materials** are critical to reducing energy use, enhancing occupant comfort, and supporting sustainability goals.

Climate-responsive design is not just good practice—it's increasingly expected by clients, mandated by code (e.g., energy codes like IECC), and emphasized on the **ARE 5.0 PPD exam**. This section focuses on **passive design strategies**—using the building's form and orientation to naturally regulate heat, light, and airflow before relying on active mechanical systems.

Key Environmental and Climate Factors:

- **Orientation**
  The position of the building relative to the sun and wind has a major impact on heating, cooling, and lighting demands.

Design Tip: Orient the long axis of the building along the **east-west** direction to minimize low-angle solar heat gain from east and west, and to control southern exposure more easily with fixed shading.

- **Massing and Building Form**
  Compact forms conserve heat in colder climates, while stretched, open forms promote ventilation in hot, humid climates. The shape and size of the building can also affect wind exposure and shading.

Design Tip: Use narrow floor plates (around 40–60 ft) in hot climates to promote cross-ventilation and daylight penetration. In cold climates, tighter massing helps retain heat.

- **Glazing Placement and Window Design**
  Strategic window placement and sizing can maximize natural daylight while minimizing glare and unwanted heat gain. Consider the **Solar Heat Gain Coefficient (SHGC)** and **U-value** of glass.

Design Tip: Place larger windows on the **north side** (in the northern hemisphere) for diffuse daylight, and limit unshaded glass on the east and west facades.

- **Shading Devices**
  Fixed or operable shading—like overhangs, louvers, or fins—can block unwanted solar gain while preserving daylight and views.

Design Tip: Horizontal overhangs are ideal for **south-facing windows**, while **vertical fins or louvers** work best for **east and west** facades where the sun is lower in the sky.

- **Material Selection and Thermal Mass**
  Materials like concrete, brick, or stone can absorb and store heat during the day and release it at night—helpful in climates with high daily temperature swings.

Design Tip: Use **thermal mass** on interior surfaces that receive direct sunlight in the winter for passive heating, and insulate from the exterior to retain heat.

- **Natural Ventilation and Airflow**
  Designing for cross-ventilation reduces reliance on mechanical cooling and improves indoor air quality. Wind patterns should be analyzed to place operable windows, vents, or atria effectively.

Design Tip: Align openings on **opposite walls** of rooms and use clerestory or stack vents to enhance the **stack effect**—where warm air rises and draws in cooler air below.

- **Renewable Energy and Climate Synergy**
  Passive strategies often pair well with active systems like **photovoltaics**, **solar hot water**, and **geothermal HVAC**. Roof orientation and configuration should accommodate these systems when feasible.

Design Tip: A south-facing sloped roof (in the northern hemisphere) can be ideal for solar panels. Keep rooftop equipment consolidated to preserve space.

Example 1: Classroom Wing Orientation
In a school design:

- Classrooms are oriented along the **east-west axis**, with most windows facing north or south.
- **South-facing windows** use horizontal overhangs to shade during summer while allowing winter sun.
- **North-facing glazing** provides soft, even daylight with minimal glare—ideal for learning environments.

Outcome: Improved thermal comfort, reduced HVAC loads, and better student focus from quality daylighting.

Example 2: Hot-Arid Climate Office Building
In a desert environment:

- The building has **deep recessed windows** and **high thermal mass** materials like concrete and masonry.
- Courtyards and shaded outdoor spaces provide passive cooling and air circulation.
- **Minimal glazing** is used on the west side to avoid afternoon overheating.

Outcome: Significant energy savings and improved occupant comfort without heavy mechanical reliance.

Example 3: Cold Climate Housing Project
In a northern climate:

- Building massing is compact to reduce exposed surface area.
- High-performance insulation and triple-glazed windows retain heat.
- **South-facing living rooms** capture low-angle winter sun; deciduous trees provide summer shading.

Outcome: Lower heating demand, increased resilience in power outages, and greater occupant comfort.

Example 4: Coastal Building with Natural Ventilation
In a temperate, coastal location:

- Operable windows and vents are aligned with **prevailing winds** to encourage cross-ventilation.
- Clerestory windows in a central corridor allow warm air to exhaust vertically.
- Shading structures allow ocean breezes to flow through without direct solar gain.

Outcome: Cooling loads are reduced, air quality is improved, and users enjoy a stronger connection to the natural environment.

Summary

Environmental and climate considerations are not just about sustainability—they are about **performance, comfort, and long-term value**. Smart early decisions can reduce mechanical loads, improve building quality, and help meet energy code or certification requirements (e.g., LEED, WELL, Passive House).

On the **ARE PPD exam**, expect to evaluate site orientation diagrams, select appropriate passive strategies based on climate, and assess how a configuration supports environmental performance.

Ultimately, **climate-responsive design** is not just better for the planet—it's better architecture.

7. **Accessibility and Universal Design**

Accessibility and Universal Design are essential considerations in all architectural projects—not only because they are legally required (per the **Americans with Disabilities Act (ADA)** or local accessibility codes), but also because they promote **equity, safety, and inclusion** in the built environment.

From the earliest stages of planning and configuration, architects must ensure that **barrier-free access** is provided to all public areas and essential functions within a building. Accessibility affects much more than just clearances—it influences how people move through, perceive, and experience a space.

Key Accessibility Principles That Impact Configuration:

- **Accessible Route Continuity**
  A continuous, unobstructed path is required from **site arrival points** (like parking, sidewalks, or transit stops) to all **accessible building entries**, and through interior circulation routes.

Design Tip: Ensure accessible paths are at least **36 inches wide**, with appropriate turning radii (usually **5 feet**) at corners and doors.

- **Entry Sequence and Site Grading**
  Main entrances must be accessible, which affects site grading, ramp placement, stair design, and entry vestibules. Changes in level (more than ½ inch) require **ramps, lifts**, or **elevators**.

Design Tip: Avoid split-level entry foyers unless an equivalent accessible route is clearly and conveniently provided.

- **Slope and Ramp Design**
  Maximum slope for an accessible ramp is **1:12 (8.33%)**, with landings every 30 feet and at changes in direction. Handrails are required for ramps over 6 inches in rise.

Design Tip: Whenever possible, use **natural topography** and gentle grading to eliminate the need for ramps altogether.

- **Elevator Placement and Reach**
  Multi-story buildings must have **elevators** (with few exceptions), and their location directly affects vertical circulation and fire egress planning. Elevator cab dimensions and door locations must accommodate wheelchair users.

Design Tip: In a central core layout, elevators should be co-located with stairs and service shafts to consolidate infrastructure and maintain compact circulation.

- **Corridor and Door Clearances**
  Corridors serving accessible routes must maintain **minimum clear widths** and include turning spaces at key junctions. Doorways must allow **32 inches clear** when open at 90°, and include **maneuvering space** (typically 18 inches latch-side on the pull side).

- **Restrooms and Toilet Room Design**
  Restroom layouts must accommodate **wheelchair turning radii**, **grab bars**, **accessible fixtures**, and **lavatory knee clearance**. At least one accessible stall is required in multi-fixture restrooms.

Design Tip: Provide a **family/unisex toilet room** as part of the public restroom strategy—it's helpful for caregivers, parents, and individuals requiring assistance.

- **Tactile Signage and Wayfinding**
  Accessible signage must include **raised characters**, **Braille**, and proper mounting height. Tactile indicators are also required for stairways, exits, and changes in floor level.

- **Floor Surfaces and Transitions**
  Surfaces must be **firm, stable, and slip-resistant**. Changes in level greater than ¼ inch must be beveled or ramped. Avoid thresholds higher than ½ inch.

Universal Design vs. Code Minimums
While ADA and accessibility codes define minimum legal requirements, **Universal Design** goes further by making environments usable by **people of all ages and abilities**—without the need for adaptation.

Examples of Universal Design strategies include:

- Providing **automatic doors** at main entrances (not required by code, but highly beneficial).
- Installing **lever handles** instead of knobs, which are easier to use for people with limited dexterity.
- Using **lower-mounted ATM screens, kiosks, or reception desks** for wheelchair users and children.
- Designing **circulation paths** wide enough to accommodate **two wheelchairs passing** (60 inches clear).

Example 1: Entry Sequence (Expanded)
In a civic building design:

- The main entrance is clearly visible from the accessible parking area.
- A **barrier-free route** leads from the parking to the entry with a gentle slope, avoiding the need for a ramp.
- Tactile warning strips are placed at the transition between sidewalk and vehicular zones.
- Entry doors include **automatic operators** and are flanked by high-contrast, Braille-labeled signage.

Result: All users, regardless of ability, enter through the **same dignified and intuitive pathway**.

Example 2: Multi-Level Retail or Educational Facility
In a two-story public building:

- The elevator is located near the main entry and clearly visible upon arrival.
- Interior corridors are a **minimum of 5 feet wide**, allowing wheelchair users to pass comfortably.
- Restrooms on both levels include **fully compliant accessible stalls**, grab bars, and lavatories with proper knee clearance.
- Emergency egress includes **Areas of Refuge** at stair landings with **two-way communication devices**.

Design Tip: When possible, **group accessible restrooms and vertical circulation** near programmatically significant zones like auditoriums or classrooms for better integration.

Common Oversights to Avoid (for ARE and Practice):

- Providing an elevator but failing to connect it to all key program areas.
- Using stairs or level changes at a main entrance without a parallel accessible route.
- Designing toilet rooms that meet fixture count but lack adequate turning space or clearances.
- Placing signage too high or without Braille/raised text.
- Including accessible parking spaces without an accessible path to the building.

Summary
Accessibility is not just a code issue—it's a core design responsibility. Architects must plan early for accessible routes, vertical circulation, and barrier-free facilities. On the **ARE PPD exam**, expect to evaluate whether a configuration meets accessibility requirements or to choose layouts that better support equitable access.

Universal Design elevates this approach by promoting inclusivity, dignity, and independence for all users.

Great design includes everyone—from a child in a stroller to an elder with a walker to a person in a wheelchair—and starts from day one of the design process.

8. **Circulation and Wayfinding**
Circulation is more than just movement—it's how people **experience** a building. Clear, efficient, and legible circulation paths are essential for **functionality, safety, and user**

comfort. Good circulation supports intuitive **wayfinding**, enhances spatial orientation, and accommodates both able-bodied and mobility-impaired users.

On the **ARE 5.0 PPD exam**, you'll be expected to evaluate circulation layouts, determine appropriate widths and flow, and assess wayfinding logic in plans and sections.

Key Circulation Planning Concepts:

- **Primary vs. Secondary Circulation**
  Primary circulation includes **main corridors, lobbies, vertical circulation (stairs/elevators)**, and major public spaces. Secondary circulation includes **branch corridors, access aisles, and internal pathways** within spaces.

Design Tip: Place major destinations (e.g., auditorium, café, restrooms) along primary paths to reduce user confusion and walking distances.

- **Efficient Flow and Travel Distances**
  Circulation paths should be **direct, continuous, and unobstructed**. Excessive dead ends, backtracking, or disjointed circulation frustrates users and wastes space.

Design Tip: Use simple, symmetrical layouts or centralized hubs to help users navigate instinctively—especially in large or complex buildings.

- **Visual Cues and Landmarks**
  Incorporate **visual orientation points**, such as atriums, stairwells, or artwork, to help users anchor themselves in the space. These cues aid wayfinding and create moments of spatial identity.

Design Tip: Provide views to the outside or interior landmarks at corridor ends to reinforce intuitive movement.

- **Hierarchy and Legibility**
  Circulation should have a clear **hierarchy of scale, lighting, and materials** to distinguish between major and minor paths. This supports both everyday users and first-time visitors.

Design Tip: Use wider, naturally lit corridors for public routes, and narrower, simpler paths for staff or service areas.

- **Accessibility and Safety**
  Circulation must meet **accessibility requirements** (e.g., minimum clear widths, turning spaces, level changes) and support **safe egress** in emergencies. Clear sightlines reduce collision risk and increase security.

Design Tip: Minimize blind corners, provide handrails where necessary, and integrate tactile wayfinding for low-vision users.

Wayfinding Strategies:
Effective wayfinding ensures that users can navigate the building **without confusion or anxiety**. This involves both spatial logic and informational design:

- **Zoning and Spatial Organization**
  Group related functions together and assign each zone a distinct identity through materials, color, or spatial form.

Example: In a healthcare facility, locate diagnostic services in one wing and treatment areas in another, with clear transitions and signage at zone thresholds.

- **Line of Sight and Visual Transparency**
  Allow long views down major corridors or between floors via atriums or interior windows to provide visual connections between destinations.

- **Signage and Environmental Graphics**
  While spatial layout should do most of the work, signage helps clarify function and orientation. Use **consistent typography, icons, and color schemes** to reinforce intuitive navigation.

- **Lighting as Wayfinding**
  Use lighting intensity, direction, or color to emphasize circulation routes—e.g., a continuous ceiling light running down a hallway or spotlighting key decision points.

- **Flooring and Material Cues**
  Subtle material transitions in flooring or wall finishes can reinforce path directions or indicate changes in program or circulation type.

Example 1: Museum with Centralized Atrium
In a cultural or museum setting:

- A large **central atrium** anchors all public circulation paths and allows visitors to orient themselves from any point.
- Upper-level galleries overlook the atrium, providing **visual connections** across floors.
- Signage and floor plans are reinforced by **physical landmarks** (e.g., a sculpture or water feature) visible from key nodes.

Result: Visitors can intuitively return to the central hub or continue forward with confidence.

Example 2: School with Ring Circulation
In a K–12 school:

- Classrooms are arranged around a **continuous corridor loop**, with centrally located administration and communal spaces.
- Each grade level is zoned into its own quadrant, distinguished by colors and signage.
- Students can walk in one direction to reach any classroom, reducing cross traffic and confusion.

Result: Young students navigate easily, and supervision is simplified through clear sightlines.

Example 3: Healthcare Facility with Dual Circulation
In a hospital:

- **Public corridors** are separated from **staff/service circulation**, ensuring patient privacy and operational efficiency.
- Service corridors allow staff to move equipment, food, and waste without disturbing patient zones.
- Public circulation is designed with generous widths, natural light, and comfortable waiting areas at key junctions.

Result: The experience is more human-centered for patients and families, while staff movement remains efficient and unobtrusive.

Example 4: Office Tower with Core-Centric Circulation
In a multi-story office building:

- Elevators and stairs are located in a central core, with office spaces arranged around the perimeter.
- A main corridor on each floor loops around the core, providing **short and direct access** to every suite.
- Exit stairs are placed at core corners to meet egress requirements and minimize travel distances.

Result: Circulation is efficient, code-compliant, and easily scalable across multiple floors.

Summary
Thoughtful circulation and wayfinding strategies make buildings **intuitive, inclusive, and pleasant** to navigate. Whether the goal is speed, clarity, or comfort, circulation is at the heart of the user experience.

On the **PPD exam**, expect to evaluate plan layouts, circulation flow diagrams, or section cuts that test your ability to spot confusing or inefficient routes and recommend improvements. Ultimately, circulation is more than a code issue—it's an architectural tool for creating **clarity, comfort, and connection** within a building.

9. **Historic and Contextual Considerations**
When working in **historic districts** or with **adaptive reuse projects**, architects face unique challenges and opportunities. The building configuration must balance respect for the **existing urban fabric**, **architectural heritage**, and **cultural significance** with the needs of modern functionality, safety, and sustainability.

Understanding the historical and contextual context is vital—not just for preserving architectural character, but also for navigating **regulatory frameworks**, such as local historic preservation ordinances and **Secretary of the Interior's Standards for Rehabilitation**.

Key Factors Affecting Configuration in Historic and Contextual Projects:

- **Preservation of Significant Elements**
  Identify which features are character-defining—such as façades, cornices, window patterns, and materials—and plan interventions to **retain, repair, or restore** these elements.

Design Tip: Often the façade or street-facing elevation is protected, while the interior can be more flexible for redesign and system integration.

- **Adaptive Reuse vs. New Construction**
  Adaptive reuse projects may require inserting **new structural cores**, **mechanical systems**, or **vertical circulation** in ways that minimize impact on historic fabric.

Design Tip: Locate new cores in less significant areas, or within interior courtyards, to preserve historic spatial arrangements.

- **Urban Context and Scale**
  Building massing and height often need to **respect the scale and rhythm** of neighboring historic buildings, especially in historic districts.

Design Tip: Step back upper floors, use complementary materials, and align cornice lines to maintain contextual harmony.

- **Modern Code and Performance Upgrades**
  Historic buildings often require significant upgrades to meet **life safety**, **accessibility**, **energy codes**, and **structural standards**.

Design Tip: Innovative strategies like adding **sprinkler systems**, inserting **fire-rated partitions**, or upgrading windows with **high-performance glazing** can improve performance without compromising historic character.

- **Material Compatibility and Detailing**
  New additions or repairs should use materials and detailing that complement but do not mimic historic fabric, maintaining a **clear distinction between old and new**.

- **Community and Regulatory Coordination**
  Engage early with historic preservation boards, planning commissions, and community stakeholders to understand constraints and gain approvals.

Example 1: Warehouse to Office Conversion
A mid-century warehouse in a historic district is converted into a modern office:

- The **original brick façade** and large industrial windows are preserved and repaired.
- A **new structural core** with elevators and stairs is inserted internally, using a steel-framed infill that minimizes load on the original masonry walls.
- Modern **MEP systems** are concealed in raised floors and dropped ceilings, preserving the open interior volume.
- Glass partitions create flexible office spaces without compromising daylight.

Outcome: The building maintains its historic character externally while supporting modern workplace needs internally.

Example 2: Historic Rowhouse Adaptive Reuse
An early 20th-century rowhouse is transformed into a boutique hotel:

- The street-facing façade and original wood windows are restored per preservation guidelines.
- Interior walls are reconfigured to accommodate guest rooms and en-suite bathrooms.
- A rear addition is carefully massed to be subordinate to the main block, using materials that differentiate it from the original structure.
- Accessibility is achieved via a discreetly located elevator shaft at the rear.

Outcome: The hotel offers modern comfort while celebrating the building's historic identity.

Example 3: Infill Construction in a Historic District
A new mixed-use building is proposed between two historic structures:

- The new building's height is set to align with adjacent buildings to maintain the streetscape rhythm.
- Façade materials reflect the texture and scale of neighboring buildings but use contemporary detailing.
- Street-level retail is emphasized to support pedestrian activity and enliven the historic corridor.
- Sustainable features like green roofs and solar shading are integrated discreetly.

Outcome: The infill building respects historic context while asserting a contemporary architectural language.

ARE 5.0 Exam Relevance
On the **PPD exam**, you may encounter scenarios requiring you to balance historic preservation with code compliance, program needs, or sustainability. Understanding how to maintain **historic character** while integrating **new systems, accessibility, and safety features** is key. You might be asked to:

- Identify appropriate locations for new mechanical cores or vertical circulation.
- Propose interventions that minimize damage to character-defining elements.
- Determine compatible materials or massing strategies for additions.

Summary
Historic and contextual considerations require a **sensitive, informed approach** to design—one that respects the past while accommodating the present and future. The architect's role includes stewarding cultural heritage, creatively adapting old buildings, and ensuring compatibility with their surroundings.

This balance of preservation and innovation embodies the challenge—and reward—of working within historic or contextual frameworks.

10. Sustainability and Energy Performance Goals
Sustainability is no longer an optional aspect of architectural design—it is a core responsibility that directly influences building configuration, material choices, and systems integration. Today's projects frequently align with recognized frameworks such as **LEED (Leadership in Energy and Environmental Design)**, **WELL Building Standard**, **Living Building Challenge**, or other regional sustainability certifications.

Design decisions around **orientation, massing, envelope performance**, and **site strategies** are guided by goals to reduce environmental impact, enhance occupant health, and optimize energy efficiency.

Key Sustainability Considerations Affecting Configuration:

- **Solar Orientation and Passive Design**
  Optimizing building orientation is critical for harnessing natural light and solar energy while minimizing heat gain or loss.
    - **South-facing façades** (in the Northern Hemisphere) maximize winter solar gain.

- Use **overhangs, louvers, or shading devices** to reduce summer overheating.
- Orient narrow floor plates east-west to allow deeper daylight penetration.

- **Daylight Autonomy and Visual Comfort**
  Maximizing natural daylight reduces reliance on electric lighting and improves occupant wellbeing.

Design Tip: Incorporate **open atriums, light wells, and interior glazing** to distribute daylight to core spaces. Use **high-performance glazing** with appropriate visible light transmittance (VLT) and solar heat gain coefficients (SHGC).

- **Envelope Performance**
  High-performance building envelopes with effective insulation, air sealing, and thermal breaks reduce heating and cooling loads.

Design Tip: Select materials with good thermal mass to stabilize interior temperatures, such as concrete or masonry, especially in climates with significant temperature swings.

- **Renewable Energy Integration**
  Solar panels (photovoltaics) and solar thermal collectors impact roof design and orientation.

Design Tip: Design roofs with appropriate pitch and orientation to maximize solar gain; consider structural loading and maintenance access.

- **Green Roofs and Vegetated Surfaces**
  Green roofs improve stormwater management, reduce urban heat island effect, and provide insulation.

Design Tip: Roof structure must accommodate additional load; design roof layout to balance accessible space with green zones.

- **Water Efficiency and Site Ecology**
  Sustainable projects incorporate rainwater harvesting, low-flow fixtures, and native landscaping to reduce potable water use.

- **Indoor Environmental Quality (IEQ)**
  WELL standards emphasize ventilation, air quality, acoustics, and occupant comfort, which influence spatial layout and system design.

Example 1: Narrow Floor Plate with Open Atrium
A corporate office building is designed to optimize daylight autonomy:

- Floor plates are limited to **30-40 feet in depth** to allow daylight penetration.
- A central atrium extends through multiple floors, bringing daylight into interior corridors and shared spaces.
- Operable windows and automated shading devices allow occupants to control daylight and glare.
- High-efficiency LED lighting supplements daylight, with occupancy sensors to reduce energy use.

Outcome: The building achieves significant reductions in electric lighting loads while enhancing occupant satisfaction.

Example 2: Solar-Oriented Residential Development
A multi-family housing project integrates passive solar design and renewable energy:
- Units are oriented so living spaces face south to maximize winter solar gain.
- Roofs are angled to optimize solar panel efficiency, with integrated inverters and battery storage in mechanical rooms.
- Vegetated balconies and green roofs contribute to stormwater management and microclimate regulation.
- High-performance windows and insulation reduce heating and cooling demands.

Outcome: The development meets a high LEED certification level and reduces utility costs for residents.

Example 3: Adaptive Reuse with Energy Upgrades
An old industrial building converted to mixed-use includes sustainability retrofits:
- Insulated glazing units replace original single-pane windows, preserving historic sightlines while improving thermal performance.
- A new HVAC system with heat recovery ventilators (HRVs) reduces energy consumption.
- Rooftop solar panels and a green roof enhance energy production and stormwater control.
- Building orientation and massing remain unchanged, but interior layout is adjusted for daylight and ventilation optimization.

ARE 5.0 Exam Relevance
In the **PPD exam**, you may need to:
- Evaluate how building orientation impacts solar heat gain and daylighting.
- Propose massing solutions that improve energy performance.
- Choose envelope strategies that balance daylight, thermal comfort, and energy use.
- Integrate renewable energy technologies or green infrastructure into site and roof design.

Understanding the **trade-offs between sustainability goals and programmatic or code constraints** is critical.

Summary
Sustainability goals shape **building form, orientation, and systems integration**. Early decisions about site response, daylight strategies, and renewable energy set the stage for achieving high-performance, healthy buildings.

A well-integrated sustainable design reduces operating costs, supports occupant wellbeing, and minimizes environmental footprint—all key responsibilities of today's architect.

11. Design Logic and Aesthetics

At the core of architectural design lies **design logic**—a coherent set of principles and formal decisions that guide the building's organization, proportions, and spatial relationships. A building's configuration should not only meet functional and technical requirements but also **express a clear conceptual intent** and create meaningful spatial experiences.

Design logic ensures that every element, from massing to circulation to façade articulation, contributes to a **unified whole**, reinforcing the project's programmatic goals and cultural context.

Key Aspects of Design Logic and Aesthetics:

- **Formal Clarity and Order**
  The arrangement of spaces and forms should follow a clear logic—whether through symmetry, hierarchy, modularity, or rhythm—to communicate organization and legibility.

Design Tip: Use axial alignments, grids, or geometric patterns to organize massing and circulation. This helps users intuitively understand space and flow.

- **Proportion and Scale**
  Respecting classical or contemporary proportions enhances visual harmony and human comfort. Scale must relate both to the human body and the urban context.

Design Tip: Apply the Golden Ratio, the Modulor, or contextual scale cues to achieve balanced and pleasing proportions.

- **Spatial Quality and Experience**
  Beyond function, spaces should evoke desired emotions—whether grandeur, intimacy, openness, or enclosure—through volume, light, texture, and materiality.

Design Tip: Manipulate ceiling heights, natural light, and material contrasts to enrich spatial experiences and support programmatic needs.

- **Cultural and Contextual Expression**
  Design logic can embody cultural values or historic references, contributing to a building's identity and sense of place.

Design Tip: Integrate local materials, vernacular motifs, or symbolic forms to reinforce contextual relevance.

Example 1: Civic Building with Symmetry and Axial Organization
A courthouse or government building might employ:

- A strong **central axis** leading from the main entrance through a formal lobby to a symbolic chamber or courtroom.
- Symmetrical wings and balanced massing convey **stability, order, and authority**.
- Proportional façade elements and monumental steps enhance presence and approachability.

Outcome: The building's formality communicates its civic importance and instills public confidence.

Example 2: Cultural Center with Fluid Spatial Connections
A museum or performing arts center might feature:

- An **asymmetrical, dynamic layout** reflecting the fluidity of creative expression.
- Interconnected volumes, curved walls, and varied ceiling heights create a sense of movement and discovery.
- Transparency and openness invite public engagement and blur boundaries between interior and exterior.

Outcome: The spatial complexity inspires curiosity and reflects the institution's innovative mission.

ARE 5.0 Exam Relevance
The exam may ask you to assess whether a building's configuration aligns with its intended use and design concept, or to identify formal strategies that enhance user experience and identity.
You might analyze:

- How massing choices reflect program hierarchy.
- Whether spatial proportions promote comfort and legibility.
- If circulation and vistas reinforce the design narrative.

Summary
Design logic and aesthetics transform architecture from mere shelter into **meaningful space**. They ensure the building is **not only functional but expressive**, embodying the client's vision and the architect's conceptual framework.

Mastering these principles equips you to create architecture that resonates deeply with users and the broader community.

Final Thoughts
Determining building configuration is fundamentally an exercise in **integration and balance**. As an architect, you are tasked with synthesizing a wide spectrum of often competing requirements—ranging from programmatic needs and environmental factors to code mandates and technical system constraints—into a single, clear, and cohesive design solution.

The **Project Planning & Design (PPD) exam** will challenge your ability to **make informed trade-offs**. You'll need to weigh the benefits and drawbacks of various options, such as:

- Balancing **daylight penetration** with potential **solar heat gain** to achieve occupant comfort and energy efficiency.
- Aligning **structural systems** to support flexible spatial layouts without compromising architectural intent.
- Navigating **code compliance** requirements while meeting client expectations for usability, aesthetics, and budget.
- Integrating **MEP systems** in ways that optimize building performance yet maintain spatial quality.

Effective Strategies for Exam Preparation:

- **Study Real-World Precedents:**
  Review case studies and precedent projects that demonstrate successful integration of multiple design factors. Pay attention to how architects resolve conflicts and prioritize goals in diverse contexts.
- **Master Code Tables and Requirements:**
  Become fluent in interpreting local and international codes (e.g., IBC, ADA) and standards. Practice applying these to sample building programs to anticipate constraints and necessary adaptations.

- **Practice Sketching and Spatial Planning:**
  Develop the habit of sketching building plans, sections, and site layouts early in your preparation. Visualizing the interaction between program, structure, and systems enhances your spatial reasoning and design fluency.
- **Integrate Systems Thoughtfully:**
  Include HVAC, plumbing, electrical, and structural elements in your sketches and diagrams. Consider how shaft locations, equipment rooms, and structural grids influence and are influenced by the overall building configuration.
- **Reflect on Design Intent:**
  Always connect your technical decisions back to overarching design goals—whether they be sustainability, accessibility, aesthetics, or user experience. This holistic mindset will serve you well both in the exam and professional practice.

Closing Encouragement
Remember, the PPD exam tests not just your technical knowledge but your **architectural judgment**—the ability to synthesize complex variables into elegant, practical, and innovative solutions. Approach your study with curiosity and critical thinking, and you'll develop the confidence to tackle any design integration challenge the exam presents.
You've got this!

B. Integrate building systems in the project design (A/E) (10-14%)

NCARB Requirements
Within this division is the beginning of systems coordination. You must consider how the selected systems fit together, both spatially and functionally, and how they are coordinated into the architectural design. You will also need to evaluate how changes in one building system impact another system as well as the overall project design.

This section of the **Project Planning & Design (PPD)** division marks a critical shift—from identifying individual building systems to **coordinating and integrating them** within the architectural design. At this stage, architects must begin to think like systems integrators, ensuring that **mechanical, electrical, plumbing (MEP), structural, fire protection, envelope, and life safety systems** all work together in harmony.

You are expected to:

- Coordinate systems both **spatially** (e.g., ductwork routing, ceiling space, shafts, floor-to-floor heights) and **functionally** (e.g., HVAC supply/return, electrical zoning, plumbing stacks).
- Understand the **interdependencies** of building systems and how decisions in one area (e.g., structure) can directly affect others (e.g., lighting layout or duct placement).
- Optimize system selection and layout to align with **project goals** such as energy performance, cost, user comfort, and ease of maintenance.
- Resolve conflicts in design through **iterative coordination** with engineers and consultants.

Real-World Relevance

In practice, system integration often determines the success of a project in terms of performance, budget, and constructability. Improper coordination can result in expensive change orders, poor building performance, or even code violations. The PPD exam expects you to understand these implications early in the design process, not just during construction documents.

Topics We'll Cover in This Section

We'll elaborate on each of the following eight key integration areas, aligned with NCARB exam expectations:

1. **Vertical and Horizontal Shaft Planning**

Proper shaft planning is a foundational step in integrating mechanical, electrical, and plumbing (MEP) systems into a building's design. Shafts serve as **vertical and horizontal conduits** that allow MEP systems to traverse between floors and zones while maintaining organization, code compliance, and accessibility.

At the PPD level, architects must anticipate how system distribution will **impact spatial organization, structural coordination, and program layout.** Effective shaft planning during early design stages helps avoid costly rework, spatial conflicts, and system inefficiencies during later phases.

Key Shaft Planning Considerations:

- **Location & Accessibility:**
  Vertical shafts should be centrally located where they can serve **multiple stacked spaces** efficiently (e.g., hotel rooms, bathrooms, mechanical risers). Avoid locating shafts at building corners or isolated locations unless required by program or layout.

- **System Stacking:**
  Align wet areas like **restrooms, kitchens, and janitor closets** to minimize plumbing riser lengths and reduce slab penetrations. Similarly, stack mechanical rooms and electrical closets for easier vertical distribution of systems.

- **Size and Clearance:**
  Sizing must accommodate the **combined routing of ductwork, piping, conduit**, and required insulation. Include maintenance clearances, access panels, and code-required fire separations.

- **Coordination with Structure:**
  Shafts should be located to avoid **conflicts with beams, columns, and load-bearing walls**. Consider floor framing direction and whether shafts penetrate through one-way or two-way slabs.

- **Code and Fire Separation:**
  Many shafts (especially those serving different occupancies or fire zones) require **fire-rated enclosures**, typically with 1- or 2-hour fire-resistance ratings. Ensure shaft walls are continuous through floors and terminate correctly at penetrations.

- **Integration with Architecture:**
  Shaft locations should not compromise the architectural design. When possible, incorporate shafts into wall thicknesses, closets, or secondary spaces to avoid disrupting primary areas or circulation.

Types of Shafts:

- **Vertical Shafts:**
  Include **mechanical risers**, **plumbing stacks**, **electrical conduits**, **fire protection risers**, and **communication cabling**. Typically run from basement/mechanical levels to roof.
- **Horizontal Distribution (Plenums, Ceiling Chases):**
  Horizontal shafts, especially in the ceiling plenum, distribute services across floors. Require coordination of **ceiling heights**, lighting, and structural depths.
- **Hybrid Shafts / Interstitial Spaces:**
  In complex buildings like hospitals, a full interstitial floor may be used to house MEP systems, providing flexible and serviceable access between occupied floors.

Example 1: Multi-Story Office Building

In a high-rise office:

- Vertical shafts are located at the **core** of the building, adjacent to elevators and stairs.
- A shared shaft serves HVAC duct risers, sprinkler risers, and electrical conduit.
- Plumbing risers are placed within **stacked restrooms** on each floor.
- Horizontal ducts branch out into tenant spaces via ceiling plenums, coordinated with beam depths and lighting layout.
  Result: Centralized shaft planning supports flexible tenant fit-outs and efficient vertical system runs.

Example 2: Mixed-Use Building

In a mixed-use project (retail below, residential above):

- A vertical shaft is located at the rear of each residential stack to serve kitchens and bathrooms.
- The shaft must penetrate the **fire-rated separation** between commercial and residential occupancies.
- Electrical and HVAC shafts are separated to reduce electromagnetic interference and improve maintenance access.
  Result: Shaft placement supports both uses while maintaining code compliance and minimizing mechanical room footprints.

Example 3: Educational Building with Atrium

In a university science building with an atrium:

- Shafts are positioned along **lab/service corridors**, behind chase walls, and in mechanical zones.
- Fire-rated shafts traverse multiple floors while respecting **atrium separation requirements**.
- Horizontal distribution is done via **bulkheads** and **dropped ceilings** in secondary spaces, leaving public zones clean and expressive.

Result: Strategic shaft placement ensures code compliance while preserving architectural clarity and openness.

PPD Exam Tips

On the exam, expect to:

- Identify **logical shaft locations** based on stacked programming.
- Recognize improper shaft planning that may violate code or disrupt layout.
- Choose between design options that reflect better system coordination.

  Practice strategy: Review typical shaft dimensions and clearances for ductwork, piping, and risers. Understand how they influence floor-to-floor heights and spatial efficiency.

Summary

Effective vertical and horizontal shaft planning is essential for integrating building systems into your design. It demands **foresight, coordination, and understanding of systems behavior**—skills that are foundational not just for passing the PPD exam, but for real-world project success.

2. **Mechanical System Coordination**

Mechanical system coordination—especially for HVAC—is one of the most **spatially demanding** and **technically complex** aspects of early design. It requires architects to plan how heating, cooling, and ventilation systems are routed and housed while maintaining **spatial efficiency, architectural intent, and code compliance**.

At the **PPD stage**, your role is to ensure that HVAC systems can be efficiently integrated into the structural and architectural framework without compromising room functions, ceiling heights, or design quality.

Key Coordination Issues:

1) Structural Integration

- **Duct clearances** must be coordinated with beam depths and joist spacing. Oversized ducts in small spaces may require structural beams to be penetrated or deepened, affecting cost and layout.
- **Open web steel joists** (OWSJ) are often used to allow ductwork to pass through rather than under beams—but must be planned for early.

  Design Tip: Choose structural systems that accommodate mechanical routing (e.g., double-T precast planks for long spans and service integration in institutional buildings).

2) Ceiling Height and Plenum Space

- Ceiling plenums must provide space for **ductwork, lighting, fire protection, sprinkler piping**, and access to equipment. Inadequate plenum space can lead to compromised ceiling heights or the need for dropped ceilings.
- Floor-to-floor heights are often driven by the **depth of HVAC systems**, especially in office, academic, or healthcare facilities.

  Design Tip: Use **zoned mechanical systems** to reduce duct sizes and routing complexity. Consider **underfloor air distribution** where ceiling space is limited.

3) Room Function and Mechanical Zoning

- HVAC system design should reflect **zoning based on usage patterns**. For example, classrooms, auditoriums, and kitchens all have unique ventilation and cooling demands.
- Mechanical rooms must be accessible and strategically placed to reduce long duct runs and energy loss.

Design Tip: Group rooms with similar HVAC needs (e.g., labs, offices, classrooms) to simplify duct routing and improve control efficiency.

4) Duct Routing and Diffuser Placement

- Avoid running large ducts across main circulation paths or central gathering spaces unless concealed within architectural elements like bulkheads or soffits.
- Diffuser and return placement should support comfort and indoor air quality while integrating seamlessly into the interior design.

5) Equipment and Shaft Coordination

- Air handlers, VAV boxes, and terminal units must fit within ceilings or mechanical closets without disrupting other systems.
- HVAC shafts must be aligned floor-to-floor and appropriately sized to accommodate supply/return ducts, fire-rated enclosures, and access points.

Example 1: University Classroom Building

- Large lecture halls require **dedicated air handling units (AHUs)** due to high occupancy and variable load.
- Ducts are routed through corridors with **bulkheads** above to avoid reducing ceiling height in classrooms.
- Ceiling heights are maintained using **low-profile ductwork** and **displacement ventilation** under raised floors.

  Result: The mechanical system supports high air quality and comfort while preserving architectural clarity.

Example 2: Commercial Office Tower

- A **central mechanical penthouse** houses AHUs, with vertical shafts running through service cores.
- Open office floors are served by VAV boxes distributed evenly, with ducts routed through the ceiling plenum.
- Ceiling height and lighting layout are coordinated to conceal ductwork above acoustic ceiling tiles (ACT).

  Result: Clean interior expression and flexible tenant fit-outs are maintained through well-integrated systems.

Example 3: Performing Arts Center

- Displacement ventilation is used in audience spaces, with **air supplied under seating platforms** to minimize noise and visual impact.
- Structural systems are coordinated to allow for unobstructed duct routing in backstage and support areas.

- Acoustic separation is integrated between mechanical equipment and performance spaces.
  Result: Quiet, thermally comfortable environments are created without compromising the building's expressive form.

PPD Exam Tips
On the exam, you may be asked to:
- Identify appropriate locations for mechanical rooms or shafts.
- Choose layouts that allow for proper system routing without reducing headroom or usable space.
- Evaluate trade-offs between system type (e.g., central vs. distributed) based on program and configuration.

Know This: Typical duct depths range from 12"–36" depending on airflow demand. Factor in insulation and structural beam depth when estimating plenum space.

Summary
Mechanical system coordination requires balancing **engineering needs with architectural intent**. Early alignment of structure, ceiling systems, and HVAC planning ensures efficient system performance, spatial quality, and cost-effective construction. The architect's ability to anticipate and integrate these systems is essential for holistic design solutions—and a key competency on the **PPD exam**.

3. **Electrical and Lighting Integration**
Integrating electrical and lighting systems into architectural design is about more than just power and illumination—it's about enhancing user experience, supporting building performance, and ensuring spatial and code compliance. As the architect, you're responsible for ensuring that electrical components (e.g., outlets, panels, lighting fixtures) and lighting systems are **coordinated with the building's layout, structure, and mechanical systems** while also reinforcing the overall design concept.

Electrical and lighting coordination must occur early in design, especially when ceiling types, structural systems, or daylighting strategies are involved.

Key Considerations in Electrical and Lighting Integration
1) Power Distribution Coordination
- Electrical panels, transformers, and conduit runs must be coordinated with **structural systems and program adjacencies**. Locate panels in secure yet accessible areas, often near utility entries or in designated electrical rooms.
- Avoid placing electrical rooms adjacent to **wet spaces** like bathrooms or janitor closets to comply with code and safety protocols.
- Conduits and raceways must be planned to avoid conflict with **plumbing, ducts, and structure**, especially in high-density ceiling plenums.
  Design Tip: Group electrical panels and equipment rooms near vertical shafts to simplify riser layouts and minimize disruption to occupiable spaces.

2) Lighting Fixture Integration

- Lighting layout should enhance the **architecture**—not fight it. Use fixture types that reinforce the spatial character (e.g., linear fixtures in corridors, uplights in vaulted ceilings, recessed cans in clean ceiling designs).
- Fixture placement must account for **ceiling type** (e.g., ACT, gyp board), **structure above**, and **duct routing**.
- In public or high-design areas, fixture alignment and spacing should follow architectural grids or geometry to ensure visual harmony.

  Design Tip: Work closely with the lighting designer or engineer to integrate fixture placement with architectural elements like beams, ceiling coffers, and wall features.

3) Daylighting and Electric Lighting Balance

- Coordinate electric lighting with **daylighting strategies** such as skylights, clerestories, and window placement. Overlighting daylit spaces wastes energy and can create glare.
- Use **photocell sensors** and **dimming systems** to automatically adjust electric lighting based on daylight availability.
- Spaces like classrooms, open offices, and lobbies benefit from **balanced daylighting** supplemented by tunable white or circadian-friendly electric lighting.

  Design Tip: In LEED or energy-efficient projects, daylight modeling is used to inform both electric lighting levels and fixture layout. Design narrow floor plates or use atriums to maximize daylight penetration.

4) Emergency Lighting and Code Compliance

- Emergency and egress lighting must comply with **life safety codes** (e.g., IBC, NFPA 101). These include illuminated exit signs, path-of-travel lighting, and battery backup or generator systems.
- Emergency lighting is often integrated subtly into ceiling systems or wall sconces in high-finish environments but must remain **clearly visible and functional**.

  Design Tip: Use low-profile or dual-purpose fixtures in high-visibility areas to integrate emergency lighting without compromising aesthetics.

5) Controls, Zoning, and User Comfort

- Spaces should be divided into **lighting zones** to allow for flexible use, energy conservation, and user control. For example, conference rooms, classrooms, or auditoriums should allow scenes or preset controls.
- Consider ADA accessibility when placing wall-mounted controls, and coordinate locations with finishes and furniture layouts.

Example 1: Art Museum

- Gallery lighting uses a combination of **track fixtures and recessed lighting** that integrate with the ceiling's coffered pattern.
- Dimmable controls allow for curatorial flexibility, and blackout zones are coordinated with daylight openings.
- Power distribution is routed through **raised access flooring** in exhibit areas to avoid visible conduits.

Result: The lighting enhances the art experience and preserves the architectural purity of the galleries.

Example 2: Elementary School

- Classrooms use a mix of **daylight-responsive linear LED fixtures** and high clerestory glazing.
- Electrical outlets are located along perimeter walls and floors based on student and technology needs.
- Occupancy sensors and daylight sensors help the building meet **energy code requirements (e.g., ASHRAE 90.1 or IECC).**
  Result: A bright, comfortable, and energy-efficient learning environment is created through careful lighting and power coordination.

Example 3: Mixed-Use Residential/Retail Project

- Electrical distribution is separated by use type (commercial vs. residential) with isolated panels, meters, and risers.
- Ambient and decorative lighting is coordinated with soffits and bulkheads to create a cohesive look in residential lobbies and amenity areas.
- Lighting in retail spaces is kept flexible for tenant customization.
  Result: Electrical systems are cleanly integrated across occupancies while allowing tenant adaptability.

PPD Exam Tips

You may be asked to:

- Identify proper locations for electrical rooms and panels based on program layout.
- Coordinate lighting fixture placement with structure or ceiling elements in plans and reflected ceiling plans (RCPs).
- Choose lighting strategies that align with daylighting, code compliance, and spatial function.

Know This: Understand how lighting types (e.g., direct/indirect, recessed, pendant) affect ceiling coordination. Be familiar with emergency lighting requirements and daylight zones per energy code.

Summary

Electrical and lighting integration is not only about function and compliance—it's about enhancing the design's intent, usability, and comfort. Thoughtful coordination across disciplines ensures lighting systems are well-integrated with architecture and structure while supporting energy goals and user needs. Mastery of this integration is key for both **exam success** and real-world design leadership.

4. **Plumbing System Layout**

   Effective plumbing system layout is critical to the function, code compliance, and long-term maintenance of a building. At the **Project Planning & Design (PPD)** level, you must demonstrate the ability to align plumbing systems—including water supply, drainage, venting, and risers—with the architectural plan in a way that is efficient, accessible, and serviceable across multiple floors.

Good plumbing layout reduces cost, improves energy and water efficiency, minimizes disruptions to other systems, and ensures occupant health and safety.

Key Plumbing Coordination Concepts

1) Wet Wall Alignment

- Grouping and **stacking wet spaces** (restrooms, kitchens, janitor closets, utility rooms) simplifies piping runs and reduces the number of vertical plumbing shafts.
- **Wet walls** should contain all necessary piping (supply, waste, vent) and should be located in interior partitions to protect from temperature extremes and freezing.
  Design Tip: Align wet walls vertically through the building whenever possible to reduce the need for offsets, elbows, and additional penetrations.

2) Riser Planning

- Plumbing **risers** are vertical pipes that carry supply water, waste, or vents from floor to floor. They are typically located in chases or shafts.
- For multi-story buildings, risers must be accessible via **pipe chases** or service shafts and must meet fire-rated enclosure requirements where applicable.
  Design Tip: Coordinate plumbing risers with mechanical and electrical shafts early to avoid spatial conflicts. Label plumbing zones clearly on floor plans and RCPs.

3) Drainage Path and Gravity Flow

- Plumbing systems depend on **gravity for drainage**, which affects slope, pipe diameter, and routing. Horizontal drain lines must slope appropriately (typically 1/4" per foot) toward vertical stacks or main lines.
- Long drainage runs with excessive offsets can lead to blockages, venting issues, or floor elevation changes.
  Design Tip: Minimize horizontal waste pipe runs by placing restrooms or kitchens near vertical stacks. Keep pipe slopes in mind when designing floor-to-floor heights.

4) Venting Requirements

- Every plumbing fixture requires venting to allow air into the drainage system and prevent siphoning of traps.
- Vent stacks must be coordinated with architecture to avoid unsightly roof penetrations and must be protected from backflow or water intrusion.
  Design Tip: Combine vent stacks where possible and coordinate with the roof plan to minimize visual clutter and potential weatherproofing issues.

5) Accessibility and Code Compliance

- Plumbing fixtures must comply with **ADA (or local accessibility codes)**. This affects fixture height, clear floor space, knee clearance, reach ranges, and controls.
- Restroom configurations must provide turning radii, grab bar locations, and accessible lavatories, all coordinated with plumbing connections.

Design Tip: Verify fixture locations and dimensions with code diagrams. Ensure trap and piping placements do not conflict with knee space or accessibility requirements.

6) Serviceability and Access

- Access panels must be provided for **valves, cleanouts, and major connections** in concealed spaces (e.g., above ceilings or behind walls).
- Avoid routing piping through occupied or high-finish spaces where future leaks or maintenance could cause disruption.
  Design Tip: Centralize plumbing systems near service corridors, janitor closets, or vertical shafts to enable quick access and minimize risk in sensitive areas.

Example 1: Mid-Rise Residential Building

- Bathrooms and kitchens in stacked units share a common plumbing wall.
- Plumbing risers are located in a central chase accessible from a corridor.
- Wet walls include recessed cleanouts and are framed to allow venting and insulation.
  Result: Reduced piping runs, easier maintenance access, and efficient use of floor-to-floor space.

Example 2: K-12 School

- Restrooms are placed back-to-back across a corridor to share plumbing infrastructure.
- Custodial sinks and drinking fountains are aligned with restroom plumbing.
- Roof drain lines and stormwater leaders are routed through designated chases to avoid cutting through structural members.
  Result: Efficient layout reduces cost, avoids structural conflicts, and simplifies future maintenance.

Example 3: Restaurant Fit-Out in a Commercial Shell

- Kitchen and bar plumbing are consolidated along a central service wall.
- Floor drains, grease traps, and backflow preventers are located in accessible mechanical closets.
- Plumbing layout is coordinated with equipment, refrigeration, and fire suppression systems.
  Result: Compact, code-compliant system supports food service operations and tenant flexibility.

PPD Exam Tips

You may be asked to:

- Identify optimal restroom or plumbing chase locations on a plan.
- Evaluate layouts for code compliance or maintenance concerns.
- Choose between design options based on stack alignment or drainage efficiency.

Know This:

- Minimum slope for horizontal drainage is typically **¼" per foot**.
- Fixtures must be vented and served by a minimum pipe size, often **1½"–4"** depending on fixture type.

- ADA-compliant toilets must have centerlines 16"–18" from adjacent walls, with grab bars and clearances maintained.

Summary
A thoughtful plumbing system layout is essential for a functional, code-compliant, and efficient building. Aligning wet walls, risers, and drainage paths early in the design process reduces conflicts, supports long-term maintenance, and ensures a smooth transition from design to construction. On the **PPD exam**, expect to demonstrate a strong understanding of plumbing coordination across multiple levels and occupancy types.

5. **Fire Protection System Coordination**
Fire protection system coordination is essential to ensure life safety, code compliance, and insurance requirements—without compromising spatial quality, aesthetics, or circulation. As an architect, your role at the **PPD level** is to plan how sprinkler systems, fire alarms, and fire-rated construction are integrated into the building layout in a way that supports both **function and form**.

Fire protection systems include **active systems** like sprinklers and alarms, and **passive systems** like fire-rated walls, doors, and floor assemblies. Coordination requires working closely with engineers, consultants, and code officials early in design.

Key Fire Protection Integration Concepts

1) **Sprinkler System Layout**
   - Sprinkler heads must be spaced and located according to **NFPA 13**, considering ceiling geometry, room type, and obstructions (beams, light fixtures, soffits).
   - Wet, dry, or pre-action systems may be required based on building use, height, climate, or hazard classification.
   - Sprinkler piping should be routed above ceilings or within chases to remain concealed in finished areas.
   Design Tip: In high-finish spaces, use **concealed or recessed sprinkler heads** that align with lighting and ceiling grids to preserve aesthetics.

2) **Fire-Rated Assemblies**
   - Building codes (IBC) require fire-rated separations between **different occupancies**, **exit paths**, **vertical shafts**, and **certain structural components**.
   - These assemblies include fire-rated walls, floors, ceilings, and doors—with continuity through concealed spaces and proper firestopping at penetrations.
   Design Tip: Align rated assemblies with structural grids and partitions to simplify detailing and material transitions. Clearly identify rated walls in construction documents to avoid conflicts.

3) **Vertical Shaft and Stairwell Protection**
   - Vertical elements like stairwells, elevators, and mechanical shafts require **2-hour fire-rated enclosures** in most multi-story buildings.
   - Penetrations for ducts, conduits, or piping must include **listed fire dampers, collars, or sleeves** to maintain the rating.

Design Tip: Consolidate vertical services and circulation paths within fire-rated zones to reduce the number of required protected assemblies.

4) **Fire Alarm and Detection Systems**
   - Systems include **smoke detectors, pull stations, strobes, and speakers**, all of which must be visible, accessible, and integrated into the space.
   - Placement is governed by NFPA 72 and ADA requirements, which affect mounting heights and spacing.
     Design Tip: Coordinate fire alarm device placement with finish materials and ceiling types. Use matching finishes or integrated cover plates in visually sensitive areas.

5) **Egress Path Protection**
   - Corridors, exit stairs, and areas of refuge may require fire-rated enclosures depending on the occupancy and travel distances.
   - Doors in rated corridors and stairs must be **self-closing**, fire-rated, and equipped with appropriate **hardware and vision panels**.
     Design Tip: Use glazing with rated framing in key locations (e.g., stair lobbies or corridors) to bring in daylight while preserving protection and visibility.

6) **Concealment and Aesthetics**
   - In exposed ceilings or architectural ceilings (e.g., wood slats, coffered ceilings), coordinate sprinkler heads and detectors so they are **flush-mounted, aligned**, or placed inconspicuously.
   - Minimize visual clutter by aligning sprinklers and fire alarm devices with lighting or ceiling modules.
     Design Tip: Early coordination with MEP consultants helps ensure sprinkler piping does not conflict with architectural elements like clerestories, skylights, or decorative ceilings.

Example 1: High-Rise Office Building

- A **fully sprinklered building** allows for larger open areas and relaxed fire-resistance requirements in certain zones (per IBC trade-offs).
- Fire-rated shafts enclose elevators and stairs, with fire-resistive glazing at stair landings to allow visibility.
- Sprinklers are aligned with ACT ceiling grid and coordinated with HVAC diffusers and lighting fixtures.
  Result: Life safety is maintained while preserving a clean, modern aesthetic in tenant spaces.

Example 2: Cultural Center with Exposed Ceilings

- Performance halls have **concealed sidewall sprinklers** to maintain architectural ceilings and acoustics.
- Fire alarm strobes and pull stations are placed at transitions or recessed into walls with coordinated finishes.
- Rated egress corridors are designed with clerestory glazing above the fire-rated wall line to bring in daylight without breaking code.

Result: Fire protection systems are subtly integrated, supporting both safety and design excellence.

Example 3: Mixed-Use Residential with Retail Podium

- Fire separation is provided between commercial and residential occupancies using 2-hour horizontal assemblies.
- Sprinklers and fire alarms are zoned by occupancy type and floor level, with tamper and flow switches monitored at the fire command center.
- Common areas feature decorative ceilings with carefully coordinated sprinkler head placement.
  Result: All code-mandated systems are included with minimal compromise to user experience or spatial clarity.

PPD Exam Tips

On the exam, you may be asked to:

- Identify locations where fire separation is required between occupancies or uses.
- Choose plan configurations that allow for proper egress protection and rated shaft enclosures.
- Evaluate the integration of fire protection systems in ceilings, walls, or circulation spaces.

Know This:

- Sprinkler head spacing and mounting depends on ceiling type, slope, and occupancy hazard (e.g., light hazard vs. ordinary hazard).
- Fire-resistance ratings must be maintained at all penetrations and junctions.
- Not all buildings require sprinklers—but when they do, the sprinkler system influences allowable floor area, egress path lengths, and fire-rated construction requirements.

Summary

Effective fire protection system coordination ensures that life safety measures are not only code-compliant, but also gracefully integrated into the building's architecture. Architects must consider both active and passive systems early in design to maintain aesthetics, preserve spatial intent, and meet code requirements. Mastery of these coordination principles is essential for passing the **PPD exam**—and for delivering responsible, beautiful buildings in practice.

6. **Structural and MEP Coordination**

One of the most critical aspects of successful building planning is the early and thoughtful coordination between the **structural system** and the **mechanical, electrical, and plumbing (MEP)** systems. Poor coordination can lead to cost overruns, construction delays, field conflicts, and even functional failures.

At the **PPD stage**, architects are expected to recognize potential clashes between structural and MEP elements and guide design decisions that facilitate integrated, buildable solutions. This involves aligning **ductwork, piping, conduit, cable trays, and equipment** with structural grids, beams, joists, and load-bearing elements.

Key Concepts in Structural and MEP Coordination

1) Understanding Structural Grids and Load Paths

- Structural grids define the **primary organization** of a building's load-bearing system, often using steel, concrete, or wood framing.
- Major MEP runs should be aligned with **grid lines or bays** to reduce the need for structure penetrations or awkward rerouting.
- Misalignment between MEP systems and grid spacing often leads to **clashes with beams or load-bearing walls**.

  Design Tip: Work within consistent bay sizes and structural module logic to support efficient duct routing and MEP zone distribution.

2) Duct Routing and Beam Conflicts

- Ductwork—especially large supply or return trunks—requires clear vertical space. Ducts often conflict with **beams, joists, or structural braces**, particularly in tight plenums or drop ceilings.
- To resolve this, consider **beam penetration strategies** (web openings), **raised structure**, or **alternative routing paths** like corridors or soffits.

  Design Tip: Use **open-web steel joists (OWSJ)** or castellated beams when feasible—they allow ducts to pass through the structure rather than under it, saving headroom and reducing plenum congestion.

3) Floor-to-Floor Height Considerations

- A building's floor-to-floor height is driven largely by the **depth of structure and MEP systems**. Insufficient space leads to lowered ceilings, system clashes, or design compromises.
- Structural systems such as **flat slabs or post-tensioned concrete** allow for tighter floor assemblies, freeing up vertical space for MEP.

  Design Tip: Coordinate structural and MEP sections early to test plenum depth and adjust building height accordingly. Use BIM models when available to assess spatial conflicts.

4) Shaft and Core Integration

- MEP risers (e.g., duct shafts, plumbing stacks, electrical risers) must pass vertically through multiple floors. Structural coordination is essential to allow for **penetrations through slabs**, **beam framing around shafts**, and **fire-rated enclosures**.
- Misaligned shafts create costly framing adjustments or call for mezzanines, offsets, or transfer systems.

  Design Tip: Group MEP risers into consolidated cores where possible, and align them with structural grids to simplify framing and vertical service distribution.

5) Penetrations Through Structural Members

- Routing pipes, ducts, or conduit through structural elements requires careful planning. Certain penetrations are acceptable (e.g., through beam webs), but must maintain the **integrity of the structural member**.
- Engineers may require **sleeves, knockouts, or reinforced openings**, especially in concrete structures.

Design Tip: Avoid penetrating beam flanges or post-tensioned concrete slabs unless absolutely necessary. Always coordinate location, size, and fireproofing of penetrations with structural engineers.

6) Equipment Loads and Vibration Isolation

- Heavy MEP equipment such as air handling units (AHUs), chillers, or generators must be supported structurally and often require **pads, mounts, or curbs** designed for point loads and vibration.
- Equipment-induced vibration may affect building performance (especially in hospitals, labs, or performing arts spaces), so **isolation mounts** and stiffening strategies must be coordinated.
  Design Tip: Locate large mechanical equipment over **columns or load-bearing walls** when possible, and isolate vibration-sensitive areas (e.g., recording studios) with floating floors or resilient structural designs.

Example 1: University Laboratory Building

- Deep steel beams were coordinated to run perpendicular to primary ductwork to allow cross-runs of exhaust and supply air.
- Mechanical penthouse sits atop the core zone to minimize duct runs and avoid long horizontal spans.
- Plumbing chases were pre-planned to align with structural bays, eliminating the need for structural slab cuts.
  Result: A highly serviceable, conflict-free system supports dense lab infrastructure within a tight floor-to-floor constraint.

Example 2: Corporate Office Tower

- OWSJ framing was selected to support open office spans and permit mechanical distribution in the ceiling zone.
- Electrical conduit and cable trays run parallel to structural framing and are mounted below beams where needed.
- Vertical shafts are centrally located within the core to minimize slab penetrations and improve service access.
  Result: Clean integration between structure and MEP systems allows for tenant flexibility and uncluttered ceilings.

Example 3: Performing Arts Center

- Displacement ventilation was routed through a raised floor system to avoid ceiling conflicts with trusses and rigging.
- Mechanical rooms and heavy air-handling units are located directly above structural transfer beams for support.
- Lighting catwalks and acoustical panels were carefully coordinated with duct runs and sprinkler systems to avoid interference.
  Result: All systems are fully integrated without visual intrusion or functional compromise.

PPD Exam Tips
You may be asked to:

- Evaluate plan or section views to detect potential structure–MEP conflicts.
- Recommend structural systems that support specific MEP layouts.
- Identify ceiling zones that allow full MEP system integration without reducing spatial quality.

Know This:

- Ducts often require **12–36 inches of plenum space**, depending on system size and routing complexity.
- Open-web joists or slab systems with dropped beams offer different advantages for MEP integration.
- Coordination between systems is easier when **bays are repetitive**, **floor heights are generous**, and **shaft locations are centralized**.

Summary
MEP and structural coordination is a foundational part of building design—and a key focus of the PPD exam. By aligning systems spatially, understanding material and framing constraints, and resolving potential conflicts in early design, you'll produce buildings that are not only efficient to construct but also adaptable, maintainable, and architecturally coherent.

7. **Envelope and System Performance Integration**
The building envelope is a critical interface between the interior and exterior environment. Its performance must support **thermal insulation, air and moisture control, vapor management, and system penetrations**—all while maintaining aesthetic goals and constructibility. As an architect in the **Project Planning & Design (PPD)** phase, you must coordinate envelope elements with mechanical systems and structural components to ensure **energy efficiency, durability, and occupant comfort**.

Effective envelope-system integration helps achieve code compliance (e.g., IECC), meet performance standards (e.g., LEED, Passive House), and minimize long-term operational costs.

Key Elements of Envelope-System Integration
1) Continuous Insulation and Thermal Bridging

- Insulation must be continuous across the envelope to maintain **thermal performance** and eliminate cold spots or **thermal bridges** that reduce energy efficiency and create condensation risks.
- Structural elements, slab edges, canopies, and parapets often interrupt insulation continuity.
  Design Tip: Use **thermal breaks** in structural attachments (e.g., balcony slabs, roof penetrations) and maintain insulation alignment across wall-roof-floor interfaces.

2) Vapor and Air Barrier Continuity

- The **vapor barrier** controls moisture diffusion, while the **air barrier** controls unintended air leakage. In cold climates, vapor barriers are often located inside the insulation; in hot/humid climates, they may be on the exterior side.
- Penetrations from mechanical systems (e.g., ducts, exhausts, conduits) can compromise these barriers if not properly detailed and sealed.
  Design Tip: Plan MEP penetrations early and specify **prefabricated sleeves, gaskets, or boots** to maintain barrier integrity. Coordinate with consultants to minimize unnecessary penetrations.

3) Mechanical Penetration Detailing

- HVAC systems often require penetrations for intakes, exhausts, refrigerant lines, and flues. These must be **flashed, sealed, and insulated** to maintain envelope integrity.
- Poor detailing can lead to **air leaks, thermal loss, and water intrusion**.
  Design Tip: Consolidate mechanical penetrations on **less visually sensitive façades** or rooftop areas. Ensure appropriate flashing and **backer rods + sealants** are specified.

4) HVAC System and Envelope Interdependence

- Envelope performance directly affects mechanical loads—**high-performance envelopes** reduce heating and cooling demand, potentially allowing for smaller HVAC systems.
- Systems such as **displacement ventilation**, **radiant heating**, or **ERVs** (Energy Recovery Ventilators) rely on a tight, well-insulated envelope to function effectively.
  Design Tip: Work with mechanical engineers to model energy loads based on envelope U-values, SHGCs, and air infiltration rates. Use performance specs, not just prescriptive tables.

5) Roof System Coordination

- Roofs often include significant HVAC equipment. Penetrations for ductwork, conduits, and piping must be flashed and insulated properly to avoid compromising **waterproofing and vapor control**.
- HVAC units on flat roofs should be placed on **curbs with appropriate slope and membrane detailing**.
  Design Tip: Minimize rooftop penetrations, or group them into shared curbed zones. Maintain **insulation and air barrier continuity** through parapets and over roof transitions.

6) Window and Curtain Wall Integration

- Windows and curtain walls must be integrated with surrounding walls for **thermal continuity, moisture management, and air barrier connection**.
- Mechanical louvers, PTACs, or HVAC terminal units in curtain walls require precise detailing to avoid failures at the envelope interface.
  Design Tip: Use **window head/jamb/sill details** that show exact transitions between insulation, flashing, and barriers. Ensure mechanical louvers include drip edges and sealed perimeters.

Example 1: Passive House–Inspired School Building

- Exterior insulation wraps the entire envelope, including foundation walls, with **thermal breaks at all penetrations**.
- ERVs and ground-source heat pumps are used with a high-performance envelope that limits heating/cooling demand.
- Penetrations for exhaust and intake are carefully sealed with grommets and tapes rated for airtightness.
  Result: The envelope enables ultra-low energy use while supporting indoor air quality and comfort.

Example 2: Mixed-Use Midrise with Curtain Wall

- HVAC terminal units are integrated below operable windows in residential units.
- Penetrations for refrigerant lines are routed through **preplanned cores** in floor slabs and sealed at the curtain wall interface.
- The vapor barrier transitions cleanly from opaque wall to glazing system using a **continuous sealant bead and flashing system**.
  Result: All systems are concealed, well-integrated, and energy-efficient while maintaining visual design goals.

Example 3: Commercial Office Building with Green Roof

- Rooftop HVAC units are elevated on curbs integrated into the tapered insulation layout of the green roof.
- Roof penetrations are coordinated with **membrane flashing and root barriers** to prevent water or root intrusion.
- Vapor barriers and air barriers continue unbroken beneath units and up parapets.
  Result: Durable, high-performance roof assembly supports mechanical systems and sustainable strategies.

PPD Exam Tips

You may be asked to:

- Identify detailing problems in wall sections (e.g., missing insulation at penetrations).
- Select an envelope strategy that best supports mechanical system performance.
- Recognize where the air or vapor barrier may be compromised by system layout.

Know This:

- **Continuous insulation (CI)** means insulation uninterrupted by studs, floor lines, or other elements.
- **Air barriers** must be continuous and durable—often membrane-based or fluid-applied.
- Infiltration significantly impacts HVAC sizing—**tight buildings use smaller, more efficient systems.**

Summary

Envelope and system performance integration is about more than sealing gaps—it's about designing a building that performs as a whole. Coordinating mechanical penetrations, maintaining thermal and moisture control, and detailing for constructibility all help ensure a durable, efficient, and high-performing design. On the **PPD exam**, your ability to recognize weak points and make holistic design decisions will be critical to your success.

8. System Impacts on Spatial and Programmatic Design

Building systems—mechanical, electrical, plumbing (MEP), structural, life safety, and acoustical—do not operate in isolation. Their selection and layout can significantly influence **room dimensions**, **floor-to-floor heights**, **finish choices**, **acoustical performance**, and **the character of interior spaces**. Architects must understand how these systems shape spatial planning and respond accordingly to deliver buildings that meet programmatic needs, code requirements, and user expectations.

At the **PPD level**, you are expected to demonstrate how integrated systems impact space usage and design logic—not just whether they "fit" in a section, but whether they **enhance or hinder** function, comfort, and aesthetics.

Key Areas Where Systems Influence Design

1) Room Sizes and Ceiling Heights

- Large ductwork, piping, and cable trays can require **increased ceiling plenums**, which in turn affects **floor-to-floor heights** and **overall building massing**.
- Equipment like mechanical units, fire suppression systems, and lighting fixtures also need vertical and horizontal clearance, often influencing room size and layout.
  Design Tip: Use corridor spaces or utility zones as distribution pathways for systems to preserve ceiling heights in primary program areas like classrooms, lobbies, or auditoriums.

2) Finishes and Material Selections

- MEP systems influence **interior finish durability** and compatibility. For example, areas with high humidity (labs, restrooms, kitchens) require moisture-resistant finishes.
- Fire protection systems may require rated walls or ceilings, impacting finish type and detailing.
  Design Tip: Choose ceiling and wall systems that allow for **easy access to concealed systems** while maintaining aesthetic goals—for example, use removable ACT panels in service zones, and drywall in public-facing areas.

3) Acoustics

- HVAC noise, plumbing flow, and structural vibration can affect acoustic comfort—critical in spaces like schools, healthcare facilities, or performance halls.
- Mechanical systems may require **duct lining, vibration isolators, or sound baffles** to control noise transmission between rooms.
  Design Tip: Coordinate MEP layouts with acoustical consultants to avoid placing noisy equipment above or adjacent to sensitive spaces (e.g., exam rooms, conference rooms).

4) Ceiling Systems and Lighting Design

- The size and layout of ductwork, sprinkler systems, and lighting fixtures influence the **type of ceiling system** used (ACT, open structure, stretched fabric, etc.).

- Integrating systems within decorative ceilings (e.g., wood slats or coffered ceilings) requires close coordination to **avoid visual clutter** or uneven lighting.
  Design Tip: Plan ceiling zones with clear hierarchies—use serviceable, flexible ceiling types in back-of-house areas, and refined ceiling finishes in high-visibility spaces.

5) **Programmatic Function and Adjacency**
   - Equipment noise, vibration, and maintenance access must be considered when locating mechanical rooms or utility shafts. Avoid placing them adjacent to noise-sensitive rooms unless proper mitigation is provided.
   - MEP systems can limit flexibility in space planning—for instance, plumbing stacks restrict where restrooms, kitchens, or labs can be placed.
     Design Tip: Group service-intensive functions (toilets, janitor closets, IT rooms) in central cores or stacking zones to limit impact on primary program spaces and maximize layout efficiency.

6) **User Comfort and Experience**
   - Systems affect air quality, temperature, lighting levels, and acoustics—each of which influences **user satisfaction, wellness, and productivity**.
   - Glare from poorly placed lighting or uncomfortable temperatures due to oversized ducts can undermine an otherwise elegant space.
     Design Tip: Work with MEP consultants to simulate lighting and HVAC performance during design development. Adjust layouts to reduce hot spots, glare, and poor airflow distribution.

Example 1: K–12 School

- Classrooms require **quiet HVAC** systems with concealed ducts and return grilles placed away from teaching zones.
- Corridor ceilings are lower and used as distribution zones for MEP services.
- Durable, low-maintenance finishes are chosen in areas with high foot traffic and cleaning demands.
  Result: A learning environment that supports focus, comfort, and safety through thoughtful integration of systems.

Example 2: Ambulatory Health Clinic

- Exam rooms require sound-rated partitions and sealed penetrations to ensure **speech privacy**.
- Ceiling heights are kept consistent to reduce institutional feel, while mechanical services are routed through back corridors and above procedure rooms.
- Systems are concealed behind resilient finishes, with access panels placed in staff-only areas.
  Result: A patient-centered facility where technical requirements are hidden, but performance is seamless.

Example 3: Corporate Headquarters

- Open office areas use **exposed structure ceilings** with suspended light fixtures and diffusers, requiring precise coordination of MEP systems.

- Conference rooms include integrated AV, lighting, and acoustics to support hybrid meetings.
- Finishes and lighting systems reinforce the brand's identity and workplace culture.
  Result: A high-performance, flexible, and visually cohesive environment shaped by system-smart design.

PPD Exam Tips

Expect questions asking you to:

- Choose ceiling systems that accommodate all necessary MEP services while preserving design intent.
- Analyze section cuts to evaluate whether ductwork or fire-rated assemblies affect headroom or layout.
- Determine the best room arrangement to accommodate system access while preserving acoustical separation.

Know This:

- HVAC return grilles should be placed away from doors or noise-sensitive areas.
- Wall ratings may require special finish considerations (e.g., no combustible finishes on rated walls).
- Plumbing stacks and mechanical chases often dictate core layouts and room adjacencies in multi-story buildings.

Summary

Systems shape the way space is experienced, used, and maintained. Good architects understand not only how to fit systems into their plans, but how to **leverage them to enhance performance, comfort, and spatial quality**. For the PPD exam, you must demonstrate your ability to recognize system impacts on spatial organization and respond with integrated design solutions that balance function and aesthetics.

C. Integrate program requirements into a project design (A/E) (6-10%)

NCARB Requirements

Architects must be able to reconcile the relationships between various program requirements, evaluate the program's impact on cost and building systems, and assess how the program is affected by environmental factors. It's important that you are able to create a building layout that incorporates program requirements for a project.

In this section, architects must demonstrate the ability to:

- Translate and reconcile **varied and sometimes conflicting programmatic requirements**.
- Understand how programming impacts **cost**, **building systems**, **sustainability goals**, and **spatial organization**.
- Create a project layout that **logically supports the client's goals**, environmental context, and regulatory requirements.

The Project Planning & Design (PPD) exam will test your ability to **critically analyze a program**, evaluate **trade-offs**, and implement a **design solution that balances all constraints**—functionally, technically, and aesthetically.

1. **Prioritize Functional Relationships**
A successful architectural layout begins with understanding the **functional relationships** between program elements. The goal is to create a logical, efficient arrangement that enhances usability, circulation, and the overall performance of the building. Poor adjacencies can create conflicts, inefficiencies, or safety concerns, while well-planned relationships support intuitive navigation, zoning, and operational success.

Start with Diagrams
Use graphic tools during early design phases to visualize and test spatial relationships:
- **Bubble diagrams** to illustrate conceptual adjacencies.
- **Adjacency matrixes** to score and prioritize spatial relationships (e.g., must-have, should-have, avoid).
- **Zoning diagrams** to map out vertical and horizontal program zones.

These tools allow you to organize spaces before committing to walls or dimensions.

Key Strategies for Prioritizing Relationships
- **Group public, private, and service zones** to streamline circulation and improve security.
- **Separate incompatible uses**, such as:
  - Loud vs. quiet zones (e.g., mechanical room vs. meditation room)
  - Wet vs. dry zones (e.g., locker room vs. archives)
  - High-traffic vs. private spaces (e.g., lobby vs. executive suite)
- **Ensure logical flow**, especially for sequential processes:
  - **Back-of-house logistics:** loading → storage → kitchen → dining
  - **Healthcare:** check-in → waiting → exam → procedure → recovery
  - **Education:** entry → administration → classrooms → gymnasium

Support Visual and Physical Connections
- Use **visual transparency** to support wayfinding or supervision, particularly in schools, offices, or retail.
- Maintain **short travel distances** between high-use or related spaces.
- Ensure **controlled access** where security or privacy is needed (e.g., between public lobbies and restricted lab spaces).

Example: Community Center
In a community center, the program might include a gymnasium, locker rooms, classrooms, a childcare center, offices, and a multipurpose event space. You would likely:
- Place the **gymnasium adjacent to locker rooms** and close to **parking** for evening and weekend access.
- Position the **childcare center** near the main entrance, but **away from noisy areas**.
- Separate **meeting rooms and offices** from high-activity zones to ensure quiet and privacy.

- Connect **event spaces** to both service areas (kitchen, storage) and public zones for flexible use.

PPD Exam Relevance
The exam may test your understanding of:
- Functional zoning diagrams
- Space adjacencies based on program descriptions
- Identifying problematic layouts
- Recognizing logical sequencing of space based on a given use case

You may be asked to evaluate layouts for efficiency, identify missing adjacencies, or determine which of several diagrams best reflects the client's operational needs.

Summary: Prioritizing functional relationships means solving adjacency puzzles in a way that supports the building's use, improves circulation, and enhances occupant comfort. This forms the backbone of architectural planning, particularly in complex programs like healthcare, education, or civic buildings.

2. **Assess Impact on Building Systems**
Every programmatic decision you make—how much space is needed, what it's used for, how it relates to other spaces—will directly influence the design and performance of the building's systems. At the PPD level, you're expected to anticipate how programming affects **mechanical, electrical, plumbing (MEP)** systems, **structural systems**, and **material selections**—and to design accordingly.

As you plan, consider how **intensity of use**, **environmental loads**, and **system compatibility** vary by program type. The program drives system sizing, zoning, floor-to-floor heights, and even how materials perform under long-term use conditions.

Key Considerations
- **MEP Load Intensity Varies by Space Function:**
  - Kitchens, laboratories, locker rooms, and medical suites generate high heat, moisture, and odor loads.
  - These spaces require **specialized ventilation, exhaust systems**, and **increased water supply and drainage**.
  - The mechanical systems serving them are often **separately zoned**, and located near service areas for maintenance access.
- **Structural Grids Respond to Program Density:**
  - Open-plan spaces (like modern offices or exhibit halls) often demand **larger structural spans**, allowing unobstructed layouts.
  - Conversely, programs with repetitive partitions (e.g., hotels, hospitals) may benefit from **tighter column grids** to align with room modules and reduce structure-to-wall conflicts.
  - Heavy-load areas (e.g., archives, equipment rooms) require **increased floor loading capacity**, which may affect slab thickness or material selection.

- **Environmental Conditions Affect System Design:**
  - Programs with high temperature or humidity sensitivity (e.g., archives, data centers) will require **specialized HVAC control**.
  - Some uses (like natatoriums or animal care facilities) require corrosion-resistant materials and robust exhaust strategies.
  - The **durability and resilience** of finishes, assemblies, and even mechanical equipment placement are impacted by program demands.

Example: Healthcare Facility
A healthcare facility typically includes exam rooms, procedure suites, waiting areas, administrative offices, and mechanical rooms. This program:

- Requires **increased ventilation rates**, often between **6–12 ACH (Air Changes per Hour)** in treatment areas, compared to 2–4 in office spaces.
- May require **HEPA filtration**, separate exhaust for isolation rooms, and **pressurization zoning** (e.g., positive pressure in clean areas, negative pressure in isolation rooms).
- Impacts the structural layout, as heavy equipment like imaging machines or surgical lights may need **increased structural support** or vibration isolation.
- Increases the **floor-to-floor height** to accommodate large ductwork, medical gas piping, and technology infrastructure—all while preserving clear ceiling heights in patient and procedure areas.

Design Strategies

- **Plan for system zones** early in schematic design. Knowing where high-load or high-maintenance areas are will guide your mechanical and structural layout.
- **Coordinate structure with program use**: a tighter grid works well for repeated rooms; a wider span is ideal for flexible or open layouts.
- **Allocate space for service areas** like mechanical rooms, chases, or shaft walls, especially for high-intensity uses.
- **Avoid retrofit traps** by verifying that system capacities meet or exceed the operational requirements of each program zone.

PPD Exam Tips
You may be asked to:

- Choose an appropriate structural grid or mechanical zoning strategy for a given program.
- Identify design conflicts between systems and spatial needs.
- Understand how a program impacts system layout, HVAC zoning, or structural module.
- Evaluate plans or sections for system feasibility (e.g., adequate ceiling space for ducts in a lab wing).

Summary
Programmatic decisions have a ripple effect on all building systems. As you translate program data into design, be aware of how space use influences the size, zoning, and layout of mechanical, electrical, plumbing, and structural systems. Thoughtful coordination early in the design process avoids costly revisions and ensures that performance goals are met without compromising spatial quality.

3. Coordinate Program with Environmental Conditions

A thoughtful architectural design considers how **site-specific environmental conditions**—sun path, wind patterns, views, topography, and climate—can enhance or hinder the performance and comfort of programmed spaces. Proper environmental coordination leads to buildings that are more energy efficient, more enjoyable to use, and better integrated with their surroundings.

The PPD exam expects candidates to demonstrate the ability to **match interior functions with exterior opportunities or constraints**—and to do so strategically to support both **passive design principles** and **user comfort**.

Key Strategies for Environmental Coordination

- **Maximize Daylighting for Occupied Spaces:**
  - Orient **regularly occupied rooms** (offices, classrooms, reading areas) toward **south or east façades** to capitalize on consistent daylight and reduce the need for artificial lighting.
  - Use **north-facing orientation** for art studios, laboratories, or workspaces that benefit from diffused light and minimal glare.
  - Integrate **light shelves**, **overhangs**, and **glazing selection** to control solar gain and distribute light effectively.
- **Use Less-Desirable Zones for Support Spaces:**
  - Allocate **mechanical rooms**, **electrical closets**, **storage areas**, and **janitorial spaces** to **areas with no views**, poor solar orientation, or minimal access needs.
  - This protects prime real estate for user-facing functions and avoids waste of valuable natural light and views.
- **Respond to Topography:**
  - On sloped sites, locate **basement-level or service-intensive functions** (e.g., mechanical spaces, loading docks, storage) at the **lower grade** to minimize excavation and improve delivery access.
  - Take advantage of stepped sections to create **split-level configurations**, sunken courtyards, or green roofs with occupiable space above service areas.
- **Orient Based on Climate:**
  - In hot climates, **minimize west-facing glass** to reduce afternoon heat gain.
  - In colder climates, **maximize southern exposure** to capture winter solar radiation and reduce heating loads.
  - Consider **prevailing wind direction** for natural ventilation strategies—program courtyards and operable windows accordingly.

Example: Library Design

In a library program:

- **Reading rooms** and **study areas** should face **east** to provide gentle morning light that supports quiet, focused activity and reduces afternoon glare.
- **Stacks**—which require less daylight and may benefit from controlled humidity—can be located in the **building's interior core** or **north-facing walls**.
- **Mechanical rooms**, staff service zones, and book processing areas can be placed in **basement levels** or shaded zones to keep prime daylight access for patrons.

This not only improves energy performance and occupant well-being, but also aligns the architecture with the surrounding site context.

Design Integration Tips

- Use **site analysis diagrams** early in design to inform layout decisions.
- Integrate **passive solar strategies**: thermal mass near south-facing windows, buffer zones (like corridors) on the west side, and shaded courtyards in hot climates.
- Employ **landscape features** (e.g., berms, tree canopies, green walls) to screen undesirable environmental conditions like wind or noise.

PPD Exam Relevance

You may be asked to:

- Evaluate if a proposed plan layout is responsive to site conditions.
- Identify optimal orientations for specific program elements based on solar or wind data.
- Recommend program adjustments to improve energy efficiency or user comfort.

Look out for prompts that mention **sun paths, prevailing winds, or site slope**—they're cues to adjust program placement for optimal performance.

Summary

Coordinating your program with environmental conditions ensures that your building design is not just functional but also climate-responsive and sustainable. By aligning key program elements with solar access, topography, and wind exposure, you can enhance energy performance, occupant well-being, and long-term operational success—core goals of integrated architectural planning and design.

4. **Translate Area Requirements into Scalable Layouts**

Understanding how individual **programmatic space requirements** translate into a comprehensive building layout is essential to effective architectural planning. Each room or function is typically defined by its **Net Square Footage (NSF)**, which reflects the usable area needed for that function—but the total building size must also include additional space for circulation, structure, building systems, and walls. This expanded area is known as **Gross Square Footage (GSF)**.

At the PPD level, architects are expected to synthesize program data into a **scalable, efficient layout**, while applying appropriate **net-to-gross ratios** and planning for realistic building massing and organization.

Key Concepts and Strategies

- **Understand Net vs. Gross Square Footage:**
  - **Net Square Footage (NSF):** The actual usable space assigned to a specific function (e.g., a classroom, exam room, or office).
  - **Gross Square Footage (GSF):** The total building area, including NSF **plus circulation, mechanical spaces, structural systems, partitions, and other non-assignable areas**.
- **Apply Appropriate Net-to-Gross Ratios:**
  - Net-to-gross ratios vary significantly by building type. For example:

        - **Office Buildings:** 75%–85% net efficiency
        - **Hospitals or Labs:** 60%–70% net efficiency (due to corridors, MEP spaces, and specialized zones)
        - **Schools:** Around 68%–75% efficiency depending on layout complexity
    - Use realistic ratios in early planning to avoid underestimating the building size and cost.
- **Stack and Repeat Efficiently:**
    - Stack **repetitive modular spaces** (e.g., classrooms, hotel rooms, patient rooms) vertically to simplify structural, plumbing, and MEP layouts.
    - This reduces material waste, shortens duct/plumbing runs, and creates uniform floor plates that are easier to build and adapt in the future.
- **Plan for Non-Assignable Space Early:**
    - Include space for:
        - **Circulation corridors**
        - **Elevator and stair cores**
        - **Mechanical/electrical rooms**
        - **Toilet groups and shafts**
        - **Structure (columns, shear walls, etc.)**
    - These spaces can consume **25%–40%** of the total building area, especially in complex or high-rise projects.

Example: Classroom Design in a School

Suppose a school program calls for **900 NSF per classroom**. However, when accounting for:

- Corridors and circulation
- Wall thicknesses
- Structural grid spacing
- Mechanical chases and service areas

…the **total gross area per classroom** might increase to approximately **1,200 GSF**. That's a **grossing factor of 1.33**, or **75% net efficiency**.

Scaling this to 20 classrooms:

- **NSF:** 900 × 20 = 18,000 NSF
- **GSF:** 1,200 × 20 = 24,000 GSF
  This additional 6,000 square feet must be planned for in the overall site massing, budget, and systems coordination.

Design Integration Tips

- Use **program area spreadsheets** that include both NSF and calculated GSF to inform early massing models.
- Adjust net-to-gross assumptions as the design evolves and spatial efficiency improves or declines.
- Consider circulation logic when determining grossing factors—double-loaded corridors are typically more efficient than single-loaded ones.
- Use **modular planning** to simplify stacking and reduce variability in room sizes across floors.

PPD Exam Relevance

You may be asked to:

- Estimate the total building area based on program requirements and efficiency assumptions.
- Compare layouts for spatial efficiency or planning logic.
- Identify whether program elements are missing key support spaces (e.g., restrooms, circulation, mechanical).
- Evaluate if a design meets both functional and dimensional program requirements.

Expect to encounter **scenario-based questions** that ask you to synthesize planning data, translate it to building scale, or assess the viability of a proposed space plan.

Summary

Translating program area requirements into a workable building layout involves more than just stacking spaces together—it requires a deep understanding of how circulation, systems, and structure interact. For the PPD exam and real-world projects, mastering net-to-gross planning ensures your designs are not only functional but also buildable, code-compliant, and budget-conscious.

5. **Balance Flexibility with Specificity**

One of the central challenges in architectural programming and design is striking the right balance between spaces that require **specific configurations** and those that benefit from **long-term adaptability**. Buildings are dynamic—users, technologies, and organizations evolve over time. While some program elements must meet precise dimensional and operational needs, others can and should be designed for **future flexibility**.

The PPD exam evaluates your ability to identify where design specificity is critical and where flexibility should be prioritized—both in the layout of spaces and the selection of systems.

Key Design Strategies

- **Incorporate Flexible Planning Concepts:**
  - Use **open plans** and **clear structural spans** to avoid rigid layouts and support future changes.
  - Design **neutral or multi-use spaces** that can shift function without significant renovation (e.g., converting a breakout space into meeting rooms).
  - Include **swing spaces** for short-term reconfiguration or future growth.
- **Support Flexibility with Systems and Infrastructure:**
  - Plan **adaptable MEP zoning** to serve areas that may change function (e.g., using underfloor air distribution in open office spaces).
  - Allow for **horizontal and vertical expansion** by designing core systems, structure, and egress to accommodate future floors or additions.
  - Use **movable or demountable partitions** and **modular furniture systems** in areas of evolving use.
- **Provide Specificity Where Performance is Critical:**
  - Highly technical or specialized rooms (e.g., commercial kitchens, operating rooms, laboratories) require tightly controlled conditions and should be **precisely sized and located**.

    - Consider code-required dimensions for accessibility, egress, and equipment clearances that cannot change over time.
    - Use precise coordination in **vertical service areas** (e.g., shafts, elevator cores) where relocation is cost-prohibitive.
- **Design Shared and Multi-Use Spaces:**
    - Oversize **corridors**, **lobbies**, or **waiting areas** to accommodate informal gathering or future programming shifts.
    - Create **shared breakout zones** between program areas that can flex to accommodate growth or seasonal demand.

Example: Flexible Office for a Tech Company
A tech company's needs can shift rapidly as teams grow, shrink, or reorganize. A well-designed office might include:

- **Open-plan work areas** with modular desks and minimal fixed partitions.
- **Demountable glass walls** for conference rooms that can expand or contract.
- **Shared lounges or breakout spaces** that can serve as meeting zones, event spaces, or team collaboration areas.
- A **core infrastructure** (e.g., raised flooring, accessible ceiling plenums) that allows for quick rerouting of power and data.

This approach gives the company room to evolve while minimizing renovation costs and disruption.

PPD Exam Relevance
Expect questions that test your ability to:

- Identify which spaces should be fixed versus flexible.
- Evaluate design layouts for adaptability.
- Recommend planning strategies for long-term use and operational change.
- Apply knowledge of structural and MEP systems to support modular or expandable space design.

You may also be asked to recognize when specificity is necessary due to **life safety**, **code**, or **technical equipment constraints**, versus when flexibility can add long-term value.

Summary
Balancing flexibility with specificity is about understanding both **current program needs** and **future unknowns**. A well-designed building is resilient—able to evolve without expensive reconstruction. As an architect, you must know where to lock in dimensions and systems, and where to leave space and infrastructure for growth, reprogramming, and change. This skill is especially important in institutional, educational, and commercial buildings where adaptability is key to long-term success.

6. **Align Program with Budget Constraints**
Not all program spaces are created equal when it comes to cost. Certain functions carry significantly higher construction and operational costs due to **system complexity**, **finish requirements**, **equipment needs**, or **specialized code compliance**. As an architect, part of

your responsibility during the planning and design phases is to **balance program needs with financial realities**, ensuring that the design vision is delivered within budgetary limits.

The PPD exam will assess your ability to recognize **cost drivers**, make trade-offs between program intensity and resource availability, and apply strategies to control construction and lifecycle costs while maintaining functional integrity.

High-Cost Program Elements
Certain types of spaces significantly increase project cost due to their infrastructure needs, specialized finishes, or high structural demands. Examples include:

- **Laboratories, healthcare suites, and imaging rooms:**
    - Require high HVAC zoning, medical gas systems, advanced MEP coordination, and strict environmental controls.
- **Commercial kitchens and clean rooms:**
    - Demand code-compliant finishes (e.g., stainless steel, seamless floors), complex ventilation systems (e.g., hoods, grease ducts, fire suppression), and specialized plumbing and drainage.
- **Large-span or high-volume assembly areas (e.g., gyms, theaters, auditoriums):**
    - Require robust structural systems (long-span beams, trusses), advanced acoustic treatments, and high-capacity HVAC systems to maintain comfort in large volumes.
- **Security-intensive or technology-rich areas (e.g., control rooms, data centers, detention facilities):**
    - Feature high-density infrastructure and redundancy, often exceeding typical cost per square foot.

Cost-Conscious Design Strategies
To keep the overall project budget within bounds, use the following approaches when integrating program into your design:

- **Group High-Cost Spaces Strategically:**
    - Locate expensive spaces **close to existing or centralized utilities** to reduce distribution costs.
    - Avoid scattering specialized program spaces across the building—this increases MEP complexity and infrastructure redundancy.
- **Offset Expensive Spaces with Simpler Areas:**
    - Balance the budget by pairing **cost-intensive areas** with **lower-cost functions** such as:
        - Open-plan workspaces instead of private offices
        - Multipurpose rooms instead of multiple single-use rooms
        - Standardized classroom layouts instead of custom-designed learning environments
- **Design for Efficiency and Modularity:**
    - Use **modular layouts** for repetitive programs to reduce construction variability.
    - Avoid excessive customization where it doesn't add functional or aesthetic value.
- **Consider Phasing or Shelling:**
    - For budget-limited projects, consider **shelling out future high-cost program areas** (e.g., a lab wing or theater) for later completion.

 - Prioritize essential functions in **Phase 1** while deferring less critical elements until funding is secured.
- **Select Alternate Construction Methods:**
  - Use **prefabrication**, **modular construction**, or **standardized detailing** to reduce cost in large or repetitive program areas.

Example: Civic Building Trade-Offs
In a civic center project:
- The design team prioritizes **public-facing areas** like the **entry lobby**, **community meeting hall**, and **event space**, allocating more budget to durable, high-end finishes and architectural detailing in these zones.
- To maintain the overall project budget, **back-of-house areas** (e.g., staff break rooms, storage, utility spaces) are specified with cost-effective materials, minimal detailing, and standard systems.
- A future **conference wing** is planned as a **shell space**, allowing for future build-out when additional funding becomes available.

PPD Exam Relevance
On the PPD exam, you may be asked to:
- Identify cost-intensive program spaces within a scenario.
- Evaluate design layouts for cost efficiency.
- Recommend layout or program changes to align with budget constraints.
- Recognize where phasing or system simplification is appropriate.

Understanding **cost drivers** and making logical, informed decisions around program placement and system selection is essential.

Summary
Aligning the architectural program with budget constraints requires strategic thinking and realistic prioritization. Not every space can be treated equally—some require significant investment, while others can be designed efficiently. Successful architects manage these trade-offs while maintaining design intent, functional integrity, and code compliance. Balancing these priorities is central to your role—and a critical part of your PPD exam preparation.

7. **Integrate Code and Accessibility into the Program**
Incorporating **building code** and **accessibility compliance** from the outset of program integration is not optional—it is essential. As you develop layouts and spatial relationships, program elements must comply with the **International Building Code (IBC)** and the **Americans with Disabilities Act (ADA)** or applicable local accessibility regulations.
Failure to address these requirements early can result in costly redesigns, user discomfort, or even legal liabilities. The PPD exam expects you to recognize how code and accessibility standards shape layout decisions and core building organization.

Key Code & Accessibility Considerations
- **Restroom Quantity and Size:**
  - Determine the required number of plumbing fixtures based on **occupancy classification**, **occupant load**, and **use group** (refer to IBC Chapter 29).

  - Provide **accessible fixtures** per ADA and IBC Section 1109—typically at least one accessible toilet and lavatory per restroom group.
  - Factor in **gender-neutral** or **family restrooms**, where applicable, especially in educational or healthcare settings.
- **Accessible Routes:**
  - All **program elements must be served by an accessible route**, including circulation paths, doorways, corridors, and room entrances.
  - Routes must comply with **maximum slopes (1:20)**, **clear width (minimum 36")**, and include required **maneuvering clearances**.
  - **Changes in level** must be addressed via **elevators**, **platform lifts**, or **ADA-compliant ramps**—stairs alone are not sufficient.
- **Egress and Vertical Circulation:**
  - Comply with **egress path requirements** including width, travel distance, and door swing (IBC Chapter 10).
  - **Fire-rated stairs**, **area of refuge**, and **exit separation distances** must be coordinated with program zoning.
  - Elevators are required in multi-story buildings serving **public or assembly functions** unless exempted by IBC/ADA (e.g., small multi-level buildings under specific thresholds).
- **Accessible Seating, Counters, and Service Areas:**
  - Provide accessible options at **dining tables**, **transaction counters**, **reception desks**, and **assembly seating**.
  - Ensure visibility and reach ranges for signage, controls, and alarm systems (per ADAAG).

Example: Two-Story Daycare Facility
In a two-story daycare center:
- The **presence of children with mobility impairments** requires full accessibility to both levels.
- An **elevator or ADA-compliant ramp** must be included in the core to allow full access to classrooms, restrooms, and play areas on both floors.
- The need for **diaper changing facilities**, child-height fixtures, and compliant restroom layouts must also be factored in.
- Corridors must provide **sufficient turning radii** for strollers and wheelchairs, while egress stairwells must meet fire-rated enclosure requirements with appropriate landings and handrails.

This affects the location of vertical circulation cores, restroom placement, and mechanical space allocation, all of which must be coordinated in the early design stages.

Design Integration Tips
- Use **occupancy calculations** early to size restrooms and egress systems appropriately.
- Develop a **compliance checklist** for program-specific ADA requirements.
- Model **accessible paths of travel** in your schematic plans and ensure full coverage across all functional zones.
- Coordinate closely with structural and MEP teams to place elevators, ramps, and stair cores efficiently.

- Review IBC and ADA guidelines for **special occupancy types** (e.g., daycare, healthcare, assembly, educational).

PPD Exam Relevance

On the exam, you may be asked to:

- Identify missing accessibility features in a floor plan.
- Determine whether egress and restroom counts meet code for a given occupancy.
- Propose changes to layouts that improve accessibility or life safety compliance.
- Interpret relevant tables and provisions from IBC Chapters 10 (Means of Egress), 11 (Accessibility), and 29 (Plumbing).

Expect scenario-based questions that blend **programmatic layout decisions** with **code interpretation**.

You'll be asked to **interpret program data**, **identify layout or system conflicts**, and **resolve them through sound design judgment**. Expect to work through scenarios that demand not just knowledge, but the ability to apply that knowledge across disciplines—architecture, structure, MEP, code, and sustainability.

PPD Exam Tips

☑ **Expect to:**

- **Analyze program-to-plan test questions** where you're asked to identify **inconsistencies or omissions**, such as:
  - Missing required adjacencies (e.g., restrooms too far from public spaces)
  - Improper access or circulation paths (e.g., no accessible route between related program areas)
- **Assess how program changes impact system and layout decisions**:
  - Example: Adding a **wet lab** to an upper floor may require additional plumbing risers, dedicated exhaust systems, reinforced structure, and zoning adjustments.
- **Make informed trade-offs** between:
  - **Area**: balancing net-to-gross ratios
  - **Cost**: prioritizing budget-critical functions
  - **Performance**: ensuring spaces meet functional and environmental expectations

💡 **Know This:**

- **Net area ≠ usable area**:
  - Always account for **circulation, structure, shafts, and mechanical systems**.
  - Grossing factors vary by building type—apply realistic assumptions based on the program and use.
- **Stacking diagrams are essential** for efficient planning in multi-story buildings:
  - They help visualize **vertical adjacencies**, simplify MEP routing, and support modular layouts.
- **Program affects every system**:
  - HVAC zoning is impacted by room function and occupancy.
  - Structural grids must support program-driven wall and furniture layouts.
  - Circulation logic is shaped by user flows dictated by the program.

  - Even **envelope design** (e.g., window placement, shading) may be dictated by space use inside.

Summary
Successful program integration requires you to embed code and accessibility requirements into the design from the beginning—not as an afterthought. Restrooms, corridors, vertical circulation, and access routes all impact the spatial organization of your building. A well-informed architect ensures that **functional intent** aligns seamlessly with **legal and ethical obligations**, producing spaces that are inclusive, safe, and fully code-compliant.

Successfully integrating program requirements into your design goes far beyond checking boxes—it involves synthesizing **functional goals, spatial relationships, technical constraints, and long-term flexibility** into a cohesive architectural response.

You must deliver a solution that:
- Supports the **client's mission and organizational needs**
- **Fits the site context**—topographically, environmentally, and socially
- Meets **code, accessibility, and budgetary limitations**
- Prepares the building for **future adaptability**

D. Integrate environmental and contextual conditions in the project design (A/E) (3-7%)

NCARB Requirements
You will need to incorporate various design strategies, including those assessed in Section 1, into the project. This may include elements like view corridors, shading elements, building materials, and landscape design.

In this domain, candidates must demonstrate the ability to integrate a variety of environmental, climatic, and contextual factors into their architectural solutions. These include physical site conditions, solar orientation, local climate, surrounding structures, landscape elements, and cultural context. These strategies must support both **building performance** and **architectural coherence**.

This section builds upon concepts from **Environmental & Contextual Conditions** and tests your ability to **synthesize those ideas into a holistic design**.

1. Respond to Solar Orientation
Understanding the **solar path** is essential in creating buildings that maximize comfort and energy efficiency. Solar orientation directly influences **thermal gain**, **natural lighting**, and **glare control**—all of which impact both the user experience and long-term building performance.

Design strategies should be tailored to the **building's latitude, climate zone**, and **use type**, ensuring spaces are well-lit and thermally stable without relying excessively on mechanical systems.

Key Strategies:

- **Orient major program spaces**—such as offices, classrooms, or living areas—toward **east or north** for stable daylight and reduced solar heat gain.
- **South-facing glazing** can be highly effective in **temperate or cold climates**, where low-angle winter sun can be welcomed with proper shading.
- **Use architectural shading devices** to respond to solar angles:
  - **Horizontal overhangs** on south façades (best for blocking high summer sun)
  - **Vertical fins** on east and west façades (for low-angle morning/evening sun)
  - **Light shelves** to bounce daylight deeper into interior spaces
- **Avoid unshaded west-facing windows**, which receive intense, low-angle afternoon sun that contributes to overheating and discomfort—especially in warmer climates.

Example: Desert Climate Strategy
In a hot-arid desert climate (e.g., Phoenix, AZ), solar control is critical:

- A building might include **recessed windows** with deep horizontal overhangs on the south elevation.
- West façades are kept solid or are protected with **vertical fins** or **architectural screens** to reduce heat gain.
- North façades are prioritized for diffuse daylight in offices and classrooms, minimizing reliance on electric lighting.

PPD Exam Connection:
You may be asked to:

- Adjust a building's layout or massing to optimize solar orientation
- Identify poor solar design choices (e.g., unshaded glazing in hot climates)
- Recommend shading strategies for specific façade orientations
- Apply sun-path logic to a given site plan or elevation

Takeaway:
Responding to solar orientation isn't just about comfort—it's about integrating **environmental intelligence** into your design from the beginning. Strategic orientation and shading reduce HVAC loads, improve indoor environmental quality, and demonstrate a core competency in sustainable, climate-responsive design—a key focus of the PPD exam.

2. **Optimize Views and Visual Connectivity**
An effective architectural configuration thoughtfully integrates **views** to enhance occupant experience, promote well-being, and connect the building to its context. Optimizing views means **highlighting positive visual connections**—such as natural landscapes, cityscapes, or public spaces—while **minimizing exposure to undesirable or distracting elements**, such as service areas, parking lots, or noisy streets.

Key Strategies:

- **Place frequently occupied or social spaces**—like offices, lobbies, lounges, or break rooms—toward windows with desirable views. These spaces benefit most from daylight and visual connection.

- **Frame views with glazing** that balances transparency with solar and thermal control. Consider the use of **low-e coatings, shading devices, or fritted glass** to manage glare and heat gain without sacrificing outlook.
- **Manage privacy and sightlines** by orienting spaces and windows away from neighboring buildings or public thoroughfares where privacy is a concern.
- **Analyze building heights and setbacks** to ensure views aren't obstructed by adjacent structures or mature vegetation.
- **Incorporate landscape features** like courtyards, gardens, or water elements as focal points visible from inside.
- Use architectural elements such as **screening, louvers, or vegetation** to mask less desirable views or create visual buffers.

Example: Coastal Hotel
A coastal hotel might **angle its guest room wings** to maximize ocean views for as many rooms as possible, creating a fan shape that opens toward the water. Balconies could be designed with **wind screens or recessed terraces** to protect guests from strong prevailing winds, while still maintaining open sightlines. Public areas such as the lobby or dining spaces might be placed where panoramic views can enhance the ambiance and marketability.

PPD Exam Connection:
Exam questions may require you to:
- Identify the best placement of program spaces based on view and privacy considerations.
- Recommend glazing strategies that balance views with solar and glare control.
- Analyze site or massing diagrams to optimize or protect important sightlines.
- Suggest mitigation methods for undesirable visual conditions.

Takeaway:
Optimizing views and visual connectivity is a critical design consideration that bridges **user comfort, environmental responsiveness, and site integration**. On the PPD exam, demonstrating your ability to thoughtfully arrange spaces to capitalize on views while controlling privacy and environmental factors is key to showing your comprehensive design understanding.

3. **Respond to Topography and Site Features**
Successful architectural design embraces the **unique physical characteristics of a site**, using topography, existing vegetation, drainage patterns, and soil conditions to inform the building's configuration and placement. Responding thoughtfully to these elements can reduce environmental impact, lower construction costs, and create a harmonious relationship between the building and its landscape.

Key Strategies:
- **Adapt building forms to sloped or uneven terrain** by employing **split-level designs, stepped floor plates, or terraced foundations**. This minimizes the need for extensive excavation, reduces soil disturbance, and preserves natural drainage patterns.

- **Align the building along site contours** rather than against them to limit grading and earthwork. This approach helps maintain the natural hydrology of the site, reducing erosion risks and stormwater runoff issues.
- **Preserve existing vegetation, especially mature trees and native plants**, which serve as critical buffers for erosion control, stormwater absorption, and microclimate regulation. Vegetation can also enhance privacy and contribute to a building's aesthetic integration into the landscape.
- **Analyze geotechnical conditions** to avoid unstable soils, landslide-prone areas, or flood zones. The structural design and foundation type may need to adjust accordingly.
- Use site features creatively as **natural shading, windbreaks, or viewsheds**, capitalizing on microclimates and enhancing energy efficiency.

Examples:

- A **hillside library** might be designed with a **partially buried lower level**, reducing its visual impact and providing thermal insulation through earth sheltering. The upper floors open toward sweeping views, creating a dynamic connection between interior spaces and the natural panorama.
- On a **steeply sloped residential site**, a **multi-level home with staggered terraces** allows each floor to access outdoor space directly, while minimizing the need for retaining walls and heavy earthmoving.
- For a **campus project on rolling terrain**, buildings could be clustered along ridgelines with pathways and roads following natural valleys, preserving drainage courses and creating intuitive circulation.

PPD Exam Connection:

You may encounter questions that ask you to:

- Select appropriate building types or configurations based on slope conditions.
- Propose site planning strategies that minimize grading and protect natural features.
- Identify potential risks related to topography such as erosion, flooding, or unstable soils.
- Recommend foundation types or structural systems suited to specific site conditions.

Takeaway:

Designing with topography and site features in mind requires balancing **engineering considerations with environmental sensitivity**. It's a key skill tested in the PPD exam to assess your ability to integrate a project within its natural context, optimize resource use, and enhance occupant experience through a site-responsive design.

4. **Address Local Climate Conditions**

Architectural design must respond thoughtfully to the **local climate** to enhance occupant comfort, reduce reliance on mechanical systems, and improve overall building performance. Each climate zone demands specific strategies that address temperature extremes, humidity, solar exposure, and prevailing winds.

Climate-Responsive Design Strategies:

- **Cold Climates:**
  - Emphasize **compact building forms** with minimized surface area to reduce heat loss.

    - Use **high-performance insulation** and airtight construction to maintain interior warmth.
    - Maximize **south-facing glazing** to harness passive solar heat gain during winter months.
    - Incorporate **thermal mass** inside to store and slowly release heat.
    - Minimize large openings on north and east façades to limit heat loss and cold drafts.
- **Hot and Humid Climates:**
    - Design for **natural ventilation**, such as cross-ventilation, to cool interior spaces without energy use.
    - Use **deep overhangs, verandas, and shading devices** to protect glazed areas from direct sun.
    - Specify **light-colored or reflective roofing materials** to reduce heat absorption.
    - Encourage **raised floor plates or permeable surfaces** to enhance airflow and reduce heat retention.
    - Avoid airtight construction that traps humidity; instead, allow for controlled moisture management.
- **Mixed or Temperate Climates:**
    - Design for **seasonal adaptability** with operable windows, adjustable shading, and flexible HVAC zoning.
    - Use **dynamic façade elements** like louvers or blinds to control solar gain based on season.
    - Balance insulation and ventilation strategies to optimize comfort year-round.
    - Integrate **landscaping for wind buffering in winter and shading in summer**.

Example: Pacific Northwest Building
In the **Pacific Northwest**, characterized by a **wet, temperate climate**, a building might include **operable clerestory windows** to promote stack ventilation during warmer months while facilitating daylight penetration. Deep roof overhangs help protect façades and entryways from heavy rain, while large glazed areas are carefully shaded to prevent overheating. Moisture-resistant materials and thoughtful site drainage further respond to the local environment.

PPD Exam Connection:
Exam questions may require you to:
- Identify appropriate design responses for specific climate zones.
- Recommend passive design strategies that reduce mechanical energy loads.
- Analyze how façade treatments and window placement affect thermal comfort.
- Consider how local climate influences building form, materials, and systems integration.

Takeaway:
Aligning architectural strategies with local climate conditions is essential for creating **resilient, energy-efficient, and comfortable buildings**. In the PPD exam, demonstrating your understanding of climate-responsive design principles shows your ability to integrate environmental factors seamlessly into project planning and design.

5. **Respond to Cultural and Urban Context**
Architectural design must thoughtfully engage with the **cultural identity and urban fabric** of its location. This means respecting local traditions, historic character, and the scale of neighboring buildings while contributing a meaningful and contemporary interpretation. Successful contextual design strengthens community connections, supports local heritage, and enhances the public realm.

Key Principles:

- **Maintain Scale and Massing:**
    - Analyze the height, bulk, and rhythm of adjacent buildings to ensure your design fits harmoniously within the streetscape.
    - Avoid overpowering neighboring structures; use setbacks, step-backs, or modulated façades to transition scale sensitively.
- **Reflect Local Architectural Vocabulary:**
    - Incorporate materials, colors, textures, and architectural details common to the area without resorting to pastiche.
    - Use familiar forms or motifs as inspiration, but reinterpret them to express a contemporary design language.
- **Enhance the Public Realm:**
    - Design ground floors that engage pedestrians with transparent façades, active uses, and inviting entrances.
    - Create or improve pedestrian connections such as pathways, plazas, or arcades that promote walkability.
    - Use landscaping thoughtfully to reinforce local plant palettes and contribute to a cohesive streetscape.
- **Address Cultural Significance:**
    - Understand the cultural values and history of the site and community.
    - Avoid designs that disrupt or disrespect local traditions, and where possible, incorporate cultural symbols or narratives.

Examples:

- A **new city hall located within a historic district** might balance modern materials like glass and steel with contextual brick and stone elements. The design respects the **scale and rhythm** of neighboring buildings while introducing **contemporary detailing and transparency** that reflects civic openness.
- In a **residential infill project within a traditional neighborhood**, the architect might preserve the existing street wall and rooflines, while using modern window proportions and clean lines to create a dialogue between old and new.
- A **cultural center in an indigenous community** could incorporate locally sourced natural materials and traditional patterns, while providing modern, accessible spaces that honor heritage and support contemporary uses.

PPD Exam Connection:
On the exam, you may be asked to:

- Evaluate design proposals for their sensitivity to scale and context.
- Choose appropriate materials and massing strategies for urban infill sites.

- Address how a project enhances or detracts from the pedestrian experience.
- Identify cultural considerations that influence site planning and architectural expression.

Takeaway:
Responding to cultural and urban context is about creating **architecture that belongs**, adding value to its surroundings without erasing identity. In the PPD exam, demonstrating an awareness of place and culture shows your ability to design with both respect and innovation.

6. **Integrate Landscape and Site Design**
A well-integrated architectural design works in tandem with its **landscape and site design** to create a cohesive, functional, and environmentally responsible project. The site is more than a backdrop—it shapes user experience, supports sustainability goals, and reinforces the architectural narrative. Thoughtful landscape design can mitigate environmental impacts, enhance accessibility, and create inviting, usable outdoor spaces.

Key Strategies:

- **Use Native and Adaptive Plantings:**
  - Select plant species that are well-suited to the local climate and soil conditions to reduce irrigation needs and maintenance.
  - Native plants also support local biodiversity by attracting pollinators and promoting ecological health.
  - Group vegetation into **hydrozones** (based on water needs) to improve irrigation efficiency.
- **Manage Stormwater On-Site:**
  - Incorporate **bioswales, rain gardens, green roofs**, and **permeable paving** to manage stormwater runoff and reduce the burden on municipal systems.
  - Design grading to direct water into landscaped areas for **filtration, infiltration, and reuse**, supporting sustainable site development goals (e.g., LEED Sustainable Sites credits).
  - Avoid over-reliance on underground storm sewers by using **natural drainage** wherever possible.
- **Support Architecture Through Site Features:**
  - Create **outdoor rooms, courtyards, terraces, and plazas** that reflect and extend interior functions such as dining areas, meeting spaces, or informal gathering spots.
  - Align **pedestrian circulation** with building entrances, views, and natural desire lines for intuitive and accessible movement.
  - Use landscape elements to **frame views, define edges, provide shading**, and reinforce wayfinding.
- **Integrate Grading and Topography:**
  - Coordinate site grading with building entries and paths to ensure ADA compliance and comfortable transitions.
  - Blend buildings into sloped sites using **retaining walls, terraces, or split-level plazas** that complement natural landforms.

Example: Academic Campus Design
On a university campus, shaded courtyards and tree-lined walkways serve multiple roles: they offer **comfortable outdoor gathering spaces**, **guide circulation between buildings**, and manage **stormwater runoff** through integrated planted swales and permeable pavers. The landscape supports both ecological goals and student wellness, while reinforcing the overall campus identity.

PPD Exam Connection:
Expect questions that test your ability to:

- Evaluate sustainable site planning strategies such as low-impact development (LID) techniques.
- Coordinate grading, planting, and hardscape design with architectural layout.
- Optimize outdoor space for user experience and environmental performance.
- Identify stormwater management solutions that integrate with overall site design.

In the **Project Planning & Design (PPD)** division, environmental and contextual integration is a recurring theme. Expect **scenario-based and diagrammatic questions** that test your ability to analyze environmental data and apply it to real-world design decisions. These scenarios often require prioritization, trade-offs, and synthesis of multiple variables.

You may be asked to:

- **Adjust building form and massing** in response to the **sun's path, prevailing winds**, or site slope to improve passive performance.
- **Select climate-appropriate strategies**, such as thermal mass, shading devices, or ventilation approaches tailored to cold, hot-humid, or temperate climates.
- **Interpret site plans, sun diagrams, or section cuts** to propose layouts that optimize **daylighting, ventilation, and drainage** while maintaining code compliance.
- **Resolve design conflicts** where environmental performance goals must be balanced against **budget limits, programmatic needs, historical context**, or client preferences.

Tip: Study psychrometric charts, solar orientation diagrams, and climate zone maps. Understand how these inform decisions about massing, glazing, materiality, and MEP coordination.

Takeaway:
Landscape and site design are **critical components of integrated architecture**. They support environmental performance, enhance usability, and connect the building meaningfully to its surroundings. For the PPD exam, you'll need to demonstrate your ability to blend architectural and landscape strategies into a unified, contextually responsive solution.

Summary
Integrating environmental and contextual conditions into architectural design is not just a best practice—it's a foundational skill that improves **building performance, occupant well-being, and long-term sustainability**. A design that responds to its environment can reduce energy consumption, lower operating costs, and create spaces that feel both **functional and meaningful**.

By considering **climate, site features, urban context, and cultural identity**, you strengthen the building's connection to its **place**—visually, environmentally, and socially.
In the **PPD exam**, this integration is often tested indirectly: through how you make choices about **massing, materials, systems, and spatial layout**. You are expected to **translate abstract environmental data into concrete architectural strategies**, while balancing competing design constraints.

Final Thought:
A successful candidate will demonstrate not only **technical knowledge**, but also **design judgment**—the ability to align performance goals with architectural expression and project realities.

Chapter Seven

Project Costs & Budgeting (8-14%)

NCARB Requirements
In this last section, you'll consider the bottom line: how much does this project cost.

In this final section of the PPD division, you'll focus on the project's bottom line: **how much will this project cost?** Understanding the full scope of costs — from conceptual budgeting to selecting cost-effective design alternatives — is essential to ensuring that your design aligns with the owner's financial expectations.

The exam expects candidates to understand how to evaluate and manage construction costs during early design phases and how design decisions influence the budget. You'll need to demonstrate the ability to **evaluate options, align them with the program and goals**, and guide clients toward feasible and cost-effective solutions.

A. Evaluate design alternatives based on the program (A/E) (2-6%)

NCARB Requirements
The first design is not necessarily the best, so analyzing design alternatives against the program requirements, project goals, and project budget, as well as considering the various factors that affect costs, is critical as you determine the most appropriate design option that fits within the project budget.

The first design is not necessarily the best. A strong designer must analyze and compare **multiple design alternatives** against several critical parameters:

- **Program requirements**
- **Project goals**
- **Site context**
- **Sustainability objectives**
- **Code compliance**
- **Project budget**

Evaluating alternatives through this lens ensures that the selected design is not only inspiring but also technically and financially feasible.

Key Concepts to Understand:

1. **Program Requirements Alignment**
Design excellence begins with meeting the programmatic needs of the client. No matter how aesthetically compelling a design may be, it must ultimately serve the **intended function**, **accommodate the necessary spatial relationships**, and **enhance the user experience**. On the

PPD exam — and in real-world practice — your ability to evaluate how well a design alternative aligns with the program is crucial.

Key Considerations:

- **Functionality Over Form**:
  The program defines the required spaces, their sizes, and how they should relate to one another. A design that ignores these criteria, even if it's visually striking, will not succeed in addressing the client's needs.
- **Spatial Adjacencies & Flow**:
  Efficient circulation and logical spatial relationships are critical. Key spaces should be placed in proximity based on function — for example, public areas should be easily accessible from entries, and service areas should be discreet but convenient.
- **User Experience**:
  The arrangement of spaces impacts how people use and feel in the building. Orientation, daylight, privacy, acoustics, and views all contribute to the comfort and efficiency of occupants.

Common Pitfalls:

- Oversizing or undersizing rooms due to a lack of attention to programmed square footage.
- Misaligned adjacencies (e.g., placing mechanical rooms near noise-sensitive areas).
- Ignoring accessibility requirements that affect room layout and clearances.

Examples:

- **Healthcare Clinic**:
  A design with an open, inviting lobby and flowing circulation may appear elegant, but if patient exam rooms are positioned too close together without adequate soundproofing, patient privacy is compromised. Similarly, if rooms are laid out inefficiently, staff workflows suffer, leading to operational inefficiencies.
- **Elementary School**:
  A classroom layout might be visually dynamic with angled walls and non-traditional forms, but if sightlines are poor, natural daylight is blocked, or if the proximity between restrooms and classrooms is overlooked, the design may fail to support effective learning environments.
- **Office Building**:
  In a corporate headquarters, placing executive offices far from collaborative team spaces may conflict with the company's desire for a transparent and team-oriented culture, even if the layout looks clean on paper.

Evaluation Tips for the Exam:

- Always reference the program when reviewing schematic diagrams. Ask yourself: **Does this layout serve the intended use?**
- Use bubble diagrams or adjacency matrices to track functional relationships.
- When given multiple floor plans or massing options, look for those that most closely reflect the program's priorities — not just what looks most interesting.

2. Budget Compatibility

Design must always be developed within the financial constraints of the project. A compelling concept that exceeds the owner's budget is ultimately impractical and unsustainable. On the PPD exam, you'll be expected to evaluate whether a design alternative is **financially feasible** — both in terms of **initial construction costs** and **long-term life-cycle expenses**.

Key Considerations:

- **Construction Costs**:
  The materials, structural systems, and complexity of form have direct cost implications. Even minor changes in geometry or systems can lead to significant cost differences.
- **Life-Cycle Costs**:
  While initial construction cost is critical, energy efficiency, durability, maintenance, and operational costs over time should also be considered when evaluating a design's true financial impact.
- **Cost Estimating Techniques**:
  At the schematic design phase, architects often use rough order-of-magnitude (ROM) estimates or cost-per-square-foot analysis. For more detailed comparisons, elemental or system-based estimates (e.g., by structural system, mechanical type, or envelope assembly) provide better clarity.
- **Escalation & Contingency**:
  Include a contingency for unknown conditions and escalation for anticipated inflation, especially on projects with extended schedules.

Example:

- **Structural System Comparison**:
  One massing option might call for a dramatic 30-foot cantilever supported by structural steel. While visually impressive, this introduces high material costs, specialized labor, and greater structural coordination.

  An alternative design may use CMU walls and engineered wood framing within a compact rectangular form. Though less visually dynamic, it is far more economical — particularly for projects with tight budgets, such as public schools or non-profits.

Exam Tip:
When given multiple schemes on the exam, assess:

- Structural complexity
- Envelope area-to-floor area ratio (compactness often reduces cost)
- Construction type (Type V wood frame is typically cheaper than Type I concrete or steel)
- Mechanical system implications (e.g., large open atria may increase HVAC load)

Always tie your cost evaluation back to **owner priorities and program needs**.

3. Site & Environmental Conditions

Every site introduces unique physical, environmental, and regulatory constraints that influence both **design strategy** and **project cost**. A design that fails to respond appropriately to the site can result in costly construction, zoning conflicts, or missed opportunities for sustainability.

Key Considerations:

- **Topography**:
  Steep sites may require grading, retaining walls, complex foundations, or split-level designs. Flat sites, while easier to build on, may present drainage or stormwater management challenges.
- **Soil Conditions**:
  Poor soil (e.g., expansive clay, high water tables) can necessitate deep foundations or soil remediation, significantly increasing costs.
- **Solar Orientation & Climate**:
  Site planning should optimize for daylight, passive heating/cooling, and protection from prevailing winds, reducing mechanical load and operational cost.
- **Zoning & Regulatory Constraints**:
  Setbacks, height limits, easements, parking requirements, and allowable uses all impact the feasibility and cost of a project.
- **Environmental Hazards**:
  Flood zones, seismic zones, or areas with high wind loads require specific construction techniques or additional design features that influence budget.

Example:

- **Sloped Site Excavation**:
  A design option that cuts deeply into a hillside may look clean and elegant, but requires extensive excavation, retaining structures, and drainage solutions.
  A different option that steps with the terrain or rotates the building mass to reduce grading can cut site costs significantly while also integrating more naturally with the landscape.

Exam Tip:
When presented with multiple site-sensitive options:

- Choose solutions that minimize earthwork or foundation complexity.
- Consider how orientation impacts energy use.
- Evaluate feasibility against local codes (e.g., avoid siting within setbacks or flood plains).

4. **Structural and Material Choices**
   The choice of **structural system** and **construction materials** is a major factor in determining both **initial construction costs** and **long-term performance**. On the PPD exam, you'll be expected to understand how different structural strategies and materials affect cost, constructability, environmental impact, and future maintenance.

Key Considerations:

- **Material Costs & Availability**
  Material pricing can vary by region and market conditions. Wood, concrete, steel, and engineered systems (like CLT or SIPs) all come with different cost structures. Availability and local labor expertise may also drive the feasibility of using certain systems.
- **Construction Speed & Labor**
  Some systems (e.g., prefabricated or modular assemblies) can reduce labor costs and speed up construction. Others, like cast-in-place concrete, may extend construction timelines but offer durability and fire resistance.

- **Structural System Compatibility with Design**
  The chosen system must support the architectural vision. Long spans, cantilevers, and open spaces typically require steel or engineered wood, whereas repetitive grids and load-bearing walls are suited for masonry or wood framing.
- **Durability & Maintenance**
  Long-term exposure to moisture, UV, pests, or freeze-thaw cycles should inform material selection. More durable systems may cost more initially but reduce maintenance or replacement costs over time.
- **Code Requirements & Performance**
  Fire ratings, seismic performance, and sound transmission requirements often dictate certain material choices. For example, a multifamily residential building over a podium will typically use a concrete base with wood framing above to meet code and budget.

Example:

- **Cross-Laminated Timber (CLT) vs. Concrete**
  A developer is evaluating two mid-rise office designs. One uses a CLT structural system, which offers speed of construction, reduced on-site labor, and a warm, exposed wood aesthetic. However, CLT may have higher material costs and limited supplier access in some regions.

  The alternate scheme uses cast-in-place concrete, offering high thermal mass, durability, and good acoustic separation. But it also comes with a longer construction timeline and greater formwork costs.
  In this scenario, a detailed cost-benefit analysis (including construction schedule impact and life-cycle costs) would guide the decision.

System	Pros	Cons
Wood Framing (Type V)	Low cost, easy to work with, fast	Limited height, fire resistance, and sound control
Steel Frame	Long spans, flexible layout, strong	Higher material cost, requires fireproofing
Concrete (Cast-in-Place)	Durable, fire-resistant, good acoustics	Labor-intensive, longer build time
Precast Concrete	Faster installation, durable	Transportation and crane costs, design constraints
CLT / Mass Timber	Sustainable, fast, aesthetic	Expensive, not widely available in all regions

Figure 7.1 Common Structural Systems and Cost Considerations

Exam Tip:
When comparing design schemes:

- Analyze the **structural efficiency** (Are spans optimized? Are systems overdesigned?)
- Consider the **impact of fire codes and building type** (e.g., Type I vs. Type V construction)
- Identify whether the **material choice supports budget, function, and sustainability goals**

A good architect balances **performance, aesthetics, and cost** — and the PPD exam rewards those who can justify structural choices from both a technical and financial perspective.

5. **Sustainability and Energy Performance**
Sustainable design is not just an environmental imperative — it's a financial strategy. Thoughtful integration of **energy-efficient systems**, **passive design strategies**, and **sustainable materials** can reduce operational costs, improve occupant comfort, and support long-term value. On the PPD exam, you are expected to evaluate how different design options impact both **initial costs** and **ongoing energy performance**.

Key Considerations:

- **Short-Term vs. Long-Term Cost Balance**
Many sustainable strategies involve a higher up-front investment with a longer payback period. Part of your role is to assess whether those investments align with the client's budget and priorities.
- **Passive Design Strategies**
Incorporate site-responsive elements that reduce mechanical demands:
    - Orientation for solar gain or shading
    - Cross-ventilation through operable windows
    - Thermal mass to regulate temperature swings
    - Daylighting to reduce electric lighting use
- **Efficient Mechanical and Envelope Systems**
Choosing high-efficiency HVAC systems, enhanced insulation, low-E glazing, and air sealing can significantly reduce operational energy use. These elements often have quick payback periods.
- **Material Lifecycle and Embodied Energy**
Sustainable material choices consider both durability and environmental impact. Products with low embodied carbon and high recyclability often contribute to certification goals (e.g., LEED, WELL).
- **Integration, Not Add-On**
Sustainable strategies are most cost-effective when integrated into the initial concept — not added later. For example, building orientation and massing should support passive systems from the outset.

Examples:

- **Deep Overhangs for Solar Control**
A proposed design includes deep south-facing overhangs to block summer sun while allowing winter solar gain. Although this increases the roof area and structural load — raising the initial cost — it significantly reduces cooling loads, lowering HVAC capacity needs and energy bills.
Over time, the reduced operating cost offers a favorable return on investment and improves occupant comfort — a critical factor in long-term building performance.

- **Option A** uses large west-facing glazing without shading, increasing heat gain and requiring oversized mechanical systems.

Option B uses horizontal fins and high-performance glazing on the west façade, with a modest up-front cost increase but long-term energy savings and improved daylighting.

- **Option A** uses standard batt insulation and code-minimum HVAC.
  Option B upgrades to spray-foam insulation and a variable refrigerant flow (VRF) system — higher cost, but improved envelope performance, greater zoning control, and lower utility bills.

Evaluation Tools:

- **Energy Modeling Software**: Tools like eQuest, Sefaira, or Insight help simulate energy performance across design options.
- **Life-Cycle Cost Analysis (LCCA)**: Weighs capital costs against savings in utilities, maintenance, and equipment replacement.
- **Sustainability Scorecards**: Track how options meet project goals like LEED certification, energy targets, or carbon reduction.

Exam Tip:
When reviewing design options:

- Favor schemes that take advantage of **passive strategies** and reduce long-term HVAC loads.
- Avoid designs that ignore climate data, solar exposure, or ventilation opportunities.
- Don't confuse **"green features"** (like solar panels) with **true integration** — sustainable design is about reducing demand first, then supplying it efficiently.

6. **Client Goals & Stakeholder Priorities**

At the heart of every design decision is a simple question: **What matters most to the client?** Whether it's budget control, architectural expression, operational efficiency, or long-term durability, successful architects must evaluate design alternatives through the lens of client goals. On the PPD exam, you'll be expected to prioritize options that reflect the client's stated values and align with the project's mission.

Key Considerations:

- **Understanding the Client Profile**
  Different clients have different drivers:
  - A **developer** may emphasize cost and speed of construction.
  - A **university** may value durability, low maintenance, and lifecycle cost savings.
  - A **cultural institution** may prioritize design prestige and public image.
  - A **municipality** might prioritize sustainability, accessibility, and public benefit.
- **Balancing Competing Goals**
  Projects rarely have a single priority. Your role is to **balance trade-offs** — for example, energy efficiency vs. initial cost, or iconic form vs. constructability — and communicate those trade-offs effectively to stakeholders.
- **Stakeholder Engagement**
  In addition to the client, various stakeholders may influence priorities: users, facilities managers, funders, or community members. Their input can shape what constitutes a "successful" design.

- **Project Mission Alignment**
  Design solutions should reinforce the client's broader mission — whether that's educational, commercial, civic, or environmental. This ensures long-term satisfaction and functionality.

Examples:

- **University Dormitory Project**
  A university client is evaluating two design options for a new residence hall:
  - **Option A** features a dynamic, angular façade with expensive cladding and large curtain wall areas. While architecturally bold, it increases both initial cost and long-term maintenance.
  - **Option B** uses a more subdued but highly durable façade system, with robust finishes, reduced glazing ratios, and efficient, modular room layouts. Though less flashy, it significantly reduces long-term maintenance costs and improves operational efficiency.

  Given the university's emphasis on **durability, reduced maintenance, and predictable operating costs**, Option B is more aligned with the institution's long-term goals — even if it's less visually distinctive.
- A **tech company** may prioritize open, flexible spaces that can adapt to growth and change, even if it means reducing private offices or enclosed spaces.
- A **non-profit health clinic** might prioritize value engineering and future expandability over architectural complexity.
- A **hospitality client** may emphasize design that enhances guest experience and branding, justifying higher initial finishes and features.

Exam Tip:
When choosing between design options:

- Look for cues in the scenario that hint at client preferences (e.g., "Client is concerned with maintenance," or "Client seeks LEED certification").
- Eliminate schemes that contradict the client's top priorities, even if they perform well in other areas.
- Think holistically: **A good design is not just functional or affordable — it must serve the client's mission.**

7. **Methods of Evaluation**

To make informed, balanced decisions during the early phases of design, architects rely on **structured evaluation tools**. These methods help compare design options objectively, ensuring that selected solutions align with the project's goals, constraints, and budget. Understanding these tools — and when to use them — is essential for both the PPD exam and real-world practice.

1) Matrix Analysis
A matrix analysis is a **decision-making framework** that allows design options to be compared side-by-side across multiple criteria. Each option is rated or scored based on performance in key categories such as:

- Functional fit with program
- Construction cost

- Energy performance
- Constructability
- Client preferences
- Aesthetic goals

This visual comparison allows stakeholders to see trade-offs and identify the strongest overall option.

Example:
Design Option A scores highest in energy efficiency and client preference but slightly lower in budget alignment. Option B is the least expensive but scores lower in program fit. The matrix helps clarify these trade-offs.

2) Life-Cycle Cost Analysis (LCCA)
LCCA is a method for evaluating the **total cost of ownership** over a building's lifespan, not just its initial construction cost. This includes:

- Initial capital costs
- Energy consumption
- Maintenance and repair
- Replacement cycles
- End-of-life costs (e.g., demolition or material recovery)

LCCA is critical when clients are planning for long-term occupancy or managing public funds where operational efficiency matters as much as initial investment.

Example:
A metal roof has a higher initial cost than asphalt shingles, but with a 50-year life span and minimal maintenance, it proves more cost-effective over the long term.

3) Cost per Square Foot Estimates
Used during early schematic phases, this method provides **ballpark estimates** of construction cost based on historical data from similar building types or systems.

- Especially useful for comparing massing options or material strategies
- Allows for quick budgeting feedback before committing to more detailed development

Example:
Option A is 25,000 SF at \$300/SF = \$7.5 million
Option B is 22,000 SF at \$320/SF = \$7.04 million
Although Option B has a higher per-SF cost, its compact footprint results in lower total cost.

4) SWOT Analysis
A SWOT (Strengths, Weaknesses, Opportunities, Threats) analysis is a **qualitative method** that considers internal and external factors affecting each design alternative.

- **Strengths**: What makes the design strong? (e.g., efficient layout, iconic form)
- **Weaknesses**: What are its vulnerabilities? (e.g., cost overruns, zoning issues)
- **Opportunities**: What could be leveraged in the future? (e.g., expansion, solar readiness)
- **Threats**: What could go wrong? (e.g., neighborhood opposition, material shortages)

Example:
A design with a strong sustainability strategy (strength) may face increased permitting time due to a complex envelope (threat). A SWOT analysis helps weigh such dynamics early.

Final Thoughts
The ability to critically evaluate multiple design alternatives — through both **quantitative tools** (cost estimating, LCCA) and **qualitative frameworks** (SWOT, matrix analysis) — is one of the most vital skills for any architect. In the context of the PPD exam, this skill is tested not only in multiple-choice questions but also in case studies that require logical, strategic thinking.

As an architect, your role is more than creating inspiring forms — you are a **problem solver, cost manager, advocate for the client, and long-term strategist**. Each design decision should reflect a clear understanding of trade-offs, budget implications, technical feasibility, and, most importantly, the client's vision.

Whether you're designing a public library, a healthcare clinic, or a mixed-use development, success lies in your ability to align **aspiration with affordability**, and **creativity with constraint**. Mastering these evaluation methods not only prepares you for the PPD division — it prepares you for real-world architectural practice.

B. Perform cost evaluation (A/E) (1-5%)

NCARB Requirements
It is important for you to evaluate various methods of estimating project costs, including those based on program type, square footage, or systems/assemblies. You must be able to create a preliminary cost estimate, adjust the estimate as the design develops, calculate the cost of design alternatives, and reconcile the estimate with the project budget.

As an architect, you must understand how to evaluate and track project costs throughout design development. This includes using early-phase estimating tools, comparing systems and materials, and making adjustments to keep the design aligned with the project's financial limits. The PPD exam expects you to:

- Apply different methods of cost estimation (e.g., square foot, unit cost, assemblies)
- Evaluate costs based on building type, program needs, and systems
- Compare the cost implications of design alternatives
- Reconcile estimates with the overall project budget
- Recognize when a design exceeds budget and propose corrective action

1. Preliminary Cost Estimating Techniques

In the **early phases of design**—particularly during **pre-design** and **schematic design**—architects rely on **preliminary cost estimating techniques** to provide a **Rough Order of Magnitude (ROM)** for the construction budget. These estimates are high-level and approximate, intended to give the client and project team a realistic sense of financial feasibility before design details are fully developed.

These early estimates help answer key questions:
- Is the project financially viable based on the proposed scope?
- Does the client's budget align with the desired program?
- What design decisions might significantly impact costs?

Common Preliminary Estimating Methods:
- **Cost per Square Foot**
  This is the **most widely used method** at the schematic stage. It multiplies the projected gross square footage of the building by an average cost per SF based on project type, location, and current market conditions. The rate can vary significantly based on region, construction type, and building complexity.

Example:
A 25,000 SF public library estimated at **$375/SF** = **$9.375 million** construction cost.

- **Historical Cost Data**
  Estimates can be derived by comparing the current project to recently completed projects of similar scope and type. Adjustments must be made for **inflation**, **site conditions**, and **differences in materials or systems**.

Example:
If a similar middle school was built two years ago for $310/SF, and construction costs have increased by 6% annually, a comparable project today may be closer to $350/SF.

- **Unit Cost Estimates (Functional Units)**
  This method is particularly useful for specialized building types. Rather than using overall area, cost is based on specific functional units:
  - Per **student** (schools)
  - Per **bed** (hotels, dormitories, hospitals)
  - Per **seat** (theaters, auditoriums)

Example:
A small theater with 400 seats estimated at **$1,800/seat** results in a ROM cost of **$720,000**.

Factors That Influence ROM Accuracy:
- Building type and occupancy (commercial, institutional, residential, etc.)
- Construction type (Type I–V; steel, wood, concrete, etc.)
- Location (urban vs. rural; regional cost indices)
- Schedule (faster construction may increase cost due to labor premiums)
- Level of finish and performance standards (e.g., LEED certification, specialized equipment)

Exam Tip:
⚠ On the ARE 5.0 PPD exam, you may be asked to:
- Compare two schematic options using SF-based cost data
- Identify which estimate is more appropriate at a specific project phase
- Recognize when a project's proposed scope does not match the budget

Practice Exercise:
A client wants a new 40,000 SF community center. Based on local cost data, you estimate:

- **Option A (simple design)**: $300/SF × 40,000 SF = **$12 million**
- **Option B (higher-end finishes + green roof)**: $370/SF × 40,000 SF = **$14.8 million**

This simple comparison helps guide early design discussions and sets realistic expectations.

2. **Assemblies/System-Based Estimating**
As a project moves from schematic design into **design development (DD)**, cost estimating becomes more precise. At this stage, architects and consultants begin using **assemblies or systems-based estimating**, which analyzes costs by **building components or functional systems**, rather than relying on a flat cost-per-square-foot approach.

This method provides a **more detailed view of cost implications** based on design decisions and allows for value comparison between alternative construction methods, materials, and building systems.

What is an "Assembly"?
An **assembly** is a group of related elements that work together as a system — for example, a wall assembly includes the framing, sheathing, insulation, air barrier, cladding, and finishes. This method groups cost by **major construction divisions** rather than individual materials or quantities.

Common Assemblies and Systems Evaluated:

- **Structure**
  - Options: slab-on-grade, steel framing, concrete frame, mass timber
  - Costs vary based on span length, load requirements, and complexity
- **Envelope**
  - Options: curtain wall, precast concrete panels, insulated metal panels, brick veneer
  - Consider cost per square foot, thermal performance, and maintenance needs
- **Mechanical/Electrical/Plumbing (MEP)**
  - HVAC systems: geothermal, VAV, split systems, radiant heating
  - Plumbing fixture types, water heating methods, lighting efficiency
- **Interior Finishes**
  - Options: vinyl composition tile (VCT), terrazzo, gypsum board, wood panels
  - Vary significantly in initial cost, durability, and life-cycle performance

Example:
You're comparing two façade systems for a 10,000 SF wall area:

- **Option A: Curtain Wall** = $140/SF
  Total: $1,400,000
- **Option B: Masonry Veneer** = $80/SF
  Total: $800,000

Difference: $600,000, which could be redirected to fund upgraded HVAC or reduce overall project cost.

This method helps teams weigh **cost vs. performance vs. aesthetics**, especially when options look equally viable functionally.

Advantages of Assemblies Estimating:

- More accurate than conceptual square-foot estimates
- Helps identify major cost drivers in the design
- Useful for **value engineering** and identifying cost-saving substitutions
- Aligns well with **Construction Specifications Institute (CSI) Uniformat** system, used in many estimating databases (e.g., RSMeans)

When to Use:

- Design Development and early Construction Document phases
- When comparing multiple systems that fulfill the same function (e.g., roof membrane types, HVAC systems)
- During budgeting exercises with cost consultants or Construction Managers at Risk (CMAR)

ARE 5.0 PPD Exam Insight:

🔍 Expect exam scenarios that involve:

- Evaluating the cost impact of alternative systems
- Understanding which system choices would meet performance and budget constraints
- Interpreting assemblies cost data in a case study format

Real-World Scenario:

A school design includes polished concrete floors, but preliminary system-based estimates show this adds $5/SF compared to resilient flooring. For a 50,000 SF building, this adds **$250,000**. The design team may decide to retain polished concrete in high-traffic areas only and switch to VCT in classrooms, balancing performance and cost.

3. **Adjusting Estimates as Design Evolves**

A key responsibility of the architect during design is to **continually monitor and update the project's cost estimate**. As the design becomes more defined, the level of accuracy and detail in cost estimating must increase accordingly. This process is known as **progressive estimating** or **design-phase cost reconciliation**.

At each major design milestone, cost estimates must be reviewed and adjusted to reflect:

- Evolving scope
- Material and system decisions
- Market pricing fluctuations
- Client-directed changes

Estimating by Design Phase

- **Schematic Design (SD)**

  At this early stage, the project is typically modeled using:

  - **Cost per square foot** (based on gross building area and occupancy type)
  - **Preliminary system allowances** (e.g., roofing, HVAC, structural)

Example:
A 60,000 SF office building at $325/SF = $19.5 million ROM cost.

- **Design Development (DD)**
  Now that systems, materials, and layouts are being refined, estimators use:
  - **Assemblies-based estimates** (e.g., cost per SF of curtain wall or roofing membrane)
  - Updated assumptions based on outline specifications

Example:
Switching from steel frame to mass timber may reduce structural weight and foundation requirements, impacting the overall cost model.

- **Construction Documents (CD)**
  With detailed drawings and specifications complete, estimators can generate:
  - **Quantity take-offs** (specific counts of doors, fixtures, lighting, etc.)
  - **Detailed line-item pricing** for labor, materials, equipment, and subcontractor work

These estimates are often created by:
  - A professional cost estimator
  - A Construction Manager (CM)
  - Bidders (in Design-Bid-Build projects)

Why Estimates Must Evolve:
- **Design Scope Adjustments**: Clients often request changes in area, layout, or program mid-design.
- **Material or System Changes**: Substituting a higher-end or sustainable system may affect cost.
- **Market Fluctuations**: Labor shortages, inflation, or supply chain issues can raise costs unexpectedly.
- **Code Compliance and Regulatory Costs**: Required upgrades or new standards can add scope (e.g., fireproofing, accessibility upgrades).

💡 As the design becomes more detailed, estimates shift from **approximation** to **precision** — and the risk of budget overrun decreases.

ARE 5.0 Exam Tip:
🔍 On the PPD exam, you may be asked to:
- Identify the appropriate type of cost estimate for a given design phase
- Recognize when an estimate should be revisited
- Understand the cost consequences of design changes

Real-World Example:
During SD, a high school gymnasium was estimated at $8 million using $400/SF. In DD, the team selected a long-span steel truss system and upgraded finishes, pushing the cost to $9.2 million. The architect worked with the owner to revise flooring choices and reduce skylight quantities, bringing the cost back in line with the original budget.

By treating estimating as a **living process**, architects help clients stay informed, avoid surprises, and maintain financial control from concept to construction. This proactive approach to cost reconciliation is a key skill in both exam scenarios and real-world practice.

4. **Evaluating Design Alternatives by Cost Impact**

Throughout the design process, architects must continually assess and compare the **cost implications of different design choices**. From materials and structural systems to envelope details and sustainability strategies, every decision can affect the construction budget — sometimes significantly.

It is the architect's role to guide the client in selecting options that not only meet design intent but also align with **budgetary constraints**, long-term performance goals, and project priorities.

Key Factors in Cost-Based Evaluation:

- **Initial Cost**: The up-front cost of a system or material at the time of construction
- **Life-Cycle Cost (LCC)**: Total cost of ownership, including maintenance, energy use, and replacement
- **Durability and Maintenance**: Long-term operational costs may outweigh savings in initial budget
- **Environmental Impact and Credits**: Some higher-cost systems may provide sustainability incentives or operational efficiency
- **Schedule Impact**: Some options may speed up or delay construction, affecting general conditions and labor costs

Example:
Green Roof vs. Conventional Roofing

- *Green Roof System*:
  Adds approximately **$20–30/SF** to construction cost
  Provides **stormwater retention**, reduces heat island effect
  May earn **LEED points** or comply with local regulations
  Requires **special waterproofing, structural support, and maintenance**
- *Conventional Membrane Roof*:
  Lower up-front cost (typically $10–15/SF)
  Limited ecological or energy benefits
  Shorter lifespan depending on material and exposure

Decision: If stormwater retention is a municipal requirement or LEED certification is a priority, the added cost of the green roof may be justified. Otherwise, it may not provide sufficient ROI.

Tools for Comparative Analysis:

- **Life-Cycle Costing (LCCA)**
  Calculates total cost over a system's life, including:
  - Initial construction
  - Maintenance and repairs
  - Energy use

- End-of-life replacement

Example: A geothermal HVAC system may cost $300,000 more than a standard unit, but save $40,000/year in energy. Over 15 years, that's a $600,000 return.

- **Cost-Benefit Analysis (CBA)**
  Weighs the added cost of a feature against the tangible and intangible benefits. Useful for assessing:
  - LEED credits
  - Public perception or branding
  - Performance enhancements (e.g., acoustics, daylight)

- **Value Engineering (VE)**
  A structured process of evaluating where **design alternatives** can reduce cost **without sacrificing quality or functionality**.

Example: Replacing stone cladding with precast concrete panels could save $300,000 while maintaining a similar aesthetic.

ARE 5.0 PPD Exam Tip:

🔍 On the exam, be prepared to:

- Compare two or more design options and evaluate their cost impact
- Understand when a high-cost option is justified by long-term benefits
- Use LCCA or VE concepts in case study scenarios

Real-World Scenario:

An airport terminal design team compares two glazing systems:

- **Option A**: High-performance triple-glazed curtain wall ($160/SF)
- **Option B**: Standard double-glazed unit ($110/SF)

Over 50,000 SF of glazing, the cost difference is $2.5 million. However, energy modeling shows that **Option A** could reduce annual HVAC operating costs by $200,000. If the building is meant to last 40 years, the long-term savings could exceed $8 million — making the initial investment worthwhile.

Summary:

Design is about **balancing aesthetics, performance, and cost**. Architects must be able to clearly present the financial implications of design options, helping clients make informed decisions that serve both the immediate project goals and the long-term success of the building.

This evaluative thinking is a **core competency tested on the PPD exam**, and a hallmark of responsible architectural practice.

5. **Estimating Soft Costs vs. Hard Costs**
   A comprehensive project budget includes more than just bricks and mortar. One of the most important distinctions to understand in architectural project planning is the difference between **hard costs** and **soft costs**. Together, they form the **total project cost**, which the architect must understand, even if they are not responsible for tracking every line item.

Being able to anticipate and advise on both categories is key to **budget alignment**, **client guidance**, and **realistic planning**—especially during the **Programming & Analysis (PA)** and **Project Planning & Design (PPD)** phases.

Hard Costs

Hard costs are the **direct construction costs** associated with the physical building and site. These are the tangible, "in-the-ground" or "above-ground" costs paid to contractors and vendors.

Typical examples include:

- Labor and trades (carpenters, electricians, plumbers)
- Materials (concrete, steel, lumber, finishes)
- Building systems (HVAC, electrical, plumbing)
- Equipment (cranes, scaffolding)
- Site work (grading, utilities, paving, landscaping)

Hard costs generally make up **~75% to 85%** of the **total project budget**, depending on the type and complexity of the project.

Soft Costs

Soft costs refer to **indirect, non-construction expenses** related to the planning, design, financing, and administrative aspects of the project. While not physically part of the building, they are essential to delivering the project legally, safely, and according to the client's goals.

Typical soft cost categories include:

- **Architectural and engineering (A/E) design fees**
- **Permits and inspections**
- **Surveys and geotechnical reports**
- **Legal and insurance fees**
- **Commissioning and testing**
- **Construction management fees**
- **Contingencies (design or construction)**
- **Loan interest and financing fees**
- **Furniture, Fixtures & Equipment (FF&E)** (often included in soft cost planning)

Soft costs typically account for **~15% to 25%** of the total project cost. This percentage can be **higher for institutional or complex projects**, such as hospitals or labs, which require more intensive design, permitting, and coordination.

Example:

Let's say a **public library project** has a **construction cost of $10 million.**

- **Hard Costs (80%)** = $8 million
  Includes building structure, HVAC, finishes, and site improvements.
- **Soft Costs (20%)** = $2 million
  Might include:
  - $750,000 for architectural and engineering design
  - $150,000 for permits and code review
  - $200,000 for commissioning and testing
  - $500,000 for FF&E
  - $400,000 for legal, insurance, and project management fees

Total Project Budget = $10M (Hard) + $2M (Soft) = $12M

Special Considerations:

- **Soft costs are often overlooked or underestimated** during early planning. Educating the client about their full financial responsibility is essential.
- Soft costs can **vary by region, project delivery method, or funding source** (e.g., publicly funded projects may require more legal review and insurance).
- **Contingency funds**, typically classified under soft costs, should be built into the budget to absorb unforeseen design revisions or construction issues.

ARE 5.0 PPD Exam Insight:

🔍 On the exam, you may be asked to:

- Distinguish between hard and soft costs in a project scenario
- Allocate or estimate costs based on budget breakdowns
- Identify which items belong in each category (e.g., "Which of the following is a soft cost?")

Real-World Scenario:

A university science building includes state-of-the-art lab equipment and complex MEP coordination. Although the base construction (hard) cost is $50M, the project's **total budget is $65M** due to:

- High design and engineering costs (soft)
- Specialized commissioning
- Extended permitting process
- Specialized FF&E requirements

Understanding this distinction helps the architect advise the client accurately and helps avoid unpleasant budget surprises later in the process.

Summary

Effective project planning requires not only tracking the **building costs**, but also forecasting the **hidden or less visible costs** that make up a substantial portion of the budget. By clearly distinguishing hard and soft costs, architects can:

- Help clients develop realistic project budgets
- Align design scope with available funds
- Prepare for downstream costs beyond the construction phase

6. **Applying Contingencies and Escalation**

In early design and pre-construction phases, cost estimates are inherently uncertain. Many details—materials, systems, site conditions, labor costs—are not yet fully known or finalized. To help manage this uncertainty, architects and estimators apply **contingencies** and **escalation allowances** to provide a more accurate and risk-adjusted picture of project cost.

These additions to the budget are essential for **financial resilience** and help clients plan for **potential overruns** as the project evolves or market conditions shift.

Types of Contingencies

Contingencies are added as **percentage-based allowances** to account for risks and unknowns that are not yet fully defined. They differ depending on the project phase and responsibility:

- **Design Contingency** *(typically 5–10%)*
  - Accounts for changes in scope or systems as the design is refined.
  - Managed by the architect and owner during the design phase.
  - Covers the natural progression of the design process (e.g., moving from massing to detailed layouts and systems).

Example: A project in Schematic Design might carry a 10% design contingency to account for layout refinements, material decisions, and coordination of structure and MEP systems.

- **Construction Contingency** *(typically 5–10%)*
  - Held by the owner or contractor to address unforeseen conditions during construction, such as:
    - Differing site conditions
    - Minor scope adjustments
    - Errors or omissions in the drawings
  - Particularly important in **Design-Bid-Build** projects, where the contractor may not be involved in early design.

Example: A $15 million construction budget might include a 5% construction contingency ($750,000) for field conflicts or RFI-related work.

Escalation

Escalation accounts for projected **inflation in construction costs** between the time of the estimate and the actual construction start or midpoint. Construction inflation is typically driven by:

- Increases in labor or material costs
- Supply chain volatility
- Market demand (boom or slowdown in regional construction)

The escalation rate depends on:

- The time lag before bidding or construction
- The local/regional market
- Economic forecasts

Rule of Thumb: 3–6% annual escalation is common in a stable market, but can be higher in high-demand regions or volatile economic conditions.

Example:

Let's say a project has a base estimate of **$20 million** in today's dollars, but construction won't start for **18 months**. If an **escalation rate of 6% per year** is expected:

- 18-month escalation = **(6% ÷ 12) × 18 = 9% total**
- Escalated cost = **$20M × 1.09 = $21.8 million**

That additional **$1.8 million** needs to be budgeted now to avoid future shortfalls.

ARE 5.0 PPD Exam Insight:

🔍 You may be asked to:

- Calculate total project cost with contingency or escalation

- Identify which type of contingency applies to a given situation
- Select appropriate contingency percentages based on project phase

Real-World Scenario:
A civic center project is budgeted for $30 million. The architect advises the client to include:

- 10% design contingency ($3 million)
- 5% construction contingency ($1.5 million)
- 8% escalation (estimated over 18 months = $2.4 million)

The **adjusted total budget** becomes **$36.9 million**, helping the client avoid budget gaps once real bids are received.

Summary
Contingency and escalation allowances are **not "padding" the budget**—they are smart, essential planning tools. They:

- Absorb scope evolution and design growth
- Protect against unknowns in the construction environment
- Help maintain alignment between the client's expectations and actual market conditions

Architects must advocate for realistic budgeting early in the process, even when difficult, to safeguard the integrity of the project through completion.

7. **Reconciling Cost Estimates with the Budget**
Cost reconciliation is a critical step in the design process—especially when the **estimated construction cost exceeds the client's budget**. When this happens (and it often does), the architect must **lead a structured response** to realign the project's scope, systems, or schedule with financial realities, without compromising the core goals of the project.

Being proactive, collaborative, and strategic during this phase is essential to maintain client trust, ensure project feasibility, and move the design forward.

Key Strategies to Reconcile Over-Budget Projects
If the current cost estimate exceeds the target budget, the following approaches are typically explored:

- **Value Engineering (VE)**
  Involves revisiting **materials, systems, or assemblies** to find cost savings without significantly altering function or design intent.
  Examples:
  - Swap out a stone veneer for high-quality precast concrete.
  - Replace custom curtain wall with standard window wall systems.
  - Reduce ceiling height in secondary spaces to simplify HVAC distribution.

- **Scope Reduction**
  Directly reducing the **building size** or **program elements** to lower cost. While often unpopular, this may be necessary for budget alignment.
  Examples:
  - Eliminate one multipurpose room or reduce classroom count.

- Simplify public lobbies or reduce built-in custom casework.
- Combine support functions into shared spaces (e.g., copy rooms, break areas).

- **Phasing the Project**
  Deferring non-critical program areas or site work to a future phase can help get the first phase of the project built within budget.
  Examples:
  - Build shell space now, finish interior later.
  - Delay landscaping or surface parking lot expansion.
  - Construct an auditorium wing in a separate capital phase.

Collaboration with Estimators and Contractors
Reconciling costs is not a solo effort. Architects must **work closely with professional cost estimators or contractors**, especially in delivery methods such as **CM-at-Risk** or **Design-Build**, where early cost input is part of the process.

Key activities include:
- **Aligning pricing assumptions** (e.g., labor rates, material costs, allowances)
- **Identifying major cost drivers** (e.g., structural spans, envelope systems, MEP complexity)
- **Confirming scope interpretations** (Was a roof deck counted twice? Was landscaping overestimated?)
- **Refining design details** that may be inflating cost unnecessarily

💬 This stage is an opportunity to **optimize**, not just cut. Aim for creative, intelligent adjustments that support the design's performance and experience while trimming excess.

ARE 5.0 PPD Exam Tip:
🔍 You may be tested on:
- How to respond when estimates exceed the budget
- The appropriate use of VE, phasing, or scope reduction
- Roles of the architect and estimator in cost reconciliation

Real-World Scenario:
A $40M elementary school project comes back with a DD estimate of $44M. The team takes the following steps:
- Eliminates a small standalone admin annex, combining functions into the main building.
- Substitutes a sloped metal roof with a membrane roof over a flat structure.
- Delays construction of the outdoor amphitheater until a future bond issue.

Through these changes, the project is brought back under budget without compromising educational goals or core functionality.

Summary
When budgets and estimates don't match, architects must:
- **Lead conversations** that balance design quality with fiscal reality
- **Collaborate across disciplines** to clarify and refine assumptions
- **Offer strategic options** that respect the client's vision and constraints

This is a key area of responsibility—and a critical skill tested on the **ARE 5.0 Project Planning & Design (PPD)** exam. You're not just designing buildings—you're helping clients make informed, responsible decisions that shape the success of their investment.

8. **Tools & Resources for Cost Evaluation**
Effective cost evaluation relies heavily on using the right **tools and resources** to generate accurate, timely, and regionally relevant estimates. These tools assist architects in developing budgets, comparing alternatives, and forecasting long-term financial impacts—all crucial for making informed design decisions and for the ARE 5.0 Project Planning & Design (PPD) exam.

Commonly Used Tools and Resources

- **RSMeans and Other Cost Databases**
RSMeans is one of the most widely recognized sources for detailed **regional cost data**, including labor, materials, and equipment pricing. It allows architects to access up-to-date cost information tailored to specific cities or states, improving the accuracy of preliminary and detailed estimates.
Example: Using RSMeans, an architect can find the average cost per square foot for concrete slab construction in Denver vs. Miami, adjusting budgets accordingly.

- **Pro Forma Templates**
These financial models combine estimated hard and soft costs, anticipated revenue (in commercial projects), financing costs, and contingency allowances into a comprehensive budgeting tool. Pro formas help owners and design teams understand the **total project cost and financial feasibility**.
Example: A mixed-use developer might use a pro forma to evaluate how design changes impact overall project profitability and financing terms.

- **Energy Modeling and Life-Cycle Cost Analysis (LCCA)**
These tools evaluate **long-term operational costs** of design choices, such as HVAC systems, glazing options, or insulation levels, helping prioritize investments that reduce energy consumption and maintenance over time.
Example: An energy model may show that increasing insulation thickness adds 2% to construction costs but reduces annual heating costs by 15%, paying back over 10 years.

- **Project Management Software**
Platforms like **Deltek**, **Procore**, or **Buildertrend** help track budgets, schedule milestones, change orders, and contractor payments throughout the project lifecycle. These tools improve communication and cost control during construction phases.
Example: Procore's budgeting module allows real-time tracking of expenditures against the approved budget, flagging overruns early.

ARE 5.0 Exam Insight

💡 On the exam, you may be asked to:

- Choose the most appropriate estimating method based on the project phase or information available
- Interpret data from a cost report or spreadsheet
- Understand the purpose and benefits of tools like energy modeling or pro formas

Summary

The successful architect integrates **quantitative tools** with qualitative judgment to manage project costs effectively. Familiarity with these resources supports better decision-making and is an essential skill for the **Project Planning & Design (PPD)** exam and professional practice alike.

Being able to perform **accurate, evolving cost evaluations** is essential for keeping a project **financially viable** from concept through construction. As the design progresses and decisions become more defined, architects must continually assess how those choices impact the overall budget.

While it is not the architect's job to produce detailed contractor-level estimates, the architect must be able to:

- Interpret and respond to **preliminary cost data**
- Understand how **materials, systems, and form** influence construction and life-cycle costs
- **Collaborate with cost consultants** and contractors to reconcile estimates
- **Communicate budget implications** to clients and guide them toward financially sound decisions
- Apply tools like **RSMeans, life-cycle costing (LCCA), energy modeling**, and **budget tracking software** when appropriate

You'll also be expected to apply concepts like:

- Differentiating **hard vs. soft costs**
- Using **contingencies and escalation factors**
- Performing **cost-benefit evaluations** of design alternatives
- Strategically **adjusting scope, phasing, or materials** when a project exceeds budget

On the ARE 5.0 PPD Exam:

🔍 Expect questions that test your ability to:

- Identify appropriate estimating methods at different design phases
- Interpret and analyze budget data
- Make cost-conscious design decisions that align with project goals
- Distinguish between valid and invalid cost assumptions

Final Thought

Cost awareness is not separate from design—it is an **integral part of design leadership**. By thinking strategically about cost from the earliest stages, architects can create solutions that are not only beautiful and functional, but also **buildable and financially responsible**. Mastering this mindset is essential not just for passing the PPD exam, but for thriving in real-world architectural practice.

C. Evaluate cost considerations during the design process (A/E) (2-6%)

NCARB Requirements
Architects need to consider client priorities and life cycle costs when selecting materials and systems, and they must evaluate the cost effectiveness of various design decisions. As an architect, you will need to determine if a project design must be modified based on cost evaluations and budgetary needs.

Throughout the design process, architects must consider how each design decision affects the **project's cost**, not only in terms of **initial construction**, but also in **long-term value** and **client priorities**. This includes making thoughtful selections of materials and systems, evaluating energy and maintenance costs, and knowing when to **revise the design** to stay within budget.

1. Life-Cycle Cost Analysis (LCCA)

Cost evaluation during design is not only about selecting the lowest-cost option at the time of construction—it's about making **economically responsible decisions** across the **entire life span of the building**. This approach is called **Life-Cycle Cost Analysis (LCCA)**, and it helps architects and clients weigh the long-term value of design choices beyond initial capital expenses.

What LCCA Includes:
LCCA accounts for the **Total Cost of Ownership (TCO)** over a building's useful life, often 30–50 years. This includes:

- **Initial construction cost**
  (labor, materials, installation)
- **Energy consumption**
  (utilities such as electricity, gas, water usage, etc.)
- **Maintenance and repair**
  (routine upkeep, part replacements, service intervals)
- **Replacement cycles**
  (how often a system or material will need full replacement)
- **End-of-life costs**
  (removal, recycling, or disposal of building components)

These factors give clients a more **realistic financial picture** of their investment, helping them prioritize value over time, not just up-front affordability.

Why It Matters in Design
Architects use LCCA to:

- Justify premium materials or systems with better long-term performance
- Support sustainability goals through reduced energy and waste
- Compare alternatives (e.g., flat vs. sloped roof systems, HVAC types)
- Help clients make informed decisions based on *net cost*, not just *first cost*

LCCA is particularly valuable when evaluating:

- Envelope systems (high-performance glazing, insulation)
- Lighting (LED vs. fluorescent)
- HVAC systems (geothermal vs. standard split systems)
- Roofing systems (single-ply membrane vs. multi-ply built-up roofing)
- Flooring (VCT vs. polished concrete vs. terrazzo)

☑ **Example:**

Lighting System Comparison:

- *Option A*: Fluorescent fixtures
    - Lower initial cost: ~$75 per fixture
    - Bulb life: ~15,000 hours
    - Higher energy usage
- *Option B*: LED fixtures
    - Higher initial cost: ~$120 per fixture
    - Bulb life: ~50,000 hours
    - 40–60% lower energy use

While Option B costs more initially, energy savings and longer life reduce maintenance and utility expenses, often paying back the difference within 3–5 years. Over a 20-year period, the LED system results in significantly lower total cost of ownership.

🔍 **On the ARE 5.0 PPD Exam:**

You may be tested on your ability to:

- Compare two materials or systems based on **life-cycle cost performance**
- Recommend the most **cost-effective solution over time**, even if it's not the cheapest up-front
- Identify the correct factors that make up an LCCA analysis
- Select the best system or material based on client goals (e.g., energy savings vs. low maintenance)

Real-World Scenario:

A client is choosing between:

- **Built-up roofing (BUR)** with a 20-year lifespan and moderate maintenance
- **Single-ply membrane roofing** with a 12-year lifespan but lower up-front cost

LCCA shows that BUR, though more expensive to install, requires fewer replacements and less disruption over 40 years—ultimately saving $150,000 in lifecycle cost. The architect recommends BUR to meet the client's 50-year master plan goals.

Summary

LCCA empowers architects to go beyond short-term thinking and provide clients with long-term, data-driven insights. It supports smarter, more resilient, and sustainable design decisions—and is a **critical skill on the PPD exam** and in real-world practice.

2. **Material & System Selection**

One of the architect's most powerful tools for managing project costs is **thoughtful selection of materials and building systems**. From the building structure to interior finishes, every choice impacts the **construction budget**, **operating expenses**, and **long-term performance**. Material and system selection also directly affects user experience, project durability, and aesthetic quality.

Key Considerations When Selecting Materials and Systems:
Architects must evaluate materials and systems holistically, considering not only performance and appearance but also how each choice contributes to both short- and long-term project value. Major evaluation factors include:

- **Up-Front (First) Cost**
  Initial cost to purchase, deliver, and install the material or system. Often the most visible metric, but not the only one that matters.
- **Durability and Warranty**
  Lifespan and performance reliability over time, including manufacturer-provided warranties.
- **Installation Complexity**
  Some materials or systems require skilled labor, specialized tools, or longer install times—raising labor costs.
- **Aesthetic and Design Value**
  Visual quality and its contribution to the intended architectural character or brand identity.
- **Maintenance Needs**
  Ongoing cleaning, sealing, or replacement required to maintain performance and appearance.

☑ **Example:**
Flooring Material Comparison – Public Terminal

- *Option A: Resilient Vinyl Tile (VCT)*
  - Cost: ~$5–7/SF
  - Lifespan: 10–15 years
  - Requires regular waxing and stripping
- *Option B: Terrazzo*
  - Cost: ~$30–35/SF
  - Lifespan: 30+ years
  - Minimal maintenance, high durability

In a **high-traffic airport terminal**, terrazzo's initial premium is justified by its durability, low maintenance, and aesthetic appeal. Over the building's life span, terrazzo can be **more cost-effective** and consistent with a high-performance facility's goals.

Other Examples:

- **Cladding Systems:**
  - Aluminum composite panels may be more expensive than stucco initially, but offer greater weather resistance, colorfastness, and durability in urban or coastal climates.
- **HVAC Systems:**
  - A variable refrigerant flow (VRF) system may cost more than a split-system heat pump but could reduce energy use significantly in a multi-zone building.

- **Roofing Materials:**
  - A metal standing seam roof has a higher first cost than asphalt shingles, but can last 2–3 times longer and require less maintenance.
- **Structural Systems:**
  - A steel frame allows for longer spans and faster erection time but may be more expensive than wood framing for smaller-scale projects.

Strategic Thinking for Architects:
Your job as an architect is not just to select what "looks good," but to propose solutions that:

- **Align with project goals and constraints**
- Reflect the **client's values** (e.g., low maintenance, green building, long lifespan)
- Work within the **construction and operations budget**
- Balance **performance and aesthetics** with economic practicality

Sometimes, you may need to advocate for a higher-cost option that provides better value, durability, or sustainability—and justify it with data or LCCA (Life-Cycle Cost Analysis).

ARE 5.0 PPD Exam Tip:
🔍 You may be asked to:

- Choose between competing materials or systems based on cost, maintenance, and design impact
- Identify when a specific system (e.g., mass timber, concrete slab, TPO roofing) is appropriate based on budget or climate
- Weigh aesthetics vs. performance in material selection scenarios

Summary
Material and system choices **directly shape the project's financial profile** and long-term functionality. Wise selection is a hallmark of great design and smart stewardship. By factoring in durability, maintenance, performance, and aesthetics—not just first cost—you create buildings that **last, perform, and deliver value** over time.

3. **Cost-Effective Spatial Planning**
Strategic spatial planning is a powerful tool for managing construction and operational costs—**without compromising program goals or design quality**. During early design, decisions about building layout, room adjacencies, and circulation efficiency can significantly affect a project's budget.

Poorly organized plans can result in **wasted area, higher material and labor costs, and inefficient building systems**, while smart planning leads to leaner, more constructible, and more maintainable buildings.

Key Strategies for Cost-Effective Spatial Design

- **Minimize Unused or Redundant Circulation**
Long corridors, oversized lobbies, and complex transitions add square footage (and cost) without contributing functional value. Efficient circulation can reduce total building area and improve occupant flow.

- **Stack and Cluster Plumbing-Intensive Spaces**
  Grouping restrooms, janitor's closets, kitchens, and mechanical rooms reduces the length and complexity of plumbing and HVAC runs, saving on both rough-in and maintenance costs.
- **Use Modular or Repetitive Layouts**
  Repetition of standardized room sizes (especially in housing, hospitality, and education) simplifies framing, reduces material waste, and speeds construction.
- **Plan Structural and Mechanical Grids Efficiently**
  Aligning walls, columns, and MEP zones with consistent bays reduces structural complexity and allows for cost-effective system routing.
- **Zone for HVAC Efficiency**
  Place high-load spaces (like kitchens or server rooms) near exterior walls for easier venting, and group zones by similar thermal needs to reduce mechanical zoning complexity.

☑ **Example:**
Stacking and Clustering MEP Loads
In a three-story office building, placing restrooms, mechanical shafts, and janitorial closets **vertically aligned on each floor** allows for a single plumbing riser, shared vent stacks, and shorter pipe runs.
This strategy can save:

- Thousands in plumbing and HVAC labor
- Valuable floor area that would otherwise be occupied by distributed mechanical rooms
- Future maintenance costs by improving accessibility

Additional Examples:

- **Healthcare Facility:**
  Standardizing exam rooms along a single corridor using mirrored layouts can streamline construction and reduce custom framing or finishes.
- **Educational Buildings:**
  Grouping classrooms around a centralized core of restrooms and mechanical space reduces utility runs and simplifies egress.
- **Multi-Family Residential:**
  Stacking kitchens and bathrooms in apartment units above and below each other dramatically reduces the cost of plumbing and mechanical chases.

ARE 5.0 PPD Exam Tip:
🔍 On the exam, you may need to:

- Identify more efficient layout options
- Recognize which planning decisions increase construction complexity or cost
- Choose space arrangements that improve constructability or reduce system length

Summary
Cost-effective spatial planning is about **making every square foot work harder**. It involves aligning architecture with constructability, building systems, and budget realities—**without**

sacrificing user comfort, program functionality, or aesthetics. Thoughtful organization of space early in design can lead to significant cost savings across the life of the building.

4. **Energy Efficiency and Operational Costs**
Designing for energy efficiency is not only about meeting performance targets—it's also about understanding the **financial implications of operational costs** over the building's life cycle. Decisions related to **building orientation, envelope design, mechanical systems, and daylighting** directly affect the amount of energy a building consumes—and how much it costs the owner to operate and maintain it.

Architects must evaluate the **cost trade-offs** between investing more up front for better-performing systems and the potential for **long-term savings on utilities, maintenance, and equipment life span**.

Key Design Elements Affecting Energy and Cost:

- **Orientation & Massing**
  Positioning a building to optimize solar gain (in heating climates) or reduce it (in cooling climates) can reduce the size and load of mechanical systems.
- **Envelope & Insulation**
  High-performance insulation, continuous air barriers, and proper detailing of thermal breaks can reduce heating/cooling loads dramatically.
- **Glazing & Daylighting**
  Strategic window placement, glazing type (e.g., low-E coatings), and light shelves can reduce reliance on artificial lighting while managing solar heat gain.
- **HVAC System Type & Efficiency**
  Selecting high-efficiency systems (e.g., heat pumps, VRF, radiant heating) can reduce energy consumption, but often comes at a higher first cost—making life-cycle cost analysis critical.

☑ **Example:**
Insulation Upgrade Payback

- Increasing the roof insulation from R-30 to R-40 adds **$1.50/SF** in up-front cost.
- Over a 20,000 SF roof, that's a **$30,000 premium**.
- However, the improvement reduces heating and cooling loads by **20%**, saving **$4,000–5,000 per year** in energy costs.
- Payback period: ~6–8 years, after which the savings continue to accumulate.

This type of decision may not always be obvious without evaluating **both performance metrics and cost over time**.

Additional Examples:

- **Passive Solar vs. Mechanical Heating**
  Designing south-facing glazing with thermal mass (like concrete floors) can reduce winter heating demand significantly. However, this requires careful shading design to prevent overheating in summer.
- **Efficient Lighting & Controls**
  Installing daylight sensors and LED lighting may cost more than basic fixtures and

switches, but can cut lighting energy by 50% or more—especially in commercial applications.

- **High-Efficiency HVAC Systems**
  A geothermal heat pump system may cost 30–40% more than a standard rooftop unit but could cut HVAC operating costs by 50% and qualify for incentives or LEED points.
- **Air Sealing and Thermal Bridging**
  Investing in continuous air barriers and thermally broken assemblies prevents unwanted heat loss or gain, reducing HVAC costs and improving comfort.

ARE 5.0 PPD Exam Tip:

🔍 You may be asked to:

- Evaluate trade-offs between initial cost and operational energy savings
- Choose between envelope assemblies or mechanical systems based on performance and long-term cost
- Identify strategies that reduce utility bills without overburdening construction budgets

Summary

Energy efficiency must be **measured against its cost impact**, both short- and long-term. Architects play a vital role in recommending strategies that strike the right balance between **initial investment and ongoing savings**. By doing so, you deliver projects that are **environmentally responsible, financially sustainable, and technically sound**—a critical mindset for both real-world practice and success on the **PPD exam**.

5. **Responding to Budget Revisions or Scope Changes**

No matter how well planned, most projects will at some point face a **cost overrun** or **scope adjustment**. When the estimated cost of construction exceeds the approved budget, the architect must take the lead in identifying and recommending **strategic design modifications**—a process known as **value engineering** or **scope alignment**.

This is a critical moment in the project lifecycle: how effectively you respond can determine whether a project proceeds smoothly or faces delays, redesign, or client dissatisfaction.

Common Strategies for Bringing Costs Back in Line:

- **Replace High-Cost Systems or Finishes**
  Substitute premium or specialty systems with standard-grade alternatives that offer acceptable performance and aesthetics at a lower cost.
  - Example: Swap high-end terrazzo flooring with polished concrete, or copper roofing with standing seam metal panels.
- **Simplify Structural Systems**
  Reduce long spans, eliminate cantilevers, and align structural grids with simple geometries to reduce steel tonnage or complexity of formwork.
  - Example: Replace exposed glulam beams with concealed open-web steel joists in back-of-house areas.
- **Reduce Square Footage or Program Areas**
  Cut or phase less-critical spaces—like extra meeting rooms, storage, or optional amenities—while preserving core program functions.

  - Example: Eliminate a large lobby expansion or reconfigure circulation to shrink total floor area.
- **Reconfigure Layouts for Efficiency**
  Consolidate functions, reduce corridor lengths, or combine shared spaces to make better use of every square foot.
- **Delay or Phase Construction**
  Break the project into phases to defer costs. This allows clients to proceed with essential components now and add others later when funding becomes available.

☑ **Example:**
Community Center Value Engineering
Initial design for a civic building featured:

- Exposed glulam beams for aesthetic appeal
- High ceilings in all spaces
- Extensive specialty finishes

When estimates exceeded budget by 12%, the architect:

- Replaced glulam with concealed steel joists in utility areas
- Lowered ceiling height in offices and support spaces
- Standardized finish materials across all non-public rooms

These changes maintained the core design intent in public-facing spaces while reducing cost by over $600,000.

Additional Ideas:

- **Switch Mechanical Systems**
  Instead of a VRF system, consider a packaged rooftop unit with zoning—simpler to install and maintain.
- **Reduce Custom Detailing**
  Simplify or eliminate unique millwork, custom staircases, or highly detailed facades that require skilled labor.
- **Use Design-Build Feedback Loops**
  Collaborate with contractors early to get cost input and spot overruns before construction documents are complete.

ARE 5.0 PPD Exam Tip:
🔍 Be ready to:

- Identify design modifications that reduce cost without compromising core functionality
- Recommend systems, materials, or strategies for value engineering
- Prioritize what to cut or keep based on client goals, program needs, and performance trade-offs

Summary
An architect's ability to **respond constructively to budget overruns** is essential in both professional practice and on the ARE. Your role is not just to design, but to strategically manage project scope, recommend alternatives, and **preserve the integrity of the design** while ensuring the project remains financially viable. Effective cost reconciliation is a hallmark of professional leadership.

6. **Client Priorities and Project Goals**

All cost-related decisions in a project should be **firmly rooted in the client's values, goals, and expectations**. While architects bring technical expertise, the most appropriate choices ultimately depend on what matters most to the client—whether it's durability, speed, sustainability, flexibility, aesthetics, or long-term operating costs.

The architect's responsibility is to **align design decisions with client-defined priorities**, even when budgets are tight or compromises are needed.

Common Client Prioritization Questions:

- Is **durability** more important than initial cost?
- Does the client prioritize **energy efficiency**, or is **speed of construction** a higher concern?
- Are they willing to invest in **higher-quality materials** for reduced future maintenance?
- Is **flexibility** (adaptable layouts or systems) important for future growth?
- Will **aesthetic or branding goals** outweigh certain economic trade-offs?

Understanding these preferences early—and revisiting them regularly—allows architects to make **cost-conscious decisions** that still meet or exceed expectations.

☑ **Example:**

Municipal Fire Station Design

A city is planning a new fire station and is faced with two wall system options for interior partitions:

- *Option A: Metal stud with gypsum board*
  - Lower first cost
  - Less durable under heavy use
- *Option B: Concrete masonry units (CMU)*
  - Slightly higher initial cost
  - Far greater impact resistance, lower maintenance

Given that fire stations endure constant use, heavy equipment movement, and potential abuse, the municipality prioritizes **durability and low long-term maintenance**. Despite the higher up-front cost, CMU is the **better-value option** for their needs.

More Examples of Client-Driven Decisions:

- **University Housing:**
  May prioritize **long life span and low maintenance** over sleek design, favoring hard flooring over carpet and metal over wood trims.
- **Tech Company Headquarters:**
  Might place emphasis on **rapid delivery and flexibility**, accepting modular construction methods and open-plan layouts to support growth.
- **Public Library:**
  Could prioritize **daylighting and sustainability**, justifying the cost of high-performance glazing and automated lighting controls to enhance occupant comfort and reduce operational costs.

- **Healthcare Facilities:**
  Often demand **infection control and code compliance**, leading to material and system choices based on cleanability and regulatory requirements rather than aesthetics.

Architect's Role:
The architect must not only listen to client input but also **translate those values into clear design strategies**. That means:
- Framing options in terms of **value and alignment with project goals**
- Avoiding personal bias toward design preferences that don't serve the client's priorities
- Making sure every cost-related choice supports **long-term functionality, satisfaction, and performance**

ARE 5.0 PPD Exam Tip:
🔍 Expect questions that ask you to:
- Choose between options based on stated client values
- Recommend a system, material, or layout that reflects cost alignment with project goals
- Prioritize budget trade-offs while preserving core programmatic or performance objectives

Summary
Cost-effective design isn't just about dollars—it's about **delivering value that reflects what the client truly cares about**. Architects must ensure that every recommendation—especially in budget-sensitive decisions—reinforces the client's goals. The most successful solutions are those that balance cost, performance, and mission.

Cost considerations must be woven into **every phase of the design process**, not treated as a separate task or an afterthought. Successful architects are those who can **design creatively within financial constraints**, ensuring that every decision adds value to the project without compromising functionality, aesthetics, or performance.

Architects must be able to:
- **Balance short-term vs. long-term costs**
  Weigh initial construction expenses against ongoing operational, maintenance, and replacement costs using life-cycle thinking.
- **Evaluate financial trade-offs between systems, materials, and layouts**
  Recognize where higher up-front investments lead to greater long-term savings—and where simpler, more cost-effective alternatives are sufficient.
- **Modify the design proactively** when costs begin to exceed the approved budget
  Offer informed suggestions for scope reduction, material substitution, or system changes that retain core project goals.
- **Advocate for best value**, not just the lowest initial price
  Help clients understand where strategic spending improves building performance, durability, or user satisfaction over time.

🔍 **On the ARE 5.0 PPD Exam, you should be prepared to:**

- **Choose the most cost-effective system, layout, or material** based on client goals and budget
  (e.g., selecting stacked plumbing cores or choosing between cladding systems based on cost per square foot and durability)
- **Evaluate design alternatives using both construction and life-cycle cost data**
  (e.g., analyzing if the long-term savings of high-performance glazing justify its higher up-front cost)
- **Recommend appropriate design changes when a project exceeds budget**
  (e.g., suggesting structural simplification, finish substitutions, or square footage reduction)

D. PPD Exam Strategy Tips

The **ARE 5.0 Project Planning & Design (PPD)** division tests your ability to make informed architectural decisions that balance systems, code, site, and programmatic needs. This exam is not about engineering calculations—it's about how **systems and planning concepts interact** to shape a successful building.

Ten Proven Strategies:

Below are proven strategies to help you navigate this complex exam and identify the best responses on test day.

☑ 1) Focus on System Understanding, Not Deep Engineering

- You don't need to size ductwork or calculate beam spans. Instead, understand **how systems function**, their benefits and limitations, and how they interact with building design.
- Know the **big-picture roles** of structural, HVAC, plumbing, electrical, and life safety systems.

🧠 *Example:* Be ready to recognize that **a chilled beam system** offers energy efficiency but may not be ideal in high-humidity climates due to condensation risk.

☑ 2) Make Informed Comparative Judgments

- Many questions ask you to **compare systems** based on climate, cost, sustainability, or spatial requirements.
- Use process-of-elimination to eliminate options that **conflict with program, climate, or budget**.

🧠 *Example:* You might be asked whether a **ground-source heat pump or air-source heat pump** is more appropriate. In a cold northern climate, the **ground-source system** may be more efficient despite higher upfront costs.

☑ 3) Watch for System Mismatches in Case Studies

Case studies often present **problematic or mismatched systems** that conflict with the program or site.

Look for red flags like:

- **Oversized rooftop units** blocking daylight or adding unnecessary weight

- **Electrical rooms located far from vertical risers** or lacking NEC-required clearance
- **Restrooms not vertically aligned**, complicating plumbing stacks
- **MEP equipment located in acoustically sensitive areas**, such as above classrooms or theaters

🧠 *Example:* If the case study shows **rooftop air handlers** placed above a library reading room, the best response may involve recommending **relocation or noise mitigation strategies.**

☑ **4) Use Climate and Site Context to Guide System Selection**

Many systems perform well only under the right **climatic conditions** or **site constraints**. Understand how **climate zones**, **solar orientation**, **prevailing winds**, and **soil conditions** impact decisions.

🧠 *Example:* A **passive solar heating strategy** is ineffective in a hot-humid climate; instead, cross-ventilation and shading strategies are more appropriate.

🧠 *Example:* A **slab-on-grade foundation** is unsuitable on a site with expansive clay soils—expect test questions that highlight these geotechnical red flags.

☑ **5) Apply Energy and Space Planning Principles**

Energy performance and space efficiency are commonly tested.

- Be familiar with **Energy Use Intensity (EUI)** and how mechanical system choices affect it.
- Know how structural and MEP systems influence **floor-to-floor heights**, **plenum space**, and **service coordination**.

🧠 *Example:* A **flat slab concrete structure** allows tight floor-to-floor dimensions, which is beneficial in height-restricted urban infill projects.

🧠 *Example:* A **multi-tenant office** might use **VAV systems** for zone flexibility, whereas a **warehouse** could use simple unit heaters and ventilation fans to reduce operating costs.

☑ **6) Know What's Typically Tested**

The PPD exam favors **strategic, concept-driven knowledge**. High-yield areas include:

Topic	**Typical Focus on Exam**
HVAC Systems	Efficiency, zoning, control, climate appropriateness
Structural Systems	Material selection, spans, cost, compatibility
MEP Coordination	Layout efficiency, space planning, maintenance access
Energy & Water Efficiency	Low-flow fixtures, LEED/Title 24 targets, HVAC controls
Code Requirements	IBC, NEC, accessibility (ADA reach ranges, fixture clearances)
Life Safety & Fire Protection	Sprinkler types, egress, compartmentation

Figure 7.2 High-yield areas

🧠 Tip: Use **NCARB's Handbook** and **ARE 5.0 Guidelines** to cross-check which objectives are tested most heavily. PPD overlaps significantly with **PA** and **PDD**, so make connections across topics.

☑ 7) Practice Reading Diagrams and Case Studies

PPD includes **case studies** with floor plans, mechanical layouts, schedules, or zoning information. Practice interpreting:

- RCPs and mechanical plans (e.g., duct runs, diffusers, lighting)
- Structural grids, column locations, and section cuts
- Code tables (e.g., occupancy type, construction type, allowable area)

🧠 *Example:* A plan may show a **plenum that's too shallow** for ductwork due to deep structural beams. Your task is to identify the mismatch and recommend a suitable structural change or system redesign.

☑ 8) Don't Overthink – Eliminate, Then Choose Best Fit

Many questions have **more than one seemingly acceptable answer**. Your task is to select the **most appropriate solution**—based on:

- Climate
- Occupancy
- Budget
- Code
- Energy goals

When in doubt, eliminate:

❌ Systems not allowed by code

❌ Choices that exceed budget or exceed functional needs

❌ Systems that conflict with project goals (e.g., LEED certification, fast-track schedule)

☑ 9) Use Real-World Logic

Imagine yourself as a project architect—**what would you realistically do** on the job?

🧠 *Example:* If you're designing a school and the MEP system is in an inaccessible attic with no service hatch, it's a poor design—even if it's "code compliant." Look for practical, maintainable solutions.

☑ 10) Time Management + Review

- You have **125 questions** and **4 hours** (including 2 case studies).
- Don't spend too long on any one question—**flag and return later**.
- Use your final time to **review flagged questions** and verify any **units, dimensions**, or **code values**.

📌 Final Thoughts

The PPD exam tests your ability to design smartly, holistically, and within constraints. It's about making **balanced, integrated decisions**—not being an engineer, but being a thoughtful architect.

🧠 **Summary Checklist for PPD Readiness**

✓ Understand the pros/cons of structural, HVAC, plumbing, and electrical systems
✓ Know code and fire-rating impacts on system selection
✓ Evaluate system appropriateness for climate, cost, and occupancy
✓ Recognize design coordination issues in plans and sections
✓ Practice case study skills—reading diagrams, schedules, and code tables
✓ Focus on strategy and judgment, not memorization

Appendixes

A. List of Figures

B. Official reference materials suggested by NCARB

1. Resources Available While Testing

Tips:

- *You need to read through these pages several times and become very familiar with them to save time in the real ARE exams.*

United States. American Institute of Steel Construction, Inc. *Steel Construction Manual*; 14th edition. Chicago, Illinois, 2011.

Beam Diagrams and Formulas:

- Simple Beam: Diagrams and Formulas - Conditions 1-3, page 3-213; Conditions 4-6, page 3-214; Conditions 7-9, page 3-215
- Beam Fixed at Both Ends: Diagrams and Formulas - Conditions 15-17, page 3-218
- Beam Overhanging One Support: Diagrams and Formulas - Conditions 24-28, pages 3-221 & 222

Dimensions and Properties:

- W Shapes 44 thru 27: Dimensions and Properties, pages 1-12 thru 17
- W Shapes 24 thru W14x145: Dimensions and Properties, pages 1-18 thru 23
- W Shapes 14x132 thru W4: Dimensions and Properties, pages 1-24 thru 29
- C Shapes: Dimensions and Properties, pages 1-36 & 37
- Angles: Properties, pages 1-42 thru 49
- Rectangular HSS: Dimensions and Properties, pages 1-74 thru 91
- Square HSS: Dimensions and Properties, pages 1-92 thru 95
- Round HSS: Dimensions and Properties, pages 1-96 thru 100

United States. International Code Council, Inc. *2012 International Building Code*. Country Club Hills, Illinois, 2011.

Live and Concentrated Loads:

- Uniform and Concentrated Loads: IBC Table 1607.1, pages 340-341

2. Typical Beam Nomenclature

The following typical beam nomenclature is excerpted from:
United States. American Institute of Steel Construction, Inc. *Steel Construction Manual*; 14th edition. Chicago, Illinois, 2011.

E	Modulus of Elasticity of steel at 29,000 ksi	V_2	Vertical shear at right reaction point, or to left of intermediate reaction of beam, kips
I	Moment of Inertia of beam, in^4	V_3	Vertical shear at right reaction point, or to right of intermediate reaction of beam, kips
L	Total length of beam between reaction point, ft	V_x	Vertical shear at distance x from end of beam, kips
M_{max}	Maximum moment, kip-in	W	Total load on beam, kips
M_1	Maximum moment in left section of beam, kip-in	A	Measured distance along beam, in
M_2	Maximum moment in right section of beam, kip-in	B	Measured distance along beam which may be greater or less than a, in
M_3	Maximum positive moment in beam with combined end moment conditions, kip-in	L	Total length of beam between reaction points, in
M_x	Maximum at distance x from end of beam, kip-in	W	Uniformly distributed load per unit of length, kips/in
P	Concentrated load, kips	w_1	Uniformly distributed load per unit of length nearest left reaction, kips/in
P_1	Concentrated load nearest left reaction, kips	w_2	Uniformly distributed load per unit of length nearest right reaction and of different magnitude than w1, kips/in
P_2	Concentrated load nearest right reaction and of different magnitude than P_1, kips	X	Any distance measured along beam from left reaction, in
R	End beam reaction for any condition of symmetrical loading, kips	x_1	Any distance measured along overhang section of beam from nearest reaction point, in
R_1	Left end beam reaction, kips	Δ_{max}	Maximum deflection, in
R_2	Right end or intermediate beam reaction, kips	Δa	Deflection at point of load, in
R_3	Right end beam reaction, kips	Δx	Deflection at point x distance from left reaction, in
V	Maximum vertical shear for any condition of symmetrical loading, kips	Δx_1	Deflection of overhang section of beam at any distance from nearest reaction point, in
V_1	Maximum vertical shear in left section of beam, kips		

3. Formulas Available While Testing

Tips:

- *These formulas and references will be available during the real exam. You should read through them a few times before the exam to become familiar with them. This will save you a lot of time during the real exam, and will help you solve structural calculations and other problems.*

Structural:

Flexural stress at extreme fiber

$$f = \frac{Mc}{I} = \frac{M}{S}$$

Flexural stress at any fiber

$$f = \frac{My}{I}$$

where y = distance from neutral axis to fiber

Average vertical shear

$$v = \frac{V}{A} = \frac{V}{dt}$$

for beams and girders

Horizontal shearing stress at any section A-A

$$v = \frac{VQ}{Ib}$$

where Q = statical moment about the neutral axis of the entire section of that portion of the cross-section lying outside of section A-A
b = width at section A-A

Electrical

$$Foot - candles = \frac{lumens}{area\ in\ ft^2}$$

$$Foot - candles = \frac{(lamp\ lumens)\ x\ (lamps\ per\ fixture)\ x\ (number\ of\ fixtures)\ x\ (CU)\ x\ (LLF)}{area\ in\ ft^2}$$

$$Number\ of\ luminaires = \frac{(foot - candles)\ x\ (floor\ area)}{(lumens)\ x\ (CU)\ x\ (LLF)}$$

where CU = coefficient of utilization
LLF = Light Loss Factor

$$DF_{AV} = 0.2x \frac{window\ area}{floor\ area}$$
for spaces with sidelighting or toplighting with vertical monitors

$$\textbf{watts} = \textbf{volts x amperes x power factor}$$
for AC circuits only

$$Demand\ charge = maximum\ power\ demand\ x\ demand\ tariff$$

Plumbing

$$1\ psi = 2.31\ feet\ of\ water$$

$$1\ cubic\ foot = 7.5\ U.S.\ gallons$$

HVAC

$$\frac{BTU}{year} = peak\ heat\ loss\ x\ \frac{full - load\ hours}{year}$$

$$BTU/h = (cfm)\ x\ (1.08)\ x\ (\Delta T)$$

$$1\ kWh = 3,400\ BTU/h$$

$$1\ ton\ of\ air\ conditioning = 12,000\ BTU/h$$

$BTU/h = (U)\ x\ (A)\ x\ (T_d)$ *where Td is the difference between indoor and outdoor temperatures*

$$U = 1/R_t$$

$$U_o = \frac{(U_w \times A_w) + (U_{op} \times A_{op})}{Ao}$$
where o = total wall, w = window, and op = opaque wall

$$U_o = \frac{(U_R \times A_R) + (U_S \times A_S)}{Ao}$$

where o = total roof, R = roof, and S = skylight

$$R = x/k$$

where x = thickness of material in inches

$$Heat\ required = \frac{BTU/h}{temperature\ differential} \times (24\ hours) \times (DD\ °F)$$

where DD = degree days

Acoustics

$$\lambda = \frac{c}{f}$$

where λ = wavelength of sound (ft)
c = velocity of sound (fps)
f = frequency of sound (Hz)

$$a = SAC\ x\ S$$

where a = Absorption of a material used in space (sabins)
SAC = Sound Absorption Coefficient of the material
S = Exposed surface area of the material (ft ^{2})

$$A = \Sigma a$$

Where A = Total sound absorption of a room (sabins)
$\Sigma a = (S_1\ x\ SAC_1) + (S_2\ x\ SAC_2) + \ldots$

$$T = 0.05 \times \frac{V}{A}$$

where T = Reverberation time (seconds)
V = Volume of space (ft ^{3})

$$NRC = average\ SAC\ for\ frequency\ bands\ 250, 500, 1000, and\ 2000\ Hz$$

4. Common Abbreviations

Tips:

- *You need to read through these common abbreviations several times and become very familiar with them to save time in the real ARE exams.*

Professional Organizations, Societies, and Agencies

American Concrete Institute	ACI
American Institute of Architects	AIA
American Institute of Steel Construction	AISC
American National Standards Institute	ANSI
American Society for Testing and Materials	ASTM
American Society of Civil Engineers	ASCE
American Society of Heating, Refrigerating, and Air-Conditioning Engineers	ASHRAE
American Society of Mechanical Engineers	ASME
American Society of Plumbing Engineers	ASPE
Architectural Woodwork Institute	AWI
Construction Specifications Institute	CSI
Department of Housing and Urban Development	HUD
Environmental Protection Agency	EPA
Federal Emergency Management Agency	FEMA
National Fire Protection Association	NFPA
Occupational Safety and Health Administration	OSHA
U.S. Green Building Council	USGBC

Tips:

- *You need to look through the following codes and regulations & AIA contract documents several times and become very familiar with them to save time in the real ARE exams. Read some of the important sections in details.*

AIA Contract Documents

A101-2007, Standard Form of Agreement Between Owner and Contractor - Stipulated Sum	A101
A201-2007, General Conditions of the Contract for Construction	A201
A305-1986, Contractor's Qualification Statement	A305
A701-1997, Instructions to Bidders	A701
B101-2007, Standard Form of Agreement Between Owner and Architect	B101
C401-2007, Standard Form of Agreement Between Architect and Consultant	C401
G701-2001, Change Order	G701
G702-1992, Application and Certificate for Payment	G702
G703-1992, Continuation Sheet	G703
G704-2000, Certificate of Substantial Completion	G704

Codes and Regulations

ADA Standards for Accessible Design	ADA

International Code Council	ICC
International Building Code	IBC
International Energy Conservation Code	IECC
International Existing Building Code	IEBC
International Mechanical Code	IMC
International Plumbing Code	IPC
International Residential Code	IRC
Leadership in Energy and Environmental Design	LEED
National Electrical Code	NEC

Commonly Used Terms

Air Handling Unit	AHU
Authority Having Jurisdiction	AHJ
Building Information Modeling	BIM
Concrete Masonry Unit	CMU
Contract Administration	CA
Construction Document	CD
Dead Load	DL
Design Development	DD
Exterior Insulation and Finish System	EIFS
Furniture, Furnishings & Equipment	FF&E
Floor Area Ratio	FAR
Heating, Ventilating, and Air Conditioning	HVAC
Insulating Glass Unit	IGU
Indoor Air Quality	IAQ
Indoor Environmental Quality	IEQ
Live Load	LL
Material Safety Data Sheets	MSDS
Photovoltaic	PV
Reflected Ceiling Plan	RCP
Schematic Design	SD
Variable Air Volume	VAV
Volatile Organic Compound	VOC
British Thermal Unit	btu
Cubic Feet per Minute	cfm
Cubic Feet per Second	cfs
Cubic Foot	cu. ft. ft^3
Cubic Inch	cu. in. in^3
Cubic Yard	cu. yd. yd^3
Decibel	dB
Foot	ft
Foot-candle	fc
Gross Square Feet	gsf

Impact Insulation Class	IIC
Inch	in
Net Square Feet	nsf
Noise Reduction Coefficient	NRC
Pound	lb
Pounds per Linear Foot	plf
Pounds per Square Foot	psf
Pounds per Square Inch	psi
Sound Transmission Class	STC
Square Foot	sq. ft. sf ft^2
Square Inch	sq. in. in^2
Square Yard	sq. yd.

5. **General NCARB reference materials for ARE:**

   Per NCARB, all candidates should become familiar with the latest version of the following codes:

   International Code Council, Inc. (ICC)
   International Building Code
   International Mechanical Code
   International Plumbing Code

   National Fire Protection Association (NFPA)
   Life Safety Code (NFPA 101)
   National Electrical Code (NFPA 70)

   National Research Council of Canada
   National Building Code of Canada
   National Plumbing Code of Canada
   National Fire Code of Canada

   American Institute of Architects
   AIA Documents

6. **Official NCARB reference materials matrix**
Per NCARB, all candidates should become familiar with the latest version of the following:

Reference	PcM	PjM	PA	PPD	PDD	CE
ADA Standards for Accessible Design U.S. Department of Justice, Latest Edition			■	■	■	
Code of Ethics and Professional Conduct AIA Office of General Counsel. The American Institute of Architects, latest edition	■					
The Architect's Handbook of Professional Practice The American Institute of Architects John Wiley & Sons, latest edition	■	■	■	■	■	■
The Architect's Studio Companion: Rules of Thumb for Preliminary Design Edward Allen and Joseph Iano John Wiley & Sons, 6th edition, 2017			■	■	■	
Architectural Acoustics. M. David Egan. J. Ross Publishing. Reprint. Original publication McGraw Hill, latest edition				■	■	
Architectural Graphic Standards The American Institute of Architects John Wiley & Sons, latest edition			■	■	■	
Building Codes Illustrated: A Guide to Understanding the International Building Code. Francis D.K. Ching and Steven R. Winkel, FAIA, PE. John Wiley & Sons, latest edition			■	■	■	
Building Construction Illustrated Francis D. K. Ching John Wiley & Sons, latest edition				■	■	
Building Structures James Ambrose and Patrick Tripeny John Wiley & Sons, 3rd edition, Latest Edition				■	■	
CSI MasterFormat. The Construction Specifications Institute, latest edition					■	■
Daylighting Handbook I Christoph Reinhart Building Technology Press, latest edition			■	■		
Dictionary of Architecture and Construction. Cyril M. Harris. McGraw-Hill, Latest edition			■	■	■	
Framework for Design Excellence American Institute of Architects Available Online			■	■		

Reference	PcM	PjM	PA	PPD	PDD	CE
Fundamentals of Building Construction: Materials and Methods Edward Allen and Joseph Iano John Wiley & Sons, latest edition				■	■	
Green Building Illustrated Francis D.K. Ching and Ian M. Shapiro Wiley, latest edition				■	■	
The Green Studio Handbook: Environmental Strategies for Schematic Design Alison G. Kwok and Walter Grondzik Routledge, latest edition			■	■		
Heating, Cooling, Lighting: Sustainable Design Methods for Architects. Norbert Lechner. John Wiley & Sons, latest edition				■	■	
The HOK Guidebook to Sustainable Design Sandra F. Mendler, William Odell, and Mary Ann Lazarus John Wiley & Sons, latest edition			■	■	■	
ICC A117.1-2009 Accessible and Usable Buildings and Facilities International Code Council, 2010			■	■	■	
International Building Code International Code Council, latest edition			■	■	■	
Law for Architects: What You Need to Know. Robert F. Herrmann and the Attorneys at Menaker & Herrmann LLP. W. W. Norton, latest edition	■					
Legislative Guidelines and Model Law/Model Regulations National Council of Architectural Registration Boards, latest edition	■					
Mechanical & Electrical Equipment for Buildings. Walter T. Grondzik, Alison G. Kwok, Benjamin Stein, and John S. Reynolds, Editors. John Wiley & Sons, latest edition				■	■	
Mechanical and Electrical Systems in Buildings. Richard R. Janis and William K. Y. Tao. Prentice Hall, latest edition				■	■	
Model Rules of Conduct National Council of Architectural Registration Boards, latest edition	■					

Reference	PcM	PjM	PA	PPD	PDD	CE
Olin's Construction Principles, Materials, and Methods. H. Leslie Simmons. John Wiley & Sons, latest edition				■	■	
Planning and Urban Design Standards American Planning Association John Wiley & Sons, latest edition			■	■		
Plumbing, Electricity, Acoustics: Sustainable Design Methods for Architecture. Norbert Lechner. John Wiley & Sons, latest edition				■	■	
Problem Seeking: An Architectural Programming Primer William M. Peña and Steven A. Parshall John Wiley & Sons, latest edition			■			
Professional Practice: A Guide to Turning Designs into Buildings. Paul Segal, FAIA. W. W. Norton, latest edition	■	■	■			
The Professional Practice of Architectural Working Drawings. Osamu A. Wakita, Nagy R. Bakhoum, and Richard M. Linde. John Wiley & Sons, latest edition				■	■	
The Project Resource Manual: CSI Manual of Practice. The Construction Specifications Institute. McGraw-Hill, latest edition		■			■	■
Simplified Engineering for Architects and Builders James Ambrose and Patrick Tripeny John Wiley & Sons, latest edition				■	■	
Site Planning and Design Handbook Thomas H. Russ McGraw-Hill, latest edition			■	■		
Space Planning Basics Mark Karlen and Rob Fleming John Wiley & Sons, latest edition			■			
Steel Construction Manual American Institute of Steel Construction Ingram, latest edition					■	
Structural Design: A Practical Guide for Architects James R. Underwood and Michele Chiuini John Wiley & Sons, latest edition				■	■	
Structures Daniel Schodek and Martin Bechthold Pearson/Prentice Hall, latest edition						

Reference	PcM	PjM	PA	PPD	PDD	CE
Sun, Wind, and Light: Architectural Design Strategies G.Z. Brown and Mark DeKay John Wiley & Sons, latest edition			■	■		
Sustainable Construction: Green Building Design and Delivery Charles J. Kibert. John Wiley & Sons, latest edition				■		
A Visual Dictionary of Architecture Francis D.K. Ching John Wiley & Sons, latest edition				■	■	

The following AIA Contract Documents have content covered in the of ARE 5.0 exams. Candidates can access them for free through their NCARB Record.

Document	PcM	PjM	PA	PPD	PDD	CE
A101-2017, Standard Form of Agreement Between Owner and Contractor where the basis of payment is a Stipulated Sum		■				■
A133-2019, Standard Form of Agreement Between Owner and Construction Manager as Constructor where the basis of payment is the Cost of the Work Plus a Fee with a Guaranteed Maximum Price		■				
A195-2008, Standard Form of Agreement Between Owner and Contractor for Integrated Project Delivery		■				
A201-2017, General Conditions of the Contract for Construction		■				■
A295-2008, General Conditions of the Contract for Integrated Project Delivery		■				
A305-1986, Contractor's Qualification Statement						■
A701-2018, Instructions to Bidders						■
B101-2017, Standard Form of Agreement Between Owner and Architect	■	■				■
B195-2008, Standard Form of Agreement Between Owner and Architect for Integrated Project Delivery		■				

Document	PcM	PjM	PA	PPD	PDD	CE
C401-2017, Standard Form of Agreement Between Architect and Consultant						
G701-2017, Change Order						
G702-1992, Application and Certificate for Payment						
G703-1992, Continuation Sheet						
G704-2017, Certificate of Substantial Completion						

The following are some extra study materials if you have some additional time and want to learn more. If you are tight on time, you can simply look through them and focus on the sections that cover your weakness:

ACI Code 318-05 (Building Code Requirements for Reinforced Concrete)
American Concrete Institute, 2005

OR
CAN/CSA-A23.1-94 (Concrete Materials and Methods of Concrete Construction) and CAN/CSA-A23.3-94 (Design of Concrete Structures for Buildings)
Canadian Standards Association

Design Value for Wood Construction
American Wood Council, 2005

Elementary Structures for Architects and Builders, Fourth Edition
Ronald E. Shaeffer
Prentice Hall, 2006

Introduction to Wood Design
Canadian Wood Council, 2005

Manual of Steel Construction: Allowable Stress Design; 9th Edition.
American Institute of Steel Construction, Inc. Chicago, Illinois, 1989

National Building Code of Canada, 2005
Parts 1, 3, 4, 9; Appendix A
Supplement
Chapters 1, 2, 4; Commentaries A, D, F, H, I

NEHRP (National Earthquake Hazards Reduction Program) Recommended Provisions for Seismic Regulations for New Buildings and Other Structures Parts 1 and 2
FEMA 2003

Simplified Building Design for Wind and Earthquake Forces

James Ambrose and Dimitry Vergun
John Wiley & Sons, 1997

Simplified Design of Concrete Structures,
Eighth Edition
James Ambrose, Patrick Tripeny
John Wiley & Sons, 2007

Simplified Design of Masonry Structures
James Ambrose
John Wiley & Sons, 1997

Simplified Design of Steel Structures, Eighth Edition
James Ambrose, Patrick Tripeny
John Wiley & Sons, 2007

Simplified Design of Wood Structures, Fifth Edition
James Ambrose
John Wiley & Sons, 2009

Simplified Mechanics and Strength of Materials, Fifth Edition
Harry Parker and James Ambrose
John Wiley & Sons, 2002

Standard Specifications Load Tables & Weight Tables for Steel Joists and Joist Girders
Steel Joist Institute, latest edition

Steel Construction Manual, Latest edition
American Institute of Steel Construction, 2006

OR
Handbook of Steel Construction, Latest edition; and *CAN/CSA-S16-01 and CISC Commentary*
Canadian Institute of Steel Construction

Steel Deck Institute Tables
Steel Deck Institute

OR
LSD Steel Deck Tables
Caradon Metal Building Products

Structural Concepts and Systems for Architects and Engineers, Second Edition
T.Y. Lin and Sidney D. Stotesbury
Van Nostrand Reinhold, 1988

Structural Design: A Practical Guide for Architects
James Underwood and Michele Chiuini
John Wiley & Sons, latest edition

Structure in Architecture: The Building of Buildings
Mario Salvadori with Robert Heller
Prentice-Hall, 1986

Understanding Structures
Fuller Moore
McGraw-Hill, 1999

Wood Design Manual and *CAN/CSA-086.1-94 and Commentary*
Canadian Wood Council

C. Other reference materials

Chen, Gang. *Building Construction: Project Management, Construction Administration, Drawings, Specs, Detailing Tips, Schedules, Checklists, and Secrets Others Don't Tell You (Architectural Practice Simplified, 2nd edition).* ArchiteG, Inc., A good introduction to the architectural practice and construction documents and service, including discussions of MasterSpec format and specification sections.

Chen, Gang. ***LEED v4 Green Associate Exam Guide (LEED GA):*** *Comprehensive Study Materials, Sample Questions, Mock Exam, Green Building LEED Certification, and Sustainability*, Book 2, LEED Exam Guide series, ArchiteG.com, the latest edition. ArchiteG, Inc. Latest Edition. This is a very comprehensive and concise book on the LEED Green Associate Exam. Some readers have passed the LEED Green Associate Exam by studying this book for 10 hours.

Ching, Francis. *Architecture: Form, Space, & Order.* Wiley, latest edition. It is one of the best architectural books that you can have. I still flip through it every now and then. It is a great book for inspiration.

Frampton, Kenneth. *Modern Architecture: A Critical History*. Thames and Hudson, London, latest edition. A valuable resource for architectural history.

Jarzombek, Mark M. (Author), Vikramaditya Prakash (Author), Francis D. K. Ching (Editor). *A Global History of Architecture*. Wiley, latest edition. A valuable and comprehensive resource for architectural history with 1000 b & w photos, 50 color photos, and 1500 b & w illustrations. It doesn't limit the topic on a Western perspective, but rather through a global vision.

Trachtenberg, Marvin and Isabelle Hyman. *Architecture: From Pre-history to Post-Modernism*. Prentice Hall, Englewood Cliffs, NJ latest edition. A valuable and comprehensive resource for architectural history.

D. Some Important Information about Architects and the Profession of Architecture

What Architects Do?

Architects plan and design houses, factories, office buildings, and other structures.

Duties
Architects typically do the following:

- Meet with clients to determine objectives and requirements for structures
- Give preliminary estimates on cost and construction time
- Prepare structure specifications
- Direct workers who prepare drawings and documents
- Prepare scaled drawings, either with computer software or by hand
- Prepare contract documents for building contractors
- Manage construction contracts
- Visit worksites to ensure that construction adheres to architectural plans
- Seek new work by marketing and giving presentations

People need places to live, work, play, learn, shop, and eat. Architects are responsible for designing these places. They work on public or private projects and design both indoor and outdoor spaces. Architects can be commissioned to design anything from a single room to an entire complex of buildings.

Architects discuss with clients the objectives, requirements, and budget of a project. In some cases, architects provide predesign services, such as feasibility and environmental impact studies, site selection, cost analyses, and design requirements.

Architects develop final construction plans on the initial proposal after discussing with clients. The architects' plans show the building's appearance and details of its construction. These plans include drawings of the structural system; air-conditioning, heating, and ventilating systems; electrical systems; communications systems; and plumbing. Sometimes, landscape plans are included as well. In developing designs, architects must follow state and local building codes, zoning laws, fire regulations, and other ordinances, such as those requiring reasonable access for people with disabilities.

Architects use computer-aided design and drafting (CADD) and building information modeling (BIM) for creating designs and construction drawings. However, hand-drawing skills are still required, especially during the conceptual stages of a project and when an architect is at a construction site.

As construction continues, architects may visit building sites to ensure that contractors follow the design, adhere to the schedule, use the specified materials, and meet work-quality standards. The job is not complete until all construction is finished, required tests are conducted, and construction costs are paid.

Architects may also help clients get construction bids, select contractors, and negotiate construction contracts.

Architects often collaborate with workers in related occupations, such as civil engineers, urban and regional planners, drafters, interior designers, and landscape architects.

Work Environment

Architects held about 127,300 jobs in 2023. The largest employers of architects were as follows:

Architectural, engineering, and related services	75%
Self-employed workers	13
Construction	3
Government	3

Architects spend much of their time in offices, where they meet with clients, develop reports and drawings, and work with other architects and engineers. They also visit construction sites to ensure that clients' objectives are met and to review the progress of projects. Some architects work from home offices.

Work Schedules

Most architects work full time and many work additional hours, especially when facing deadlines. Self-employed architects may have more flexible work schedules.

How to Become an Architect

There are typically three main steps to becoming a licensed architect: completing a bachelor's degree in architecture, gaining relevant experience through a paid internship, and passing the Architect Registration Examination.

Education

Architects typically need a bachelor's degree in architecture. Most architects earn their degree through a 5-year Bachelor of Architecture degree program. Many earn a master's degree in architecture, which can take 1 to 5 additional years. The time required depends on the extent of the student's previous education and training in architecture.

A typical bachelor's degree program includes courses in architectural history and theory, building design with an emphasis on computer-aided design and drafting (CADD), structures, construction methods, professional practices, math, physical sciences, and liberal arts.

About two-thirds of states require that architects hold a degree in architecture from one of more than 120 schools of architecture accredited by the National Architectural Accrediting Board (NAAB). State licensing requirements can be found at the National Council of Architectural Registration Boards (NCARB).

Training

All state architectural registration boards require architecture graduates to complete a lengthy paid internship—generally lasting 3 years—before they may sit for the Architect Registration Examination. Most new graduates complete their training period by working at architectural firms through the Architectural Experience Program (AXP), a program run by NCARB that guides students through the internship process. Some states allow a portion of the training to occur in the offices of employers in related careers, such as engineers and general contractors. Architecture students who complete internships while still in school can count some of that time toward the 3-year training period.

Interns in architectural firms may help design part of a project. They may help prepare architectural documents and drawings, build models, and prepare construction drawings on CADD. Interns may also research building codes and write specifications for building materials, installation criteria, the quality of finishes, and other related details. Licensed architects take the documents that interns produce, make edits to them, finalize plans, and then sign and seal the documents.

Licenses, Certifications, and Registrations
All states and the District of Columbia require architects to be licensed. Licensing requirements typically include completing a degree program in architecture, gaining relevant experience through a paid internship, and passing the Architect Registration Examination.

Most states also require some form of continuing education to keep a license. Continuing education requirements vary by state but usually involve additional education through workshops, university classes, conferences, self-study courses, or other sources..

Advancement
After many years of work experience, some architects advance to become architectural and engineering managers. These managers typically coordinate the activities of employees and may work on larger construction projects.

Important Qualities
Analytical skills. Architects must understand the content of designs and the context in which they were created. For example, architects must understand the locations of mechanical systems and how those systems affect building operations.

Communication skills. Architects share their ideas, both in oral presentations and in writing, with clients, other architects, and workers who help prepare drawings. Many also give presentations to explain their ideas and designs.

Creativity. Architects design the overall look of houses, buildings, and other structures. They must ensure that the final product is both attractive and functional.

Organizational skills. Architects often manage contracts. Therefore, they must keep records related to the details of a project, including total cost, materials used, and progress.

Technical skills. Architects need to use CADD technology to create plans as part of building information modeling (BIM).

Visualization skills. Architects must be able to envision how the parts of a structure relate to each other. They also must be able to visualize how the overall building will look once completed.

Pay

The median annual wage for architects was $96,690 in May 2024. The median wage is the wage at which half the workers in an occupation earned more than that amount and half earned less. The lowest 10 percent earned less than $60,510, and the highest 10 percent earned more than $159,800.

In May 2024, the median annual wages for architects in the top industries in which they worked were as follows:

Government	$113,030
Architectural, engineering, and related services	$95,850
Construction	$91,030

Most architects work full time and many work additional hours, especially when facing deadlines. Self-employed architects may have more flexible work hours.

Job Outlook

Employment of architects is projected to grow 8 percent from 2023 to 2033, faster than the average for all occupations.

About 8,500 openings for architects are projected each year, on average, over the decade. Many of those openings are expected to result from the need to replace workers who transfer to different occupations or exit the labor force, such as to retire.

Employment

Architects are expected to be needed to make plans and designs, particularly in sustainable design, for the construction and renovation of homes, schools, healthcare facilities, and other structures. Improved building information modeling (BIM) software and measuring technology are expected to allow architects to take on activities once performed by other workers, such as architectural and civil drafters, interior designers, and engineers.

State & Area Data

Occupational Employment and Wage Statistics (OEWS)

The Occupational Employment and Wage Statistics (OEWS) program produces employment and wage estimates annually for over 800 occupations. These estimates are available for the nation as a whole, for individual states, and for metropolitan and nonmetropolitan areas. The link below

goes to OEWS data maps for employment and wages by state and area. Use the dropdown boxes to select an occupation.
https://www.bls.gov/oes/current/oes171011.htm#st

Projections Central
Occupational employment projections are developed for all states by Labor Market Information (LMI) or individual state Employment Projections offices. All state projections data are available at www.projectionscentral.org. Information on this site allows projected employment growth for an occupation to be compared among states or to be compared within one state. In addition, states may produce projections for areas; there are links to each state's websites where these data may be retrieved.

CareerOneStop
CareerOneStop includes hundreds of occupational profiles with data available by state and metro area. There are links in the left-hand side menu to compare occupational employment by state and occupational wages by local area or metro area. There is also a salary info tool to search for wages by zip code.

Related Occupations
Architects design buildings and related structures. Construction managers, like architects, also plan and coordinate activities concerned with the construction and maintenance of buildings and facilities. Others who engage in similar work are landscape architects, civil engineers, urban and regional planners, and designers, including interior designers, commercial and industrial designers, and graphic designers.

Sources of Additional Information
Disclaimer:
Links to non-BLS Internet sites are provided for your convenience and do not constitute an endorsement.

Information about education and careers in architecture can be obtained from:

- The American Institute of Architects, 1735 New York Ave. NW., Washington, DC 20006. Internet: http://www.aia.org
- National Architectural Accrediting Board: http://www.naab.org/
- National Council of Architectural Registration Boards, Suite 1100K, 1801 K St. NW., Washington, D.C. 20006. Internet: http://www.ncarb.org
  OOH ONET Codes 17-1011.00"

Source: Bureau of Labor Statistics, U.S. Department of Labor, *Occupational Outlook Handbook*, Architects, on the Internet at https://www.bls.gov/ooh/architecture-and-engineering/architects.htm (visited May 24, 2025).

Last Modified Date: Friday, April 18, 2025

Note:
Please check the website above for the latest information.

E. AIA Compensation Survey Salary Calculator

This salary calculator provides compensation information for full-time architecture professionals working at AIA-member firms in the United States that employ at least three architectural staff members. The data reflects figures as of January 1, 2023: https://salarycalculator.aia.org/salary.aspx

F. So … You would Like to Study Architecture

To study architecture, you need to learn how to draft, how to understand and organize spaces and the interactions between interior and exterior spaces, how to do design, and how to communicate effectively. You also need to understand the history of architecture.

As an architect, a leader for a team of various design professionals, you not only need to know architecture, but also need to understand enough of your consultants' work to be able to coordinate them. Your consultants include soils and civil engineers, landscape architects, structural, electrical, mechanical, and plumbing engineers, interior designers, sign consultants, etc.

There are two major career paths for you in architecture: practice as an architect or teach in colleges or universities. The earlier you determine which path you are going to take, the more likely you will be successful at an early age. Some famous and well-respected architects, like my USC alumnus Frank Gehry, have combined the two paths successfully. They teach at universities and have their own architectural practices. Even as a college or university professor, people respect you more if you have actual working experience and have some built projects. If you only teach in colleges or universities but have no actual working experience and have no built projects, people will consider you as a "paper" architect, and they are not likely to take you seriously, because they will think you probably do not know how to put a real building together.

In the United States., if you want to practice architecture, you need to obtain an architect's license. It requires a combination of passing scores on the Architectural Registration Exam (ARE) and eight years of education and/or qualified working experience, including at least one year of working experience in the United States. Your working experience needs to be under the supervision of a licensed architect to be counted as qualified working experience for your architect's license.

If you work for a landscape architect or civil engineer or structural engineer, some state architectural licensing boards will count your experience at a discounted rate for the qualification of your architect's license. For example, two years of experience working for a civil engineer may be counted as one year of qualified experience for your architect's license. You need to contact your state's architectural licensing board for specific licensing requirements for your state.

If you want to teach in colleges or universities, you probably want to obtain a master's degree or a Ph.D. It is not very common for people in the architectural field to have a Ph.D. One reason is that there are few Ph.D. programs for architecture. Another reason is that architecture is considered a profession and requires a license. Many people think an architect's license is more important than a Ph.D. degree. In many states, you need to have an architect's license to even use the title "architect," or the terms "architectural" or "architecture" to advertise your service. You cannot call yourself an architect if you do not have an architect's license, even if you have a Ph.D. in architecture. Violation of these rules brings punishment.

To become a tenured professor, you need to have a certain number of publications and pass the evaluation for the tenure position. Publications are very important for tenure track positions. Some people say for the tenured track positions in universities and colleges, it is "publish or perish" environment.

The American Institute of Architects (AIA) is the national organization for the architectural profession. Membership is voluntary. There are different levels of AIA membership. Only licensed architects can be (full) AIA members. If you are an architectural student or an intern but not a licensed architect yet, you can join as an associate AIA member. Contact AIA for detailed information.

The National Council of Architectural Registration Boards (NCARB) is a nonprofit federation of architectural licensing boards. It has some very useful programs, such as IDP, to assist you in obtaining your architect's license. Contact NCARB for detailed information.

Back Page Promotion

You may be interested in some other books written by Gang Chen:

A. **ARE Mock Exam series & ARE Exam Guide series. See the following link:**
http://www.GreenExamEducation.com

B. **LEED Exam Guides series. See the following link:**
http://www.GreenExamEducation.com

C. ***Building Construction:*** *Project Management, Construction Administration, Drawings, Specs, Detailing Tips, Schedules, Checklists, and Secrets Others Don't Tell You (Architectural Practice Simplified, 2nd edition)*
http://www.GreenExamEducation.com

D. ***Planting Design Illustrated***
http://www.GreenExamEducation.com

ARE Mock Exam Series & ARE Exam Guide Series

ARE 5.0 Mock Exam Series

Practice Management (PcM) ARE 5.0 Mock Exam (Architect Registration Examination): *ARE 5.0 Overview, Exam Prep Tips, Hotspots, Case Studies, Drag-and-Place, Solutions and Explanations.* **ISBN**: 9781612650388

Project Management (PjM) ARE 5.0 Mock Exam (Architect Registration Examination): *ARE 5.0 Overview, Exam Prep Tips, Hotspots, Case Studies, Drag-and-Place, Solutions and Explanations.* **ISBN**: 9781612650371

Programming & Analysis (PA) ARE 5.0 Mock Exam (Architect Registration Exam): *ARE 5.0 Overview, Exam Prep Tips, Hotspots, Case Studies, Drag-and-Place, Solutions and Explanations.* **ISBN**: 9781612650326

Project Planning & Design (PPD) ARE 5.0 Mock Exam (Architect Registration Examination): *ARE 5.0 Overview, Exam Prep Tips, Hotspots, Case Studies, Drag-and-Place, Solutions and Explanations.* **ISBN**: 9781612650296

Project Development & Documentation (PDD) ARE 5.0 Mock Exam (Architect Registration Examination): *ARE 5.0 Overview, Exam Prep Tips, Hotspots, Case Studies, Drag-and-Place, Solutions and Explanations*
ISBN: 9781612650258

Construction & Evaluation (CE) ARE 5.0 Mock Exam (Architect Registration Examination): *ARE 5.0 Overview, Exam Prep Tips, Hotspots, Case Studies, Drag-and-Place, Solutions and Explanations*
ISBN: 9781612650241

Mock California Supplemental Exam (CSE of Architect Registration Examination)*: CSE Overview, Exam Prep Tips, General Section and Project Scenario Section, Questions, Solutions and Explanations.* **ISBN**: 9781612650159

ARE 5.0 Exam Guide Series

Practice Management (PcM) ARE 5.0 Exam Guide (Architect Registration Examination): *ARE 5.0 Overview, Exam Prep Tips, Guide, and Critical Content.* **ISBN**: 9781612650333

Project Management (PjM) ARE 5.0 Exam Guide (Architect Registration Examination): *ARE 5.0 Overview, Exam Prep Tips, Guide, and Critical Content.* **ISBN**: 9781612650418

Programming & Analysis (PA) ARE 5.0 Exam Guide (Architect Registration Examination): *ARE 5.0 Overview, Exam Prep Tips, Guide, and Critical Content.*
ISBN: 9781612650487

Construction and Evaluation (CE) ARE 5 Exam Guide (Architect Registration Exam): *ARE 5.0 Overview, Exam Prep Tips, Guide, and Critical Content*
ISBN: 9781612650432

Project Planning & Design (PPD) ARE 5.0 Exam Guide (Architect Registration Examination): *ARE 5.0 Overview, Exam Prep Tips, Guide, and Critical Content*
ISBN: 9781612650456

Project Development & Documentation (PDD) ARE 5.0 Exam Guide (Architect Registration Exam): *ARE 5.0 Overview, Exam Prep Tips, Guide, and Critical Content*
ISBN: 9781612650555

See the following link for the latest information:
http://www.GreenExamEducation.com

LEED Exam Guides series: Comprehensive Study Materials, Sample Questions, Mock Exam, Building LEED Certification and Going Green

LEED (Leadership in Energy and Environmental Design) is the most important trend of development, and it is revolutionizing the construction industry. It has gained tremendous momentum and has a profound impact on our environment.

From LEED Exam Guides series, you will learn how to

1. Pass the LEED Green Associate Exam and various LEED AP + exams (each book will help you with a specific LEED exam).

2. Register and certify a building for LEED certification.

3. Understand the intent for each LEED prerequisite and credit.

4. Calculate points for a LEED credit.

5. Identify the responsible party for each prerequisite and credit.

6. Earn extra credit (exemplary performance) for LEED.

7. Implement the local codes and building standards for prerequisites and credit.

8. Receive points for categories not yet clearly defined by USGBC.

There is currently NO official book on the LEED Green Associate Exam, and most of the existing books on LEED and LEED AP are too expensive and too complicated to be practical and helpful. The pocket guides in LEED Exam Guides series fill in the blanks, demystify LEED, and uncover the tips, codes, and jargon for LEED as well as the true meaning of "going green." They will set up a solid foundation and fundamental framework of LEED for you. Each book in the LEED Exam Guides series covers every aspect of one or more specific LEED rating system(s) in plain and concise language and makes this information understandable to all people.

These pocket guides are small and easy to carry around. You can read them whenever you have a few extra minutes. They are indispensable books for all people—administrators; developers; contractors; architects; landscape architects; civil, mechanical, electrical, and plumbing engineers; interns; drafters; designers; and other design professionals.

Why is the LEED Exam Guides series needed?

A number of books are available that you can use to prepare for the LEED exams:

1. *USGBC Reference Guides*. You need to select the correct version of the *Reference Guide* for your exam.

   The *USGBC Reference Guides* are comprehensive, but they give too much information. For example, *The LEED 2009 Reference Guide for Green Building Design and Construction (BD&C)* has about 700 oversized pages. Many of the calculations in the books are too detailed for the exam. They are also expensive (approximately $200 each, so most people may not buy them for their personal use, but instead, will seek to share an office copy).

   It is good to read a reference guide from cover to cover if you have the time. The problem is not too many people have time to read the whole reference guide. Even if you do read the whole guide, you may not remember the important issues to pass the LEED exam. You need to reread the material several times before you can remember much of it.

   Reading the reference guide from cover to cover without a guidebook is a difficult and inefficient way of preparing for the LEED AP Exam, because you do NOT know what USGBC and GBCI are looking for in the exam.

2. The USGBC workshops and related handouts are concise, but they do not cover extra credits (exemplary performance). The workshops are expensive, costing approximately $450 each.

3. Various books published by a third party are available on Amazon, bn.com and books.google.com. However, most of them are not very helpful.

   There are many books on LEED, but not all are useful.

   LEED Exam Guides series will fill in the blanks and become a valuable, reliable source:

   a. They will give you more information for your money. Each of the books in the LEED Exam Guides series has more information than the related USGBC workshops.

   b. They are exam-oriented and more effective than the USGBC reference guides.

   c. They are better than most, if not all, of the other third-party books. They give you comprehensive study materials, sample questions and answers, mock exams and answers, and critical information on building LEED certification and going green. Other third-party books only give you a fraction of the information.

   d. They are comprehensive yet concise. They are small and easy to carry around. You can read them whenever you have a few extra minutes.

   e. They are great timesavers. I have highlighted the important information that you need to understand and MEMORIZE. I also make some acronyms and short sentences to help you easily remember the credit names.

It should take you about 1 or 2 weeks of full-time study to pass each of the LEED exams. I have met people who have spent 40 hours to study and passed the exams.

You can find sample texts and other information on the LEED Exam Guides series in customer discussion sections under each of my book's listing on Amazon, bn.com and books.google.com.

What others are saying about *LEED GA Exam Guide* (Book 2, LEED Exam Guide series):

"Finally! A comprehensive study tool for LEED GA Prep!

"I took the 1-day Green LEED GA course and walked away with a power point binder printed in very small print—which was missing MUCH of the required information (although I didn't know it at the time). I studied my little heart out and took the test, only to fail it by 1 point. Turns out I did NOT study all the material I needed to in order to pass the test. I found this book, read it, marked it up, retook the test, and passed it with a 95%. Look, we all know the LEED GA exam is new and the resources for study are VERY limited. This one is the VERY best out there right now. I highly recommend it."
—ConsultantVA

"Complete overview for the LEED GA exam

"I studied this book for about 3 days and passed the exam … if you are truly interested in learning about the LEED system and green building design, this is a great place to start."
—K.A. Evans

"A Wonderful Guide for the LEED GA Exam

"After deciding to take the LEED Green Associate exam, I started to look for the best possible study materials and resources. From what I thought would be a relatively easy task, it turned into a tedious endeavor. I realized that there are vast amounts of third-party guides and handbooks. Since the official sites offer little to no help, it became clear to me that my best chance to succeed and pass this exam would be to find the most comprehensive study guide that would not only teach me the topics, but would also give me a great background and understanding of what LEED actually is. Once I stumbled upon Mr. Chen's book, all my needs were answered. This is a great study guide that will give the reader the most complete view of the LEED exam and all that it entails.

"The book is written in an easy-to-understand language and brings up great examples, tying the material to the real world. The information is presented in a coherent and logical way, which optimizes the learning process and does not go into details that will not be needed for the LEED Green Associate Exam, as many other guides do. This book stays dead on topic and keeps the reader interested in the material.

"I highly recommend this book to anyone that is considering the LEED Green Associate Exam. I learned a great deal from this guide, and I am feeling very confident about my chances for passing my upcoming exam."
—Pavel Geystrin

"Easy to read, easy to understand

"I have read through the book once and found it to be the perfect study guide for me. The author does a great job of helping you get into the right frame of mind for the content of the exam. I had started by studying the Green Building Design and Construction reference guide for LEED projects produced by the USGBC. That was the wrong approach, simply too much information with very little retention. At 636 pages in textbook format, it would have been a daunting task to get through it. Gang Chen breaks down the points, helping to minimize the amount of information but maximizing the content I was able to absorb. I plan on going through the book a few more times, and I now believe I have the right information to pass the LEED Green Associate Exam."
—Brian Hochstein

"All in one—LEED GA prep material

"Since the LEED Green Associate exam is a newer addition by USGBC, there is not much information regarding study material for this exam. When I started looking around for material, I got really confused about what material I should buy. This LEED GA guide by Gang Chen is an answer to all my worries! It is a very precise book with lots of information, like how to approach the exam, what to study and what to skip, links to online material, and tips and tricks for passing the exam. It is like the 'one stop shop' for the LEED Green Associate Exam. I think this book can also be a good reference guide for green building professionals. A must-have!"
—SwatiD

"An ESSENTIAL LEED GA Exam Reference Guide

"This book is an invaluable tool in preparation for the LEED Green Associate (GA) Exam. As a practicing professional in the consulting realm, I found this book to be all-inclusive of the preparatory material needed for sitting the exam. The information provides clarity to the fundamental and advanced concepts of what LEED aims to achieve. A tremendous benefit is the connectivity of the concepts with real-world applications.

"The author, Gang Chen, provides a vast amount of knowledge in a very clear, concise, and logical media. For those that have not picked up a textbook in a while, it is very manageable to extract the needed information from this book. If you are taking the exam, do yourself a favor and purchase a copy of this great guide. Applicable fields: Civil Engineering, Architectural Design, MEP, and General Land Development."
—Edwin L. Tamang

Note:
*Other books in the **LEED Exam Guides series** are published or in the process of being produced. At least **one book will eventually be produced for each of the LEED exams.** The series include:*

LEED v4 Green Associate Exam Guide (LEED GA): *Comprehensive Study Materials, Sample Questions, Mock Exam, Green Building LEED Certification, and Sustainability*, LEED Exam Guide series, ArchiteG.com. Latest Edition.

LEED GA MOCK EXAMS (LEED v4): *Questions, Answers, and Explanations: A Must-Have for the LEED Green Associate Exam, Green Building LEED Certification, and Sustainability*, LEED Exam Guide series, ArchiteG.com. Latest Edition

LEED v4 BD&C EXAM GUIDE: *A Must-Have for the LEED AP BD+C Exam: Comprehensive Study Materials, Sample Questions, Mock Exam, Green Building Design and Construction, LEED Certification, and Sustainability*, LEED Exam Guide series, ArchiteG.com. Latest Edition.

LEED v4 BD&C MOCK EXAMS: *Questions, Answers, and Explanations: A Must-Have for the LEED AP BD+C Exam, Green Building LEED Certification, and Sustainability*, LEED Exam Guide series, ArchiteG.com. Latest Edition.

LEED v4 ID&C Exam Guide: *A Must-Have for the LEED AP ID+C Exam: Study Materials, Sample Questions, Green Interior Design and Construction, Green Building LEED Certification, and Sustainability,* LEED Exam Guide series, ArchiteG.com. Latest Edition.

LEED v4 AP ID+C MOCK EXAM: *Questions, Answers, and Explanations: A Must-Have for the LEED AP ID+C Exam, Green Building LEED Certification, and Sustainability.* LEED Exam Guide series, ArchiteG.com. Latest Edition.

LEED v4 AP O+M MOCK EXAM: Questions, Answers, and Explanations: *A Must-Have for the LEED AP O+M Exam, Green Building LEED Certification, and Sustainability*. LEED Exam Guide series, ArchiteG.com. Latest Edition.

LEED v4 O&M EXAM GUIDE: *A Must-Have for the LEED AP O+M Exam: Comprehensive Study Materials, Sample Questions, Mock Exam, Green Building Operations and Maintenance, LEED Certification, and Sustainability*, LEED Exam Guide series, ArchiteG.com. Latest Edition.

LEED v4 HOMES EXAM GUIDE: *A Must-Have for the LEED AP Homes Exam: Comprehensive Study Materials, Sample Questions, Mock Exam, Green Building LEED Certification, and Sustainability*, LEED Exam Guide series, ArchiteG.com. Latest Edition.

LEED v4 ND EXAM GUIDE: *A Must-Have for the LEED AP Neighborhood Development Exam: Comprehensive Study Materials, Sample Questions, Mock Exam, Green Building LEED Certification, and Sustainability*, LEED Exam Guide series, ArchiteG.com. Latest Edition.

How to order these books:
You can order the books listed above at:
http://www.GreenExamEducation.com

OR
http://www.ArchiteG.com

Building Construction

Project Management, Construction Administration, Drawings, Specs, Detailing Tips, Schedules, Checklists, and Secrets Others Don't Tell You (Architectural Practice Simplified, 2nd edition)

Learn the Tips, Become One of Those Who Know Building Construction and Architectural Practice, and Thrive!

For architectural practice and building design and construction industry, there are two kinds of people: those who know, and those who don't. The tips of building design and construction and project management have been undercover—until now.

Most of the existing books on building construction and architectural practice are too expensive, too complicated, and too long to be practical and helpful. This book simplifies the process to make it easier to understand and uncovers the tips of building design and construction and project management. It sets up a solid foundation and fundamental framework for this field. It covers every aspect of building construction and architectural practice in plain and concise language and introduces it to all people. Through practical case studies, it demonstrates the efficient and proper ways to handle various issues and problems in architectural practice and building design and construction industry.

It is for ordinary people and aspiring young architects as well as seasoned professionals in the construction industry. For ordinary people, it uncovers the tips of building construction; for aspiring architects, it works as a construction industry survival guide and a guidebook to shorten the process in mastering architectural practice and climbing up the professional ladder; for seasoned architects, it has many checklists to refresh their memory. It is an indispensable reference book for ordinary people, architectural students, interns, drafters, designers, seasoned architects, engineers, construction administrators, superintendents, construction managers, contractors, and developers.

You will learn:

1. How to develop your business and work with your client.
2. The entire process of building design and construction, including programming, entitlement, schematic design, design development, construction documents, bidding, and construction administration.
3. How to coordinate with governing agencies, including a county's health department and a city's planning, building, fire, public works departments, etc.
4. How to coordinate with your consultants, including soils, civil, structural, electrical, mechanical, plumbing engineers, landscape architects, etc.
5. How to create and use your own checklists to do quality control of your construction documents.
6. How to use various logs (i.e., RFI log, submittal log, field visit log, etc.) and lists (contact list, document control list, distribution list, etc.) to organize and simplify your work.
7. How to respond to RFI, issue CCDs, review change orders, submittals, etc.
8. How to make your architectural practice a profitable and successful business.

Planting Design Illustrated

A Must-Have for Landscape Architecture: A Holistic Garden Design Guide with Architectural and Horticultural Insight, and Ideas from Famous Gardens in Major Civilizations

One of the most significant books on landscaping!

This is one of the most comprehensive books on planting design. It fills in the blanks of the field and introduces poetry, painting, and symbolism into planting design. It covers in detail the two major systems of planting design: formal planting design and naturalistic planting design. It has numerous line drawings and photos to illustrate the planting design concepts and principles. Through in-depth discussions of historical precedents and practical case studies, it uncovers the fundamental design principles and concepts, as well as the underpinning philosophy for planting design. It is an indispensable reference book for landscape architecture students, designers, architects, urban planners, and ordinary garden lovers.

What Others Are Saying about *Planting Design Illustrated* …

"I found this book to be absolutely fascinating. You will need to concentrate while reading it, but the effort will be well worth your time."
—Bobbie Schwartz, former president of APLD (Association of Professional Landscape Designers) and author of *The Design Puzzle: Putting the Pieces Together*.

"This is a book that you have to read, and it is more than well worth your time. Gang Chen takes you well beyond what you will learn in other books about basic principles like color, texture, and mass."
—Jane Berger, editor & publisher of gardendesignonline

"As a longtime consumer of gardening books, I am impressed with Gang Chen's inclusion of new information on planting design theory for Chinese and Japanese gardens. Many gardening books discuss the beauty of Japanese gardens, and a few discuss the unique charms of Chinese gardens, but this one explains how Japanese and Chinese history, as well as geography and artistic traditions, bear on the development of each country's style. The material on traditional Western garden planting is thorough and inspiring, too. *Planting Design Illustrated* definitely rewards repeated reading and study. Any garden designer will read it with profit."
—Jan Whitner, editor of the *Washington Park Arboretum Bulletin*

"Enhanced with an annotated bibliography and informative appendices, *Planting Design Illustrated* offers an especially "reader friendly" and practical guide that makes it a very strongly recommended addition to personal, professional, academic, and community library gardening & landscaping reference collection and supplemental reading list."
—Midwest Book Review

"Where to start? *Planting Design Illustrated* is, above all, fascinating and refreshing! Not something the lay reader encounters every day, the book presents an unlikely topic in an easily digestible, easy-to-follow way. It is superbly organized with a comprehensive table of contents, bibliography, and appendices. The writing, though expertly informative, maintains its accessibility throughout and is a joy to read. The detailed and beautiful illustrations expanding on the concepts presented were my favorite portion. One of the finest books I've encountered in this contest in the past 5 years."
—Writer's Digest 16th Annual International Self-Published Book Awards Judge's Commentary

"The work in my view has incredible application to planting design generally and a system approach to what is a very difficult subject to teach, at least in my experience. Also featured is a very beautiful philosophy of garden design principles bordering poetry. It's my strong conviction that this work needs to see the light of day by being published for the use of professionals, students & garden enthusiasts."
—Donald C. Brinkerhoff, FASLA, chairman and CEO of Lifescapes International, Inc.

Index

www.ingramcontent.com/pod-product-compliance
Lightning Source LLC
LaVergne TN
LVHW070847210926
848482LV00016B/887